Chris Mooders

AF769900

Coastal and Estuarine Studies

Managing Editors:
Malcolm J. Bowman   Richard T. Barber
Christopher N.K. Mooers   John A. Raven

Coastal and Estuarine Studies

43

B. Lahlou and P. Vitiello (Eds.)

Aquaculture: Fundamental and Applied Research

American Geophysical Union

Washington, DC

Managing Editors
Malcolm J. Bowman
Marine Sciences Research Center, State University of New York
Stony Brook, N.Y. 11794, USA

Richard T. Barber
Duke Marine Laboratory
Beaufort, N.C. 28516, USA

Christopher N.K. Mooers
Ocean Process Analysis Laboratory
Institute for the Study of the Earth, Oceans and Space
University of New Hampshire
Durham, N.H. 03824-3525, USA

John A. Raven
Dept. of Biological Sciences, Dundee University
Dundee, DD1 4HN, Scotland

Editors
B. LAHLOU
Laboratoire de Physiologie Cellulaire et Comparée
Faculté des Sciences
Parc Valrose
06108 NICE CEDEX 2 FRANCE

P. VITIELLO
Laboratoire de Biologie marine
Faculté des Sciences de Marseille-Luminy
13288 MARSEILLE CEDEX 9
FRANCE

Library of Congress Cataloging-in-Publication Data

Aquaculture : fundamental and applied research / B. Lahlou and P. Vitiello (eds.).
p. cm. — (Coastal and estuarine studies, ISSN 0733-9569 ; 43)
Papers based on lectures delivered at the International Congress on Research for Aquaculture: Fundamental and Applied Aspects, held Oct. 6-10, 1991, at Antibes-Juan Les Pins (Alpes-Maritimes, France).
Includes bibliographical references.
ISBN 0-87590-257-X
1. Fishes—Physiology—Congresses. 2. Hatchery fishes—Physiology—Congresses. 3. Fish-culture—Congresses. 4. Aquaculture—Congresses. I. Lahlou, B., 1936— . II. Vitiello, P. (Pierre) III. International Congress on Research for Aquaculture: Fundamental and Applied Aspects 1991 : Juan-les-Pins, France) IV. Series.
QL639.1.A68 1993
597′.01—dc20 93-38053
CIP

ISSN 0733-9569

ISBN 0-87590-257-X

Copyright 1993 by the American Geophysical Union, 2000 Florida Avenue, NW, Washington, DC 20009, U.S.A.

Figures, tables, and short excerpts may be reprinted in scientific books and journals if the source is properly cited.

Authorization to photocopy items for internal or personal use, or the internal or personal use of specific clients, is granted by the American Geophysical Union for libraries and other users registered with the Copyright Clearance Center (CCC) Transactional Reporting Service, provided that the base fee of $1.00 per copy plus $0.10 per page is paid directly to CCC, 21 Congress Street, Salem, MA 10970. 0733-9569/93/$01. + .10.

This consent does not extend to other kinds of copying, such as copying for creating new collective works or for resale. The reproduction of multiple copies and the use of full articles or the use of extracts, including figures and tables, for commercial purposes requires permission from AGU.

Printed in the United States of America.

CONTENTS

PREFACE

The aim of this volume is to present an account of recent progress in basic research on animal species (mostly fishes) currently exploited in aquaculture, insofar as this knowledge is a source of actual or potential applications in this industry. Each contribution is updated to contain the best of current knowledge in the field and to provide a large variety of readers with a valuable source of information.

Aquaculture is quickly developing worldwide and has attracted huge investments. As with any large-scale industry open to international competition, its success is now closely related to the use of advanced technologies which may concern all aspects of aquatic animal biology. For this reason, the topics presented in this comprehensive review were chosen in order to cover a broad spectrum of scientific interests and all scales of study, environmental to molecular. Environmental factors considered include salinity, specific ions, acidity, light, metal and organic pollutants. Animal physiology and biochemistry are examined, covering the functions of nutrition, growth, development and reproduction, with heavy emphasis on endocrinological controls and responses to stress. Molecular biology, which is creating new powerful tools in biotechnologies, is also presented from the point of view of its impact on aquaculture.

This volume is based on lectures delivered at the International Congress on Research for Aquaculture: Fundamental and Applied Aspects, held at Antibes-Juan les Pins (Alpes-Maritimes, France), October 6-10, 1991. The Congress represented the 13th conference of the European Society for Comparative Physiology and Biochemistry (ESCPB). It would not have been possible to organize so successfully without considerable support from local authorities in the southeast of France, especially the Conseil Régional de la Région Provence-Alpes-Côte d'Azur (Région PACA), the Conseil Général of the Département des Alpes-Maritimes and the city of Antibes-Juan les Pins. The Crédit Agricole Bank provided financial help, which is also gratefully acknowledged. The following public research institutions provided strong encouragement, subsidies and technical facilities: Centre National de la Recherche Scientifique (CNRS), Institut Français pour l'Exploitation Scientifique de la Mer (IFREMER), and the University of Nice-Sophia Antipolis (UNSA). A large scientific contribution was also offered by the laboratories of the Institut National de la Recherche Agronomique (INRA). Space does not permit us to thank each of the many persons who contributed so enthusiastically in arranging the scientific program or organizing the meeting at all stages. Wonderful help was generously provided by the staffs of the Laboratoire de Physiologie Cellulaire et Comparée (UNSA), the Office Régional de la Mer (PACA), the congress hall (Palais des Congrès) of Antibes-Juan les Pins.

B. Lahlou
P. Vitiello
Editors

1

Acid-Base Balance in Aquatic Invertebrates: The Effects of Environmental Factors

J.P. Truchot

In the steady state, endogenous production and external exchanges of acidic and basic substances contribute to the maintenance in body fluids of a given acid-base balance usually referred to as the pH value. Keeping an appropriate acid-base state within narrow limits is of prime importance for many basic living processes, especially those depending on protein conformation and electrical charge, which are in large part determined by the pH. Acid-base relevant mechanisms depend on various organ functions such as respiratory gas exchanges, ionic regulation, cell metabolism, etc.. (fig 1), that could either be involved in the generation of acid-base disturbances, or participate in acid-base regulatory responses. As many of these exchange functions are much influenced by the very variable ambient conditions in aquatic habitats, acid-base balance of water-dwelling invertebrates is continuously challenged in a very intricate way by environmental changes. For all these reasons, acid-base variables are usually very sensitive to changes in ambient conditions and appears thus ideally suited to contribute to the diagnosis of any problem originating in inadequate water quality parameters, either in the wild or in the aquaculture practice. Using this type of information may, however, not be an easy task. Numerous single factor studies have been conducted in laboratory conditions in order to describe environmentally-induced effects on extracellular acid-base state and to elucidate underlying physiological mechanisms. But it should be kept in mind that many factors simultaneously vary in natural habitats as well as in aquaculture conditions and that the resultant modifications of acid-base balance are not always easy to predict.

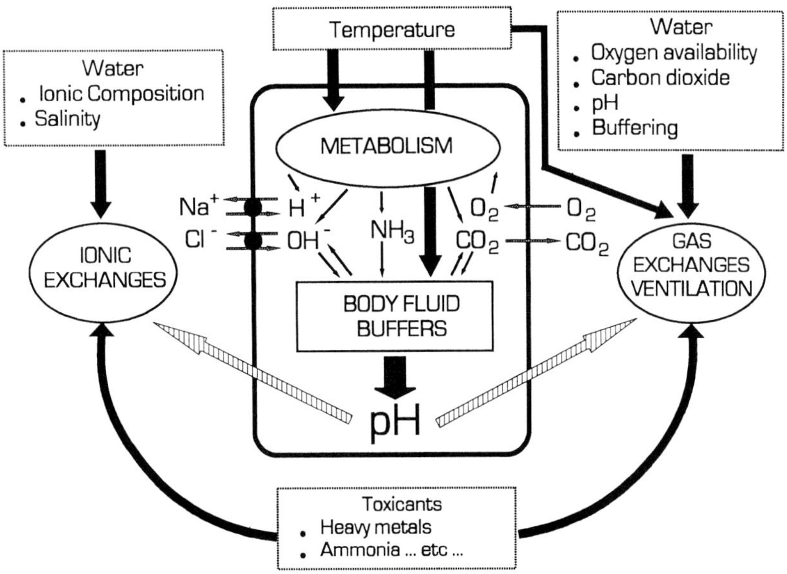

Fig. 1. Schematic diagram showing the effects of various environmental factors on blood acid-base balance in aquatic invertebrates. Blood pH is determined by the titration state of body fluid buffers which depends i) on CO_2 metabolic production and elimination by gill respiratory exchanges ; ii) on fixed acid or base (H^+, OH^-) endogenous generation and external exchanges coupled with ionic gill movements. Environmental factors affect blood pH via various functions : gas and ion exchange, metabolism.. etc. pH is a controlled variable that feeds back on gill ventilation and ion exchange to ensure acid-base regulation.

Using mainly decapod crustaceans as model invertebrates, this short review will summarize the main lines of recent laboratory work on environmentally-induced acid-base changes and show an example of how this information can be used in interpreting field observations.

Water quality parameters : laboratory studies.

Oxygen

Water as a respiratory medium is primarily characterized by a very restricted oxygen availability, due to both low O_2 solubility and poor O_2 diffusibility (Dejours, 1981). As a consequence, O_2 depletion resulting from either chemical or biological oxidations, or O_2 oversaturation caused by active photosynthesis during daytime, are of common occurrence in aquatic habitats. These changes in O_2 concentration and partial pressure may be particularly large and rapid in small water bodies such as intertidal rockpools (Truchot and Duhamel-Jouve, 1980) in which various animal species can be retained several hours at low tide.

In crustaceans as in other aquatic animals, acid-base deviations caused by changes of water oxygenation are mainly linked to large adjustments of the gill water flow rate, that contribute to the maintenance of a constant O_2 uptake through a large range of ambient Po_2. These adjustments, hyperventilation in hypoxia and hypoventilation in hyperoxia, lead to new steady states in metabolic CO_2 excretion, with modified internal CO_2 partial pressures. Thus, changes of ambient oxygen primarily induce acid-base disturbances of respiratory origin : decreased Pco_2 and increased pH, or hypocapnic alkalosis in hypoxia ; and increased Pco_2 and decreased pH, or hypercapnic acidosis in hyperoxia (Truchot, 1975, 1987 ; Dejours and Beekenkamp, 1977 ; McMahon et al., 1978 ; Toulmond and Tchernigovtzeff, 1989). At least in the last case, however, blood pH tends to progressively recover thanks to a rise in bicarbonate concentration which results in a large part from a measurable fixed acid excretion through the gills (Truchot, 1979). In freshwater fish, and probably in crustaceans also, such compensatory acid-base fluxes have been shown to be coupled to gill Na^+ and Cl^- fluxes by yet unknown mechanisms (Wood et al., 1984 ; Wood, 1991). Nevertheless, this indicates that some component of gill ion exchange is responsive to acid-base deviations and can be dynamically manipulated in order to correct acidoses of respiratory origin. By contrast, appropriate compensation of hypocapnic alkalosis is not apparent during hypoxia in crustaceans. In moderate hypoxia, the alkalosis may even be accentuated by an increase of bicarbonate concentration (Truchot, 1975a). Even in severe hypoxia, another potential source of acid-base disturbance, i.e. lactate release, surprisingly does not lead to acidosis, probably because the excess of metabolic protons appears buffered by carbonate stores in the exoskeleton or elsewhere, as suggested by a strong increase in blood calcium concentration (Lallier et al., 1987).

Carbon dioxide.

Although resulting also mainly from biological processes and thus being tightly linked, ambient changes of carbon dioxide partial pressure (Pco_2) are much less in amplitude than those of oxygen in aquatic habitats, because of a 20-30 fold higher solubility. Also carbonic acid buffering in carbonate-rich, hard waters considerably limit Pco_2 changes (Dejours, 1981). Hence, experimental data are relevant in this area only when moderate changes of Pco_2 are applied. Ambient hypercapnia usually leads to an increased blood Pco_2 and decreased pH, i.e. an hypercapnic acidosis (Truchot, 1975b ; Conti and Toulmond, 1986). Although recent evidence indicates that some crustacean species hyperventilate their gills when exposed to increased ambient CO_2 (Massabuau et al., 1984 ; Massabuau and Burtin, 1985), this response usually remains moderate (Jouve-Duhamel and Truchot, 1983) and does not result in a marked decrease in blood-water Pco_2 difference (Truchot, 1975b). The hypercapnic acidosis induced by increased ambient Pco_2 appears progressively compensated with an increase in blood bicarbonate concentration, probably by mechanisms similar to those mentioned above in hyperoxic situations.

Although ambient hypocapnia can occur naturally, its effect on acid-base balance has not been specifically studied, probably because very low water Pco_2 values are difficult to obtain experimentally. Theoretical considerations indicate that animals should become very hypocapnic and alkalotic, particularly in carbonated waters (Truchot and Duhamel-Jouve, 1980). However, in the natural setting, hypocapnia is always associated with hyperoxia which drives hypoventilatory responses mitigating the alkalosis (see below).

Water pH and buffering capacity.

Except in very acid waters, pH is not a factor influencing blood acid-base balance by itself in aquatic invertebrates. But changes of other factors associated with pH seem to act as primary causes of acid-base deviations. One such factor is the P_{CO_2} (see above), but another one relates to the water buffering capacity, and more specifically to its carbonate alkalinity. Indeed, in carbonated water at low P_{CO_2} and high pH, carbonate ions can buffer CO_2 released at the gills according to :

$$CO_2 + H_2O + CO_3^{2-} \rightleftharpoons 2\ HCO_3^-$$

To be effective, this process requires a rapid reaction for CO_2 hydration, which is made possible by externally available carbonic anhydrase in fish (Rahim et al., 1988 ; Randall et al., 1991) and most probably also crustacean gill epithelium.

As a consequence, the relationship between water CO_2 concentration and partial pressure becomes curved at low P_{CO_2} (fig 2), meaning that more CO_2 can be added or removed for a given P_{CO_2} change. Moreover, this increase in CO_2 "capacitance" is the larger, the higher the water carbonate alkalinity (Truchot, 1984). So, CO_2 excretion will cause a greater P_{CO_2} increase in gill water at low than at high carbonate alkalinity. This may be expected to result in hypercapnia and acidosis when water carbonate content is reduced and, indeed, direct measurements in crayfish *Astacus leptodactylus* (Dejours and Armand, 1980) and crab *Carcinus maenas* (Truchot, 1981 ; 1984) have shown this to be true (fig 2).

An important consequence of this effect of water CO_2 buffering capacity is that the magnitude of any acid-base disturbance of respiratory origin should depend on the water carbonate alkalinity. This is for example the case for hyperoxic exposure which induces in the crab an increase of blood Pco_2 and a decrease of blood pH which are the more marked the less the water carbonate alkalinity (Truchot, 1984).

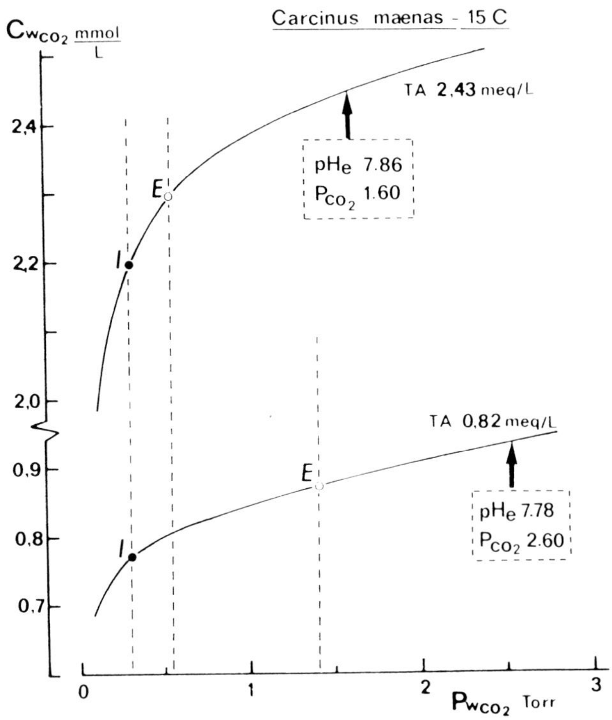

Fig. 2. The effect of titration alkalinity (TA) on hemolymph acid-base balance in the crab *Carcinus maenas*. Relationship between water total CO_2 concentration ($Cwco_2$) and CO_2 partial pressure ($Pwco_2$) are drawn for two TA values. For crabs breathing well aerated water in both cases (point I at ca 0.3 Torr $Pwco_2$) and assuming a common (expired - inspired) $Cwco_2$ difference of 0.1 mmol/L, expired water Pco_2 value (point E) is higher at low than at high TA. Accordingly, measured hemolymph pH and Pco_2 values show that crabs breathing water at low TA are acidotic and hypercapnic compared to crabs breathing at high TA (Redrawn from Truchot, 1981).

Ionic composition in freshwater.

To maintain ionic balance, freshwater animals take up Na^+ and Cl^- ions from the ambient water. Although the exact mechanisms underlying this active ion absorption are not completely elucidated, it is clear that it is at least partly coupled with the movement of acid-base equivalents, Na^+ being exchanged against one acid equivalent, H^+ or NH_4^+, and Cl^- against an alkaline equivalent, probably HCO_3^-. Thus, as shown by a number of studies, restriction of Na^+ uptake, either by water Na^+ depletion or specific inhibition of sodium transport, is associated with a reduction of acid excretion and metabolic acidosis, while conversely a metabolic alkalosis develops when Cl^- uptake is reduced (see Truchot, 1987). Exposure to very low ambient pH also affect exchange mechanisms of Na^+ more than those of Cl^-, either by reducing active uptake or increasing passive permeability, and result in metabolic acidosis (McDonald, 1983).

Salinity.

Another important determinant of acid-base balance in euryhaline invertebrates is the water salinity. Effects here appear linked not to ion availability but rather to osmotic effects. In many cases, a decrease of water salinity entails an alkalosis, the main component of which is metabolic. Conversely, increased salinity induces a metabolic acidosis (Truchot, 1987). These disturbances are long lasting, receding only after some weeks of acclimation to a new salinity. The underlying mechanisms are not elucidated but the involvement of alterations of gill ionic fluxes is doubtful. Indeed, in the crab *Carcinus maenas*, the metabolic alkalosis induced by transfer to lower salinity is accompanied by a net outflux of base, suggesting that its ultimate source resides in the tissues, perhaps in relation with processes engaged in cell volume regulation (Truchot, 1981).

Temperature.

The effects of temperature changes on acid-base balance are manifold and complex. First, temperature changes the dissociation constant of buffers and thus can alter pH independently of external acid-base exchanges. Second, temperature affects the metabolic level and can influence many processes, particularly gas exchange. The outcome of all these effects leads to a pattern of decreased blood pH with increasing body temperature which is common to many animals (Reeves, 1977 ; Truchot, 1987) The exact value of the pH/temperature slope has been much discussed but in most cases it approaches - 0.017 pH unit.°C^{-1}, the slope describing the change in neutral pH with temperature, which is also that characterizing the most common protein buffer group in the physiological range, the imidazole group of histidine. This results in a relatively constant ionization state of protein molecules and has been considered of prime importance to maintain protein function during changes of body temperature.

In water-breathers, the adjustments leading to this pH/temperature pattern appear to require large changes of body CO_2 stores and bicarbonate concentration instead of rapid ventilatory adjustments of Pco_2 as in air-breathers. For example, at increasing temperature, blood bicarbonate concentration decreases markedly in the shore crab *Carcinus maenas* (Truchot, 1973) as well as in the lugworm *Arenicola marina* (Toulmond, 1977), while Pco_2 stays constant or increases only moderately. Thus, temperature acclimation of acid-base balance is considered to require a much longer time in water-breathers than in air breathers.

Acid-base balance in natural situations.

From the bulk of data reviewed above, the effects of many environmental variables on acid-base balance are reasonably well understood from single-factor laboratory experiments. But what about acid-base changes in natural situations in which many factors often change together in a very intricate way ?

Typical and rather extreme examples of rapidly changing environmental conditions can be found in the marine intertidal zone where small water bodies remain isolated with their animal and plant inhabitants for several hours at low tide. As documented in many studies (for example Truchot and Duhamel-Jouve, 1980 ; Morris and Taylor, 1983), the main fluctuating factors in these intertidal pools are oxygen, carbon dioxide, pH and temperature. During the day, photosynthetic processes override biological oxygen consumption, resulting in hyperoxia and hypocapnia, while the converse changes, progressive hypoxia and hypercapnia take place at night. Due to a small volume and exposure to direct solar radiation, large changes of water temperature are also commonplace, but salinity

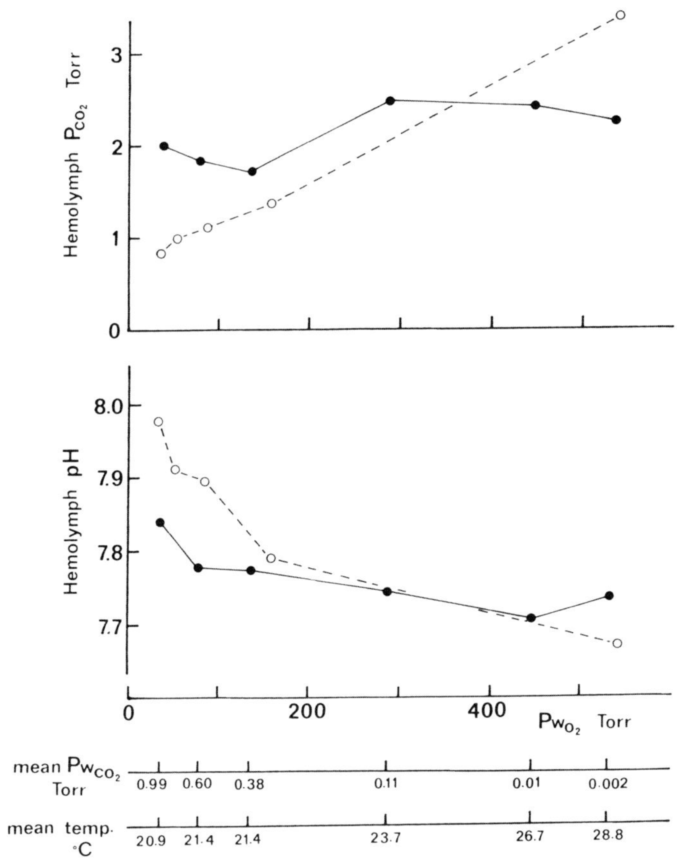

Fig. 3. Mean values of hemolymph pH and P_{CO_2} measured in crabs *Carcinus maenas* at various water P_{O_2}. Full lines and closed symbols show values recorded in an artificial tidepool with spontaneous P_{CO_2} and temperature variations (shown on the additional abscissa scales). Dashed lines and open symbols show for comparison values measured in laboratory experiments at constant water P_{CO_2} (ca 0.3 Torr) and temperature (15 °C) (from Truchot, 1986).

variations are usually moderate and less well predictable. Interestingly enough, it is relatively easy to simulate such environmental changes in medium-sized outdoor artificial enclosures. To do this, we used a shallow water tank populated with algae and animals, in which high tide was simulated during 6 h periods by a constant water flow rate from the seawater laboratory supply, alternating with 6 h low tide period without flow. In this way, we observed changes of water Po_2, Pco_2, pH and temperature that were of similar amplitude and speed as those recorded in the field at low tide (Truchot, 1986).

According to laboratory data commented above, such concomitant Po_2 and Pco_2 variations have opposite effects on the acid-base balance of water-breathers and they would thus counteract each other. Indeed, during the day, the hypercapnic acidosis induced by water hyperoxia is expected to be moderated by the effects of water hypocapnia. Similarly, at night, an increase of water Pco_2 will counteract the respiratory alkalosis resulting from the hypoxia-induced hyperventilation.

These expectations have been confirmed by acid-base measurements on blood sampled at various times of day and night from crabs *Carcinus maenas* kept in the artificial tidepool (Truchot, 1986). Fig 3 compares hemolymph pH and Pco_2 values observed at several water oxygen levels, either at constant Pco_2 and temperature in laboratory experiments, or at Pco_2 and temperature fluctuating on a diel basis as in the field. Clearly, acid-base deviations induced by changes of water oxygenation in natural conditions are damped compared to laboratory data. This probably results mainly from counteracting Pco_2 changes because the effect of temperature on blood pH is expected to act in the same direction as oxygen changes. Moreover, temperature fluctuations in the field are probably too much rapid to allow complete acclimation by changes of hemolymph bicarbonate concentration. Nevertheless, taking all day and night pH values into account and plotting them as a function of temperature shows a clear trend toward a decrease of blood pH with decreasing water temperature, with a slope amounting to -0.0157 pH unit.$^\circ C^{-1}$, a value very near that found in single factor experiments in the laboratory (fig 4). Clearly, the same pattern of changes of pH with temperature could prevail either in constant conditions and after a long temperature acclimation in the laboratory or in a natural setting with concerted and rapid variations of many environmental factors. Interestingly, in both situations, this pattern is that required to maintain a constant protein charge state, according to the imidazole alphastat concept (Reeves, 1977).

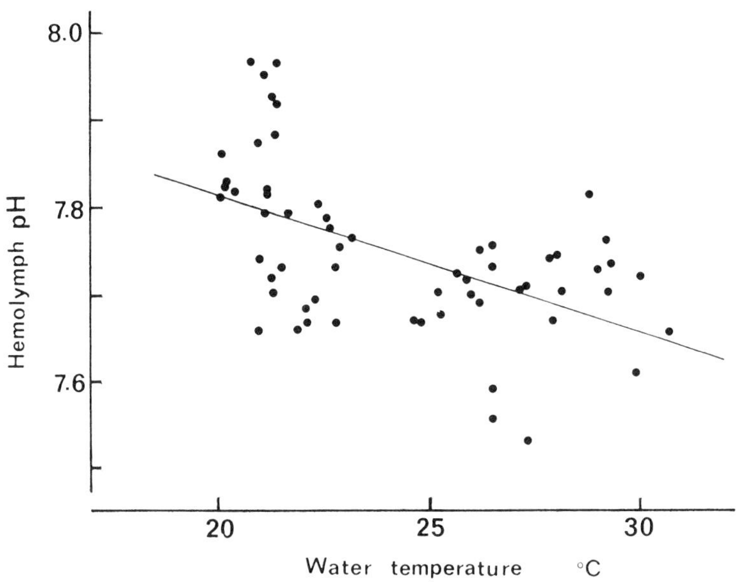

Fig. 4. Relationship between hemolymph pH and water temperature for crabs *Carcinus maenas* kept in an artificial tidepool with spontaneous, near-natural variations of water Po_2 and Pco_2. Values in the high temperature range are taken during the day and correspond to hyperoxic and hypocapnic water conditions, while values in the low temperature range are night values recorded in hypoxic and hypercapnic conditions in water. The pH/temperature slope amounts to - 0.0157 $pH.^{\circ}C^{-1}$ (from Truchot, 1986).

The physiological meaning of these findings remains unclear. But one must conclude that acid-base variations in field conditions would have been hardly predictable from laboratory data. However, such data constitute an essential basis to understand how the interplay of many environmental factors leads to the natural pattern of acid-base changes. Clearly, present level of knowledge of acid-base balance could make this very integrated and tightly regulated function very useful in assessing the suitability of water quality parameters in aquaculture conditions.

Literature references.

Conti E, Toulmond A (1986) Ventilatory response to ambient hypo- and hypercapnia in the lugworm, *Arenicola marina* (L.). J Comp Physiol B 156:797-802

Dejours P (1981) Principles of comparative respiratory physiology, 2nd edn. Elsevier/North Holland Biomedical Press, Amsterdam New York

Dejours P, Armand J (1980) Hemolymph acid-base balance of the crayfish *Astacus leptodactylus* as a function of oxygenation and the acid-base balance of the ambient water. Respir Physiol 41:1-11

Dejours P, Beekenkamp H (1977) Crayfish respiration as a function of water oxygenation. Respir Physiol 30:241-251

Jouve-Duhamel A, Truchot JP (1983) Ventilation in the shore crab *Carcinus maenas* (L.) as a function of ambient oxygen and carbon dioxide : field and laboratory studies. J Exp Mar Biol Ecol 70:281-296

Lallier F, Boitel F, Truchot JP (1987) The effect of ambient oxygen and temperature on haemolymph L-lactate and urate concentrations in the shore crab *Carcinus maenas*. Comp Biochem Physiol 86A:255-260

Massabuau JC, Dejours P, Sakakibara Y (1984) Ventilatory CO_2 drive in the crayfish : influence of oxygen consumption level and water oxygenation. J Comp Physiol B 154:65-72

Massabuau JC, Burtin B (1985) Ventilatory CO_2 reflex response in hypoxic crayfish *Astacus leptodactylus* acclimated to 20 °C. J Comp Physiol B 156:115-118

McDonald DG (1983) The effect of H^+ upon the gills of freshwater fish. Canad J Zool 61:691-703

McMahon BR, Butler PJ, Taylor EW (1978) Acid-base changes during recovery from disturbance and during long term hypoxic exposure in the lobster *Homarus vulgaris*. J Exp Zool 205:361-370

Morris S, Taylor AC (1983) Diurnal and seasonal variations in physico-chemical conditions within intertidal rockpools. Estuarine Coastal Shelf Sci 17:339-355

Rahim S, Delaunoy JP, Laurent P (1988) Identification and immunocytochemical localization of two different carbonic anhydrase isoenzymes in teleostean fish eythrocytes and gill epithelia. Histochemistry 89:451-459

Randall D, Lin H, Wright PA (1991) Gill water flow and the chemistry of the boundary layer. Physiol Zool 64:26-38

Reeves RB (1977) The interaction of body temperature and acid-base balance in ectothermic vertebrates. Ann Rev Physiol 39:559-586

Toulmond A (1977) Temperature-induced variations of blood acid-base status in the lugworm, *Arenicola marina* (L.) : II. In vivo study. Respir Physiol 31:151-160

Toulmond A, Tchernigovtzeff C (1989) Blood acid-base balance in the lugworm *Arenicola marina* ventilating in hypo- or hyperoxic sea water. J Exp Biol 142:143-153

Truchot JP (1973) Temperature and acid-base regulation in the shore crab *Carcinus maenas* (L.). Respir Physiol 17:11-20

Truchot JP (1975a) Changements de l'état acide-base du sang en fonction de l'oxygenation de l'eau chez le crabe *Carcinus maenas* (L.). J Physiol Paris 70:583-592

Truchot JP (1975b) Action de l'hypercapnie sur l'etat acide-base du sang chez le crabe *Carcinus maenas* (L.)(Crustacé Décapode). C R Acad Sci Paris 280:311-314

Truchot JP (1979) Mechanisms of the compensation of blood respiratory acid-base disturbances in the shore crab *Carcinus maenas* (L.). J Exp Zool 210:407-416

Truchot JP (1981) The effect of water salinity and acid-base state on the blood acid-base balance in the euryhaline crab, *Carcinus maenas* (L.). Comp Biochem Physiol 68A:555-561

Truchot JP (1984) Water carbonate alkalinity as a determinant of hemolymph acid-base balance in the shore crab, *Carcinus maenas* : a study at two different ambient Pco_2 and Po_2 levels. J Comp Physiol B 154:601-606

Truchot JP (1986) Changes in hemolymph acid-base state of the shore crab, *Carcinus maenas*, exposed to simulated tidepool conditions. Biol Bull 170:506-518

Truchot JP (1987) Comparative aspects of extracellular acid-base balance. Springer Verlag, Heidelberg Berlin

Truchot JP, Duhamel-Jouve A (1980) Oxygen and carbon dioxide in the marine intertidal environment : diurnal and tidal changes in rockpools. Respir Physiol 31:241-254

Wood CM, Wheatly MG, Höbe H (1984) The mechanisms of acid-base and ionoregulation in freshwater rainbow trout during environmental hyperoxia and subsequent normoxia. III. Branchial exchanges. Respir Physiol 55:175-192

Wood CM (1991) Branchial ion and acid-base transfer in freshwater teleost fish : environmental hyperoxia as a probe. Physiol Zool 64:68-102

2

Respiratory and Ionic Regulation in Fish with Changes of the Environment

N. Heisler

Utilization of water as gas exchange medium and the corresponding close contact of the body fluids with the aqueous environment via the large gill surface area presents for fish a number of limitations, but also regulatory advantages as compared to terrestrial animals. The composition of water varies to a much larger extent than that of air, due to a number of natural or contamination-related factors. The large gill surface area in combination with the typical counter current water/blood flow orientation provides a very efficient interface for respiratory gas exchange between environment and body fluids. This large interface, however, also facilitates immediate transmission of changes in environmental partial pressures of respiratory gases (O_2 and CO_2) to the body fluids, and may accordingly severely affect tissue oxygen supply, acid-base and ionic regulation, as well as respiratory control. Various, interrelated regulatory mechanisms are activated in order to cope with such challenges.

Ventilation and Pco_2

A most adverse factor for fish homoiostatic control is the much lower oxygen content of water as compared to air (~ 0.03, depending on temperature). Although the large gill surface area generally facilitates uptake of oxygen according to the metabolic demand, this is achieved only by employing a large specific ventilation (volume per unit oxygen consumption) of the viscous breathing medium water. The capacitance of pure water for CO_2 is similar to that of air, and the effective capacitance in the range of water Pco_2 of less than 1 mmHg is even larger, due to carbonate- and other non-bicarbonate buffering of CO_2. Accordingly, the high relative rate of gill ventilation (about 5 ml/µmol O_2 consumption as compared to 0.6 ml pulmonary ventilation/µmol O_2 in man) results in much lower CO_2 partial pressure differences between arterial blood and water in fish (typically 1-4 mmHg) than in terrestrial vertebrates (typically 30-45 mmHg).

Such small environmental-arterial Pco_2 differences (ΔPco_2) provide narrow limitations for compensatory modulation of Pco_2 by hyper- or hypoventilation, a mechanism frequently exploited in mammals (cf. Woodbury, 1965). According to the physical properties of water the scope for a further reduction of ΔPco_2 is generally small (Rahn, 1966). ΔPco_2 in fish is not only related to the rate of gill ventilation, but is also a function of other factors like blood or water shunting past the gas exchange surface (cf. Heisler, 1989), or incomplete CO_2 hydration and disequilibrium as a result of insufficient carbonic anhydrase activity on the mucosal side of the gill epithelium (e.g. Henry *et al.*, 1988; cf. Heisler, 1990). Hyperventilation will accordingly only reduce arterial Pco_2 significantly in fish with relatively large ΔPco_2, like carp (3-4 mmHg, cf. Claiborne and Heisler, 1984, 1986). This species is capable of reducing arterial Pco_2 during hypoxia by a factor of two, but at the same time the rate of ventilation is elevated by a factor of ~ 3 (cf. Glass *et al.*, 1990). Thus at least a fraction of ~ 1 mmHg of ΔPco_2 is related to factors like incomplete CO_2 hydration and/or blood or water shunting (cf. Heisler, 1989). In species with lower ΔPco_2 like rainbow trout (~ 1-1.5 mmHg) the effect of even excessive hyperventilation on arterial Pco_2 will be much smaller.

Hyperventilation as Regulatory Mechanism During Hypercapnia

The regulatory effect of hyperventilation has to be considered small on the background of environmental hypercapnia fish may encounter. Although surface seawater Pco_2 usually reflects equilibrium with air (0.15 - 0.3 mmHg), Pco_2 at 200-500 m depths may rise by 5-10 mmHg due to anaerobic metabolism of microorganisms. Natural freshwater Pco_2 may range from extremely low, photosynthesis-related values (less than atmospheric, < ~ 0.26 mmHg) to as high as 60 mmHg (cf. Heisler *et al.*, 1982) due to inhibition of surface gas exchange and CO_2 release from bicarbonate by anaerobic microbial metabolism. Changes in ΔPco_2 brought about by adjustment of gill ventilation will therefore be too small to effectively ameliorate the impact of hypercapnia. A rise of inspired Pco_2 by 10 mmHg will result in fish in a rise of arterial Pco_2 by a factor of 3-6 with deflections of arterial pH by ~ 0.4-0.5 pH units (cf. Heisler, 1986a), whereas the acid-base status will hardly be affected in mammals: a slight hyperventilation (+ 30%) will reduce the inspired/alveolar Pco_2 difference from 40 to 30 mmHg. Hyperventilation is therefore hardly exploited in fish during environmental hypercapnia: the response is generally small and transient without being related to the actual time course of pH normalization (e.g. Randall *et al.*, 1976; cf. Dejours, 1975, 1981).

Ventilation and Pco_2 as a Function of Oxygen Availability

The primarily oxygen-oriented regulation of ventilation in fish (cf. Dejours, 1975, 1981) is characterized by adjustment in order to provide sufficient amounts of oxygen for metabolism, whereas in terrestrial animals regulation of ventilation aims for constant arterial Pco_2. On the background of sparse availability of oxygen in the water this type of regulation is certainly advantageous for tissue oxygen supply, but carries the inherent disadvantage of endogenous respiratory acid-base disturbances. Even moderate hypoxia results in considerably elevated gill ventilation supporting the animal's oxygen supply by reducing the inspired-arterial oxygen partial pressure difference (ΔPo_2), but simultaneously effects considerable changes in arterial Pco_2 and pH (e.g. Glass *et al.*, 1990). In turn, environmental hyperoxia results in considerably reduced ventilation with simultaneously increased ΔPo_2 and ΔPco_2 (e.g. Dejours, 1973,: Heisler *et al.*, 1981, 1988; for review: Heisler, 1986a).

Hyperoxia-Induced Hypercapnia: Respiratory and Ionic Regulation

Disturbances of homoiostasis resulting from this tight coupling of ventilation to the availability of oxygen are compensated for by effective ionic regulation: changes in Pco_2 are generally compensated by equivalent changes in plasma bicarbonate. The high degree of interrelationship between respiratory and ionic regulation is evident from the regulatory pattern during environmental hyperoxia, which frequently occurs in natural freshwater habitats. As a result of the reduction in gill ventilation during hyperoxia (cf. Heisler *et al.*, 1988) arterial Pco_2 becomes elevated. The time course of Pco_2 rise, however, is much slower than expected on the basis of the water oxygen content (Fig. 1). This is partially related to conversion of endogenously produced CO_2 to HCO_3^- by non-bicarbonate buffering (cf. Fig. 2), but the main factor is that the reduction of ventilation is less than equivalent to the elevated oxygen availability. In dogfish (*Scyliorhinus stellaris*) Pco_2 rises from about

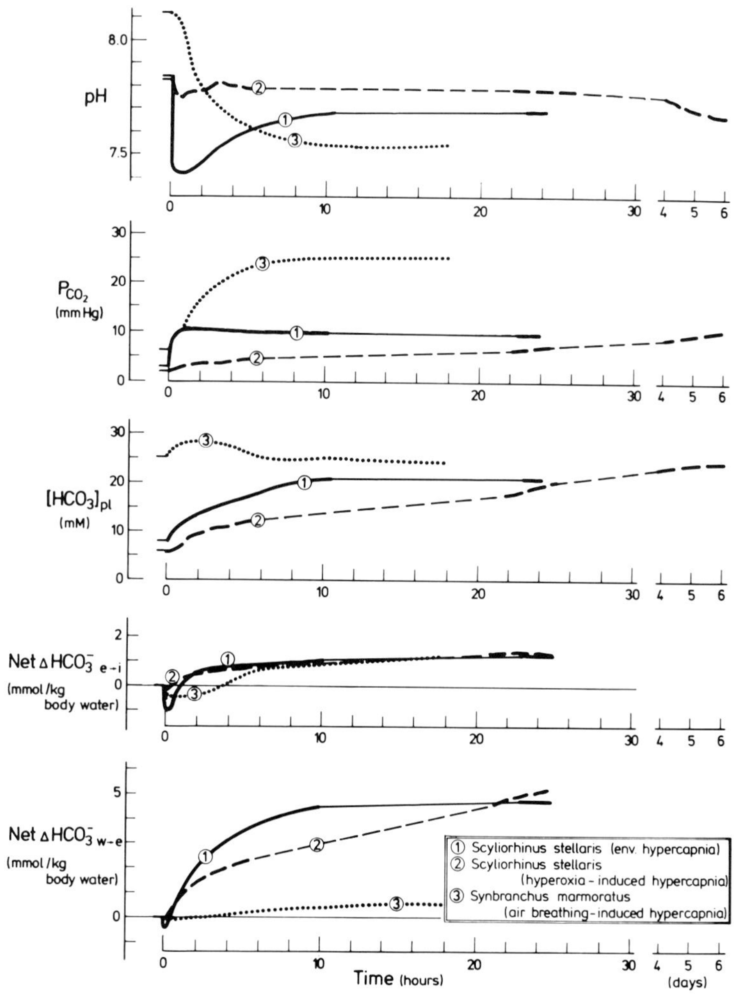

Fig. 1 Time course of pH, Pco_2 and $[HCO_3^-]$, and bicarbonate equivalent transfer between intra- and extracellular space and environmental water during environmental (1), hyperoxia induced (2) and air-breathing induced hypercapnia (3). See also text.

2 to finally 11 mmHg only after 5 days of hyperoxia, and the rise is well matched to the accumulation of plasma bicarbonate: pH is little affected during adaptation to the new environment (initially ~ - 0.08, later -0.03 units; Fig. 1; Heisler *et al.*, 1988). This indicates a switch from primarily oxygen-oriented towards pH-stat regulation (Heisler, 1988) during hyperoxia, with the adjustment of gill ventilation tracking (with some regulatory oscillations) the accumulation of bicarbonate. The disparity between oxygen demand and ventilation during this tracking period is obvious from the rise of arterial Po_2 to more than 200 mmHg during the first day and its subsequent fall to levels of finally below 100 mmHg after 5 days of hyperoxia, close to the normoxic controls. The plasma $[HCO_3^-]$ rises steadily to values of almost 25 mM, concurrent with, and almost completely compensating, the elevation of Pco_2 (Fig. 1; Heisler *et al.*, 1988). Control of ventilation is accordingly very closely related to plasma pH during this transient period of rising Pco_2. Only when the reduction in ventilation reapproaches levels just sufficient to satisfy the metabolic oxygen demand, ventilatory control is returned to the oxygen-related drive. Then deviations of pH from the controls become larger than during the initial phases of hyperoxia-induced hypercapnia (Heisler *et al.*, 1988).

Hyperoxia-Induced vs. *Environmental Hypercapnia*

The characteristics of ionic regulation during hyperoxia-induced hypercapnia are similar to the pattern observed during environmental hypercapnia of comparable degree. Differences are related to the time course of Pco_2 elevation. Hyperoxia-induced hypercapnia slowly develops in the course of several days, whereas environmental hypercapnia is almost immediately transmitted to the arterial blood by the large gill surface area (Fig. 1). The abrupt elevation of plasma Pco_2 results in a general fall in plasma and intracellular pH values. With some exceptions, pH starts to recover towards control values soon after initiation of hypercapnia by elevation of the bicarbonate concentration, partially or even completely compensating the elevation in Pco_2. During the initial phase with large deflections in pH, the rise in extracellular bicarbonate is partially due to non-bicarbonate buffering of CO_2, particularly in well buffered intracellular muscle compartments (Fig. 2). With increasing compensation and return of pH towards control values the fraction of bicarbonate supplied by this mechanism falls, approaching zero when pH compensation is complete. The response to environmental hypercapnia is accordingly characterized by large deflections in pH and the associated much larger role of non-bicarbonate buffering, due to the limited extent Pco_2 can be modulated by changes in gill ventilation (Figs. 1 and 2).

Bicarbonate-Equivalent Transfer During Hypercapnia

Bicarbonate accumulated during compensation of hypercapnia is supplied by acid-base relevant ion transfer mechanisms. After a new steady state has been attained during environmental hypercapnia of 1% CO_2, fish have gained and accumulated about 3 to 6 mmol/kg body weight of bicarbonate-equivalent ions from the environment (e.g. Heisler, Weitz and Weitz, 1976; Claiborne and Heisler, 1984, 1986; cf. Heisler 1986a), an amount much larger than the bicarbonate pool of the animals during normocapnia. Although about the same amount is finally accumulated in hyperoxia-induced and environmental hypercapnia of comparable degree (*Scyliorhinus stellaris*,

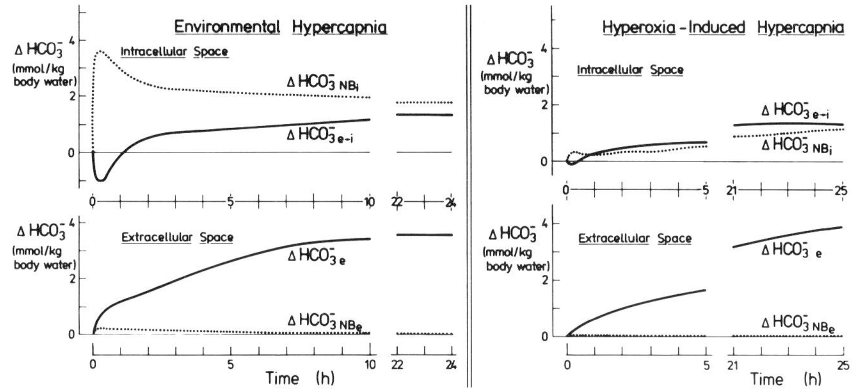

Fig. 2 Changes in extracellular bicarbonate ('e'), bicarbonate-equivalent transfer between extracellular and intracellular space ('e->i'), and contribution of nonbicarbonate buffering ('NB_i','NB_e') during environmental and hyperoxia-induced hypercapnia. See also text.

Heisler *et al.*, 1976, 1988) the rate of transfer is considerably different (Fig. 1). A high transepithelial transfer rate, close to the maximal capacity of 15 μeq/(min · kg) of bicarbonate equivalent resorption for this species (Heisler *et al.*, 1976, 1977; Holeton and Heisler, 1983; cf. Heisler, 1988) is attained only during the first hours of hyperoxia, but is reduced to ~ 5 μmoles/(min · kg) with plasma pH recovering to less than - 0.03 after about 3-5 h, and falls to about 2 μmoles/(min kg) after 21-25 h of hyperoxia. Accordingly it takes much longer to accumulate the same quantity of bicarbonate during hyperoxia-induced than during environmental hypercapnia. During this time ventilation is much higher than required on the basis of the oxygen demand, although the capacity of branchial ion transfer mechanisms would allow much faster reduction of ventilation to the final low rate, even if pH was kept completely compensated. The reason for this delayed regulation is unknown, but may be related to the involved energetics (i.e. the additional energetic cost for ventilation would be less than that of rapid accumulation of bicarbonate), to the kinetics of readjustment of the respiratory center, or to differential sensitivity of sensors for respiratory adjustment and ionic transfer.

The transepithelial ion transfer processes are accompanied by transmembrane movements of bicarbonate-equivalent ions. The initial elevation of extracellular bicarbonate after onset of environmental hypercapnia is mainly the result of a shift from the well-buffered intracellular space (Fig. 1; cf. Heisler *et al.*, 1976; Toews *et al.*, 1983), before epithelial mechanisms are activated. During this phase bicarbonate-equivalents are also released to the environment, a maladaptive response expressing still incomplete adjustment of ion retaining and resorbing structures to the elevated plasma bicarbonate levels during hypercapnia. The initial transfer of bicarbonate-equivalents from intracellular to extracellular space may represent an analogous deficiency of adjustment at the cellular level. This 'overflow' phenomenon is characteristic for environmental hypercapnia, whereas during hyperoxia-induced hypercapnia the corresponding fluxes are much smaller (Fig. 1). With full activation of the transepithelial transfer bicarbonate originally supplied by the intracellular compartments is returned, and bicarbonate gained from the environment is accumulated in both extracellular and intracellular fluid compartments. In spite of its small volume (0.2-0.25 of the body water; e.g. Heisler, 1978, 1982a) the extracellular space receives a much larger fraction than the intracellular space (0.8-0.75 of body water) (cf. Fig. 1). Nevertheless, the small fraction of bicarbonate transferred to the large volume of intracellular space results in almost complete pH compensation and even closer restoration of pH to normocapnic controls than in the extracellular fluid (cf. Heisler, 1984, 1986a). This is due to extracellular control bicarbonate levels being generally larger than intracellular bicarbonate by factors of 4 - 8 and the supportive effect of high intracellular non-bicarbonate buffering (cf. Heisler, 1984, 1986a, 1986b).

Air Breathing-Induced Hypercapnia

The switch of facultative air breathing fish from water breathing to air breathing is directly comparable to the exposure of exclusive water breathers to hyperoxia: in both cases the concentration of oxygen in the breathing medium is largely enhanced. In the Amazonian *Synbranchus marmoratus* the rate of 'gill' ventilation is reduced considerably upon air breathing, with Pco_2 rising from about 6 to 25 mmHg within 10 to 72 h (Fig. 1; Heisler, 1982a). This rise in Pco_2 effects plasma pH to fall from 8.15 to 7.5. In contrast to water breathing species pH remains deflected, with no sign of compensation. After an initial rise in plasma bicarbonate due to blood nonbicarbonate buffering, bicarbonate levels off at essentially the same value as during water breathing. Also, *Synbranchus* does not gain significant amounts of bicarbonate from the environment during hypercapnia (Fig. 1; Heisler, 1982a).

The complete lack of extracellular pH compensation may be related to a number of factors. The observed five-fold rise in Pco_2 would imply an equivalent rise in bicarbonate in order to achieve complete compensation (cf. Heisler, 1986b). On the basis of the high bicarbonate level during water breathing (24 mM), however, an equimolar reduction in extracellular $[Cl^-]$ in order to maintain electroneutrality will be impossible; even a fraction of this could hardly be tolerated (Heisler, 1982a). Also, the animals may not be able to take up sufficient amounts of bicarbonate anyway. Filling the buccal cavity with air for gas exchange prevents irrigation with water, reducing the contact time of ion transporting epithelia with water to short flush periods every 5 to 35 min (Heisler, 1982a). Additionally, the environmental conditions for these fish are quite unfavorable with respect to ion exchange with the environment (Heisler, 1982a; see below).

In contrast to extracellular pH, the intracellular pH of white and heart muscle of *Synbranchus* are almost completely compensated (Heisler, 1982a). This is achieved by transfer of bicarbonate mainly produced by blood nonbicarbonate buffering to the intracellular space. With only a limited amount of additional bicarbonate available, regulation of intracellular homoiostasis is evidently of higher priority. This type of adjustment is very efficient: the small amount of bicarbonate produced by extracellular buffering cannot effect any significant compensation on the background of very high levels of bicarbonate (24 mM) in the extracellular space, but is sufficient to serve for almost complete intracellular pH compensation on the basis of much lower intracellular bicarbonate levels (Heisler, 1982a).

Tissue Anoxia, CO_2 Production and Gill Ventilation

Incomplete or lacking supply of oxygen for tissue metabolism may frequently occur in fish, related to environmental hypoxia or anoxia, or to a discrepancy between tissue demand and supply by the oxygen transport chain, for instance during high levels of muscular activity. This results in anaerobic energy production, with lactic acid as the main metabolic intermediary product in vertebrates (only a few species such as *Carassius auratus* and *C. carassius* have adapted to long-term anoxia by switching to the acid-base neutral intermediary product ethanol; cf. Shoubridge and Hochachka, 1980; van den Thillart *et al.*, 1983). According to its low pK' value (~ 3.9) lactic acid dissociates H^+ ions which are buffered by nonbicarbonate buffers, as well as combine with bicarbonate. Although environmental hypoxia as one of the most common factors responsible for tissue anoxia results *per se* in largely elevated gill ventilation and excess elimination of CO_2, any reduction of Pco_2 for pH compensation is hardly possible. The CO_2 produced during this stage largely exceeds the amount that can be eliminated. In *Scyliorhinus stellaris*, for instance, the amount of bicarbonate combined with H^+ ions during a short period of anaerobic muscular activity is equivalent to the amount produced during 3.5 h of resting aerobic metabolism (cf. Holeton and Heisler, 1982; Randall *et al.*, 1976). Arterial Pco_2 is enhanced from about 2 to more than 5 mmHg for 3 h (Fig. 3), and remains elevated above controls for more than 10 h (Holeton and Heisler, 1983). This can certainly not be attributed to CO_2 produced from combination of H^+ and bicarbonate, but during recovery from lactacidosis the rate of aerobic CO_2 production is elevated due to the payment of the oxygen dept by reestablishment of ATP and creatine-phosphate stores, as well as aerobic processing of lactic acid. Also factors like incomplete CO_2 hydration in the gill effluent water, or larger than normal heterogeneities between blood and water flow in the gill epithelium may result in elevated ΔPco_2 between arterial blood and expired water during such periods of enhanced CO_2 elimination. The acid-base regulation is accordingly challenged by a mixed metabolic and respiratory acidosis and modulation of Pco_2 does not contribute to restoration of pH.

Ion Transfer Processes During Lactacidosis

Although lactic acid produced during environmentally-induced tissue hypoxia (water hypoxia, nitrite exposure) is more homogeneously distributed throughout the organism than during functional hypoxia (muscular activity), the general regulatory patterns are similar. After production and dissociation of lactic acid during extensive muscular activity the dissociation products H^+ and lactate$^-$ are gradually eliminated from the intracellular space (Fig. 3; Holeton and Heisler, 1983; cf. Heisler, 1984, 1986a). The time course of elimination is different for the two dissociation products and variable among fish species. Generally, the extracellular acid-base deflections peak much earlier than extracellular lactate. In turn, the acid-base variables are restored long before plasma lactate, which is achieved by transient net transfer of H^+ ions to the ambient water. In contrast, lactate is never transferred to the environment (cf. Holeton and Heisler, 1983; Holeton, Neumann and Heisler, 1983).

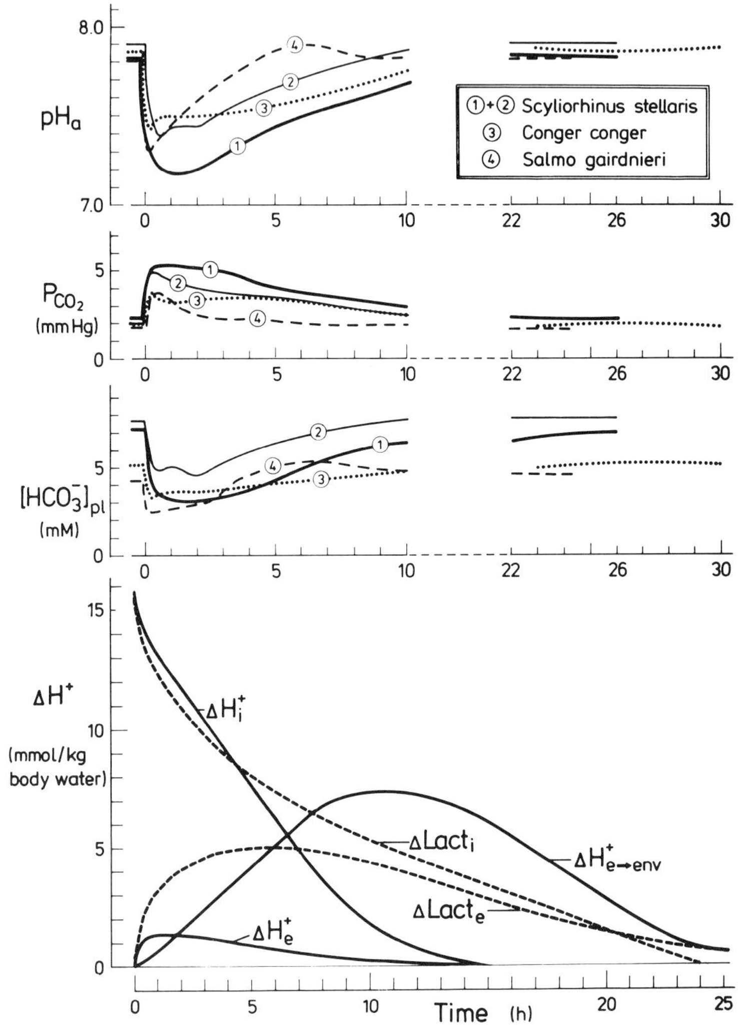

Fig. 3 Arterial plasma pH, P_{CO_2} and $[HCO_3^-]$, the amount of surplus H^+ and lactate in intra- and extracellular fluid compartments ('i', 'e') and the environmental water (e->sw) during lactacidosis in *Scyliorhinus stellaris*. See also text.

After an initial efflux period intracellular muscle pH in *Scyliorhinus stellaris* is generally less affected than extracellular pH. H^+ ions are eliminated from the cells at a tremendous rate (> 250 μmol/(min kg)), which is soon retarded due to the establishment of a transient equilibrium between intracellular and extracellular pH (Fig. 3). H^+ ions are further released from the intracellular space only at the rate of further transfer to the environment ('equilibrium limitation', Holeton and Heisler, 1983). Less than 8 % of surplus H^+ ions are buffered in blood and interstitial space.

Net H^+ transfer to the environment takes place at a rate of about 15 μmoles/(min · kg body water), similar to the maximal rate during hypercapnia. After normalization of the extracellular acid-base status H^+ ions required for aerobic processing of lactate are returned from the environment at the rate of lactic acid metabolization. This mechanism of transient storage of surplus H^+ ions in the ambient water is generally applied by all studied species (e.g. *Conger conger*, Toews *et al.*, 1983; *Salmo gairdneri*, Holeton *et al.*, 1983).

Epithelial Ion Transfer Mechanisms

As a result of the low oxygen capacitance of water the adjustment of fish gill ventilation is primarily limited to provide oxygen at a rate sufficient for aerobic metabolism, rather than to support acid-base regulation by modulating Pco_2. Accordingly, ionic regulation is much more important than in terrestrial animals, and except for very few examples such as air breathing fish, transepithelial elimination of H^+ equivalents is the leading mechanism for organismic acid-base regulation.

The epithelial ion transfer processes are generally considered to be best described as active ion exchange processes of H^+ or NH_4^+ against Na^+, and of HCO_3^- against Cl^- (cf. Evans, 1986; Heisler 1986a), although an electrogenic H^+ pump in the apical membrane of branchial chloride cells has recently also been postulated (Lin and Randall, 1991). An electrogenic H^+ pump would be linked only indirectly to movement of Na^+ or other cations (Avella and Bornancin, 1989; Lin and Randall, 1991). There is evidence for all of these mechanisms to operate under certain conditions (cf. Evans, 1986; Lin and Randall, 1991). Under physiological conditions, however, evidently not all of them are involved in the regulatory process. Recent determination of unidirectional Na^+ and Cl^- tracer ion fluxes in *Cyprinus carpio* indicated Na^+-related mechanisms to play hardly any role in transfer of acid-base relevant ions for compensation of hypercapnic acidoses (cf. Heisler, 1990), whereas the unidirectional Cl^- flux was directly related. Similar flux measurements in trout confirmed the notion that acid-base regulation under physiological conditions is to a large fraction performed by Cl^--related mechanisms (Ishimatsu *et al.*, 1991; Iwama *et al.*, 1992). The lack of relationship with Na^+ movements is in line with the proposed electrogenic proton pump (Lin and Randall, 1991); the close correlation with Cl^- together with the lack of appreciable other cation movements (required for electroneutrality), however, suggests that the main fraction of transfer is performed by electroneutral HCO_3^-/Cl^- ion exchange.

Site of ionic transfer

The ion transfer mechanisms are mainly located in the gill epithelium (for review: Heisler, 1984, 1986a). The role of the kidneys, determined by separating the urine flow from the environmental water, is small, and, at least in one species of elasmobranchs, the skin does not contribute significantly (cf. Heisler, 1984, 1986a). Recent experiments in trout, however, have indicated that the bicarbonate concentration difference between the drainage from the secondary circulatory system of the skin and the dorsal aortic plasma increased significantly during environmental hypercapnia (Ishimatsu *et al.*, 1992). Unfortunately, the quantitative contribution of this system to hypercapnic compensation cannot be estimated due to lack of data about the flow velocity in this vessel system.

Within the gills, the site of ionic transfer mechanisms is still not quite clear. On the basis of circumstantial evidence the mitochondria-rich cells (chloride cells) are generally claimed to be responsible. Recently, cannulation of the branchial vein allowed an analysis of fluid from the gill central venous sinus (CVS), draining the basal compartments of the secondary lamellae as the main location of the mitochondria-rich cells. Simultaneous determination of unidirectional Na^+ and Cl^- fluxes into systemic (apical) and CVS (basal) gill compartments indicated, that the (larger) Cl^--related response to environmental hypercapnia was primarily in the CVS compartment, whereas the (smaller) Na^+-related response was mainly in the apical (systemic) compartment (Iwama, Ishimatsu and Heisler, 1992). On the basis of these data the Na^+ fraction of ionic transfer could well be related to an electrogenic H^+ pump in the chloride cells, with simultaneous passive flux of Na^+ for electroneutrality taking place mainly in the large surface area of the apical 'respiratory' cells.

Branchial Ion Transfer Related to Water Ionic Composition

The transepithelial ion transfer mechanisms in fish are actually much more efficient than in terrestrial animals, with overall acid-base relevant transfer rates per unit standard metabolic rate in the range of 0.13-0.47 (μmol ion transfer per μmol O_2 consumption) as compared to 0.02-0.03 in mammals (Heisler, 1988). This superior efficiency is likely related to the large gill surface area and large water flow rate past the ion exchange sites, with correspondingly small inspired-expired ionic differences. Establishment of large ion gradients, like in the mammalian renal tubular system, is not required.

As a result of their exposed location, however, the ionic transfer mechanisms are subject to changes of the environment. This was apparent from comparison of literature data (Heisler, 1982b), indicating that trout in dilute water ($[Na^+] < 0.1$ mM; $[HCO_3^-]$ 0.050 mM) required much longer to achieve about the same level of pH compensation (72 h, Janssen and Randall, 1975) than trout in water of higher ionic concentration ($[Na^+]$ 0.5 mM; $[HCO_3^-]$ 3 mM, 24h, Eddy *et al.*, 1977) and *Conger* in seawater ($[Na^+]$ 480 mM; $[HCO_3^-]$ 2.5 mM, 10 h, Toews *et al.*, 1983). Closer analysis in *Scyliorhinus stellaris* indicated that bicarbonate was net gained from the environment at a constant rate in a range of $[HCO_3^-]_{pl}/[HCO_3^-]_{sw}$ = 0.3-4, but the rate fell in an apparently linear fashion at higher $[HCO_3^-]_{pl}/[HCO_3^-]_{sw}$ ratios, attaining zero at $[HCO_3^-]_{pl}/[HCO_3^-]_{sw}$= 13 (Fig. 4). At even higher ratios bicarbonate was lost to the water. Since all other ions remained constant in the environmental water these data suggest water $[HCO_3^-]$ as a limiting factor in this species (Heisler and Neumann, 1977; Heisler, 1988). Also in trout at different environmental salinities and in carp at different environmental bicarbonate concentrations environmental hypercapnia was faster compensated at higher salinities (Iwama and Heisler, 1991) or bicarbonate concentrations (N.A. Andersen and N. Heisler, unpublished data), respectively. After lactic acid infusion (Tang and Boutilier, 1988) or exhausting exercise (Tang *et al.*, 1989) genetically identical trout populations achieved much larger branchial bicarbonate-equivalent flux rates in seawater than in freshwater. The availability of

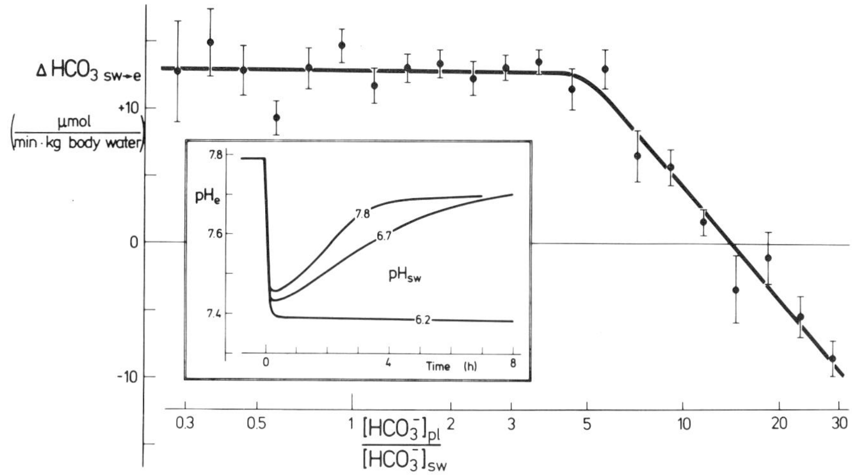

Fig. 4 Transepithelial bicarbonate-equivalent transfer rate in *Scyliorhinus stellaris* as a function of the ratio between plasma and seawater bicarbonate ($[HCO_3^-]_{pl}/[HCO_3^-]_{sw}$). Insert: kinetics of pH recovery at different seawater pH. See also text.

counter ions will accordingly limit the rate of branchial ion transfer, and may indirectly be responsible for differential utilization of epithelial ion exchange mechanisms, such as HCO_3^-/Cl^- and/or $H^+(NH_4^+)/Na^+$, in different fish species.

Maximal Plasma Bicarbonate Concentration

The degree of compensation that is achieved during hypercapnia, however, is rather independent of the environmental ionic composition, although time courses are largely different. An interesting phenomenon can be observed in *Scyliorhinus* during hyperoxia-induced hypercapnia. After some initial regulatory oscillations extracellular pH is kept compensated to within 0.03 pH below the normoxic controls, until plasma bicarbonate exceeds 20 mM. When still higher Pco_2 values are attained, the compensation is increasingly incomplete and even after 6 days of hyperoxia plasma bicarbonate does not rise above 25 mM (Heisler *et al.*, 1988). A similar value is attained also during environmental hypercapnia (Heisler *et al.*, 1976), and data on steady-state bicarbonate-equivalent resorption as a function of water bicarbonate concentration (see above) suggest as well that plasma bicarbonate cannot be elevated to levels much above 25 mM at normal seawater bicarbonate concentration.

Reviewing literature data on the compensation of hypercapnia it is apparent that quite a number of lower vertebrates never exceed bicarbonate concentrations of 23 to 33 mM (Heisler, 1988). With such a limit or threshold for plasma bicarbonate the degree of compensation that can be achieved must generally be lower in animals with high control bicarbonate concentration, due to the inability to elevate bicarbonate by the same factor as Pco_2 (Heisler, 1986b). Due to the unusually high arterial Pco_2 of about 4 mmHg, *Cyprinus carpio* maintains a comparatively high control

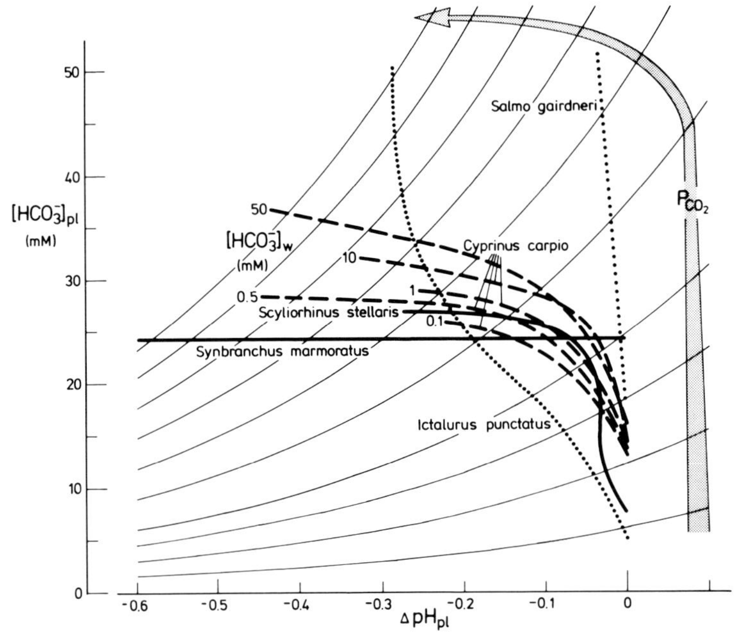

Fig. 5 Bicarbonate concentration during hypercapnia as a function of water Pco_2 and bicarbonate concentration. In most studied fish, pH compensation fails after certain plasma bicarbonate levels are attained (e.g. *Scyliorhinus*, *Cyprinus* and *Synbranchus*), whereas in some experiments an improved capacity for pH compensation and bicarbonate accumulation is observed (*Salmo* and *Ictalurus*). See also text.

bicarbonate concentration of about 13 mM, much higher than other water breathers (trout ~ 4 mM, Holeton *et al.*, 1983; *Conger* ~ 5 mM, Toews *et al.*, 1983; *Scyliorhinus* ~ 7 mM, Heisler *et al.*, 1976). Carp does in fact not raise plasma bicarbonate above 25 mM during exposure to hypercapnia of 1% or 5 % CO_2, in spite of incomplete pH compensation (80 or 45%) even after exposure for up to three weeks (Claiborne and Heisler, 1984, 1986). The lack of any compensation in the tropical air breather *Synbranchus* (Heisler, 1982a) may as well be due to a limiting bicarbonate threshold close to the control bicarbonate concentration of 24 mM, although alternative factors such as the limited exposure time of the ion-transporting epithelia to water during air breathing, or adversely low environmental ion concentrations may be involved (Heisler, 1982a). In at least two species of aquatic animals, *Cyprinus carpio* and the urodele amphibian *Siren lacertina*, bicarbonate infused into the blood stream during environmental hypercapnia is not retained, but quantitatively released to the environment (Claiborne and Heisler, 1986; Heisler *et al.*, 1982). These data as well as many literature reports support the notion of an epithelial threshold for the retention and net resorption of bicarbonate equivalents, although experiments in three fish species have demonstrated that much higher bicarbonate concentrations can be attained during compensation of hypercapnia (Fig. 5; Børjeson, 1976, 1977; Jensen and Weber, 1982; Dimberg, 1988; Cameron and Iwama, 1987).

Bicarbonate equivalent resorption is a function of water bicarbonate concentration in *Scyliorhinus stellaris* (Heisler and Neumann, 1977; Heisler, 1988; see above), and preliminary data indicate a similar relationship also for carp (cf. Fig. 5). Availability of counter ions for transfer processes, however, are hardly the only factors in determining the tremendous capability to compensate hypercapnia with plasma bicarbonate concentrations of up to 50 mM, as in the Uppsala trout (Børjeson, 1976, 1977; Dimberg, 1988), Odense tench (Jensen and Weber, 1982) and Port Aransas channel catfish (*Ictalurus punctatus*; Cameron and Iwama, 1987). Long-term acclimation may seem to be an obvious factor in the Uppsala trout, which were adapted to elevated levels of environmental Pco_2 for several months before the experiments, but plasma bicarbonate levels of 50 mM were attained in catfish already after 5 days of hypercapnia (Cameron and Iwama, 1987). A factor more likely to contribute is the environmental $[Ca^{2+}]$. All experiments, in which high plasma bicarbonate levels were achieved, have been conducted at environmental $[Ca^{2+}]$ close to or higher than 2 mM and relatively high levels of bicarbonate in the range of 3-4 mM. The capacity to achieve high levels of plasma bicarbonate during hypercapnia may accordingly be based on a combination of the advancing effect of high environmental bicarbonate with the sealing effect of Ca^{2+} on epithelial paracellular pathways, and an associated reduction of transepithelial bicarbonate leakage. Even direct effects of Ca^{2+} and/or other environmental ions on translocating structures cannot be excluded.

Conclusion

Respiratory regulation in fish is aimed primarily at satisfying the metabolic oxygen demand. The close link between environmental oxygen content and rate of ventilation prohibits any supportive readjustment of ventilation during endogenous metabolic and exogenous respiratory disturbances, but may in fact severely challenge homoiostatic regulation by hyperoxia- and hypoxia-induced changes of arterial Pco_2. Respiratory acid-base disturbances are compensated for by transfer of acid-base relevant ions among intracellular and extracellular fluid compartments, and the environmental water. These transfer processes, which are much more efficient than in terrestrial animals, are sensitive to the composition of the environmental water, in particular to the bicarbonate concentration. The success of compensatory adaptation through transepithelial acid-base relevant ion transfer is accordingly closely linked to environmental salinity and specific electrolyte composition.

References

Avella, M. and Bornancin, M. (1989). A new analysis of ammonia and sodium transport through the gills of the freshwater rainbow trout (*Salmo gairdneri*). J. exp. Biol. 142, 155-175

Børjeson, H. (1976). Some effects of high carbon dioxide tension on juvenile salmon, (*Salmo salar* L.). Acta Universitatis Upsaliensis 383, 3.35

Børjeson, H. (1977). Effects of hypercapnia on the buffer capacity and haematological values in *Salmo salar* (L.). J. Fish Biol. 11, 133-142

Claiborne, J.B. and Heisler, N. (1984). Acid-base regulation in the carp (*Cyprinus carpio*) during and after exposure to environmental hypercapnia. J. Exp. Biol. 108, 25-43.

Claiborne, J.B. and Heisler, N. (1986). Acid-base regulation and ion transfers in the carp (*Cyprinus carpio*): pH compensation during graded long- and short-term environmental hypercapnia and the effect of bicarbonate infusion. J. Exp. Biol. 126, 41-61

Cameron, J.N. and Iwama; G.K. (1987). Compensation of progressive hypercapnia in channel catfish and blue crabs. J. exp. Biol. 133, 183-197

Dejours, P. (1973). Problems of control of breathing in fishes. In: Comparative Physiology, eds. L. Bolis, K. Schmidt-Nielsen, S.H.P. Maddrell, pp. 117-133, North Holland Publ. Amsterdam.

Dejours, P. (1975, 1981). Principles of Comparative Respiratory Physiology. North-Holland Publ., Amsterdam.

Dimberg, K. (1988). High blood CO_2 levels in rainbow trout exposed to hypercapnia in bicarbonate-rich hard fresh water - A methodological verification. J. exp. Biol. 134, 463-466.

Eddy, F.B., Lomholt, J.P., Weber, R.E. and Johansen, K. (1977). Blood respiratory properties of rainbow trout (*Salmo gairdneri*) kept in water of high CO_2 tension. J. Exp. Biol. 67, 37-47.

Evans, D.H. (1986). The role of branchial and dermal epithelia in acid-base regulation in aquatic animals. In: Acid-Base Regulation in Animals (N. Heisler, ed.), pp. 139-172, Elsevier Biomedical Press, Amsterdam

Glass, M.L., Andersen, N.A., Kruhøffer, M., Williams, E.M. and Heisler, N. (1990). Combined effects of environmental Po_2 and temperature on ventilation and blood gases in the carp *Cyprinus carpio* L. J. Exp. Biol. 148, 1-17

Heisler, N. (1978). Bicarbonate exchange between body compartments after changes of temperature in the Larger Spotted Dogfish (*Scyliorhinus stellaris*). Respir. Physiol. 33, 145-160

Heisler, N. (1982a). Intracellular and extracellular acid-base regulation in the tropical freshwater teleost fish *Synbranchus marmoratus* in response to the transition from water breathing to air breathing. J. Exp. Biol. 99, 9-28.

Heisler, N. (1982b). Transepithelial ion transfer processes as mechanisms for fish acid-base regulation in hypercapnia and lactacidosis. Can. J. Zool. 60, 1108-1122

Heisler (1984). Acid-base regulation in fishes. In: Fish Physiology, Vol.X A, pp.315-401, W.S.Hoar and D.J.Randall, eds. Academic Press, New York and London

Heisler, N. (1986a). Acid-base regulation in fishes. In: Acid-Base Regulation in Animals; pp. 309-356; N. Heisler, ed.; Elsevier Science Publishers B.V., Amsterdam

Heisler, N. (1986b). Comparative aspects of acid-base regulation. In: Acid-Base Regulation in Animals; pp. 397-450; N. Heisler, ed.; Elsevier Science Publishers B.V., Amsterdam

Heisler, N. (1988). Acid-Base Regulation in Elasmobranch Fishes. In: Physiology of Elasmobranch Fishes; T.J. Shuttleworth ed., pp. 215-252; Springer, Heidelberg

Heisler, N. (1989). Acid-base regulation in fishes I. Mechanisms. In: Acid toxicity and Aquatic Animals; R. Morris, E.W. Taylor, D.J.A. Brown and J.A. Brown, eds., pp. 85-97; Society of Experimental Biology Seminar Series, Cambridge University Press

Heisler, N. (1990). Acid-base regulation: Interrelationships between gaseous and ionic exchange. In: Vertebrate Gas Exchange from Environment to Cell (R.G. Boutilier, ed.), Series: Advances in Environmental and Comparative Physiology Vol. 6 (R. Gilles, managing editor), pp. 211-251. Springer, Heidelberg, Berlin, New York, pp. 211-251.

Heisler, N., Weitz, H. and Weitz, A.M. (1976). Hypercapnia and resultant bicarbonate transfer processes in an elasmobranch fish. Bull. Eur. Physiopathol. Respir. 12, 77-85.

Heisler, N. and Neumann, P. (1977). Influence of sea water pH upon bicarbonate uptake induced by hypercapnia in an elasmobranch (*Scyliorhinus stellaris*). Pflügers Arch. 368, Suppl., R19.

Heisler, N., Holeton, G.F. and Toews, D.P. (1981). Regulation of gill ventilation and acid-base status in hyperoxia-induced hypercapnia in the larger spotted dogfish (*Scyliorhinus stellaris*). Physiologist 24, 58 (305)

Heisler, N., Forcht, G., Ultsch, G.R. and Anderson, J.F. (1982). Acid-base regulation in response to environmental hypercapnia in two aquatic salamanders, *Siren lacertina* and *Amphiuma means*. Respir. Physiol. 49, 141-158

Heisler, N., Toews, D.P., and Holeton, G.F. (1988). Regulation of ventilation and acid-base status in the elasmobranch *Scyliorhinus stellaris* during hyperoxia-induced hypercapnia. Resp. Physiol. 71, 227-246

Henry, R.P., Smatresk, N.J. and Cameron, J.N. (1988). The distribution of branchial carbonic anhydrase and the effects of gill and erythrocyte carbonic anhydrase inhibition in the channel catfish, *Ictalurus punctatus*. J. Exp. Biol. 134, 201-218

Holeton, G.F. and Heisler, N. (1983). Contribution of net ion transfer mechanisms to the acid-base regulation after exhausting activity in the larger spotted dogfish (*Scyliorhinus stellaris*). J. Exp. Biol. 103, 31-46.

Holeton, G.F., Neumann, P. and Heisler, N. (1983). Branchial ion exchange and acid-base regulation after strenuous exercise in rainbow trout (*Salmo gairdneri*). Respir. Physiol. 51, 303-318.

Iwama, G.K. and Heisler, N. (1991). Effect of environmental water salinity on the acid-base regulation during environmental hypercapnia in the rainbow trout (*Salmo gairdneri*). J. Exp. Biol. 158, 1-18

Iwama, G.K., Ishimatsu, A. and Heisler, N. (1992). Blood flow and ion flux partitioning between central venous sinus and systemic gill compartments during exposure to environmental hypercapnia in the rainbow trout (*Salmo gairdneri*). J. exp. Biol. (in press)

Ishimatsu, A., Iwama, G.K., Bentley, T. and Heisler, N. (1991). Role of the secondary circulatory system in acid-base regulation in trout. Proc. IUBS Comp. Physiol. Biochem. 3, 137 (3017).

Ishimatsu, A., Iwama, G.K., Bentley, T. and Heisler, N. (1992). Potential roles of the secondary circulatory system in acid-base regulation in rainbow trout during hypercapnia. J. exp. Biol. (in press)

Janssen, R.G. and Randall, D.J. (1975). The effect of changes in pH and Pco_2 in blood and water on breathing in rainbow trout, *Salmo gairdneri*. Respir. Physiol. 25, 235-245.

Jensen, F.B. and Weber, R.E. (1982). Respiratory properties of tench blood and hemoglobin. Adaptation to hypoxic-hypercapnic water. Molecular Physiology 2, 235-250

Lin, H. and Randall, D.J. (1991). Evidence for the presence of an electrogenic proton pump on the trout gill epithelium. J. exp. Biol. 161, 119-134

Rahn, H. (1966b). Aquatic gas exchange: theory. Respir. Physiol. 1, 1-12.

Randall, D.J., Heisler, N. and Drees, F. (1976). Ventilatory response to hypercapnia in the larger spotted dogfish *Scyliorhinus stellaris*. Am. J. Physiol. 230, 590-594.

Tang, Y., D.G. McDonald & R.G. Boutilier (1989). Acid-base regulation following exhaustive exercise: A comparison between freshwater- and seawater- adapted rainbow trout (*Salmo gairdneri*). J. Exp. Biol. 141, 407-418.

Tang, Y. & R.G. Boutilier (1988). Clearance of lactate and protons following acute lactacidosis: a comparison between seawater- and freshwater-adapted rainbow trout (*Salmo gairdneri*). Exp. Biol. 48, 41-44.

Toews, D.P., Holeton, G.F. and Heisler, N. (1983). Regulation of the acid-base status during environmental hypercapnia in the marine teleost fish *Conger conger*. J. Exp. Biol. 107, 9-20.

Wood, C.M. and Jackson, E.B. (1980). Blood acid-base regulation during environmental hyperoxia in the rainbow trout (*Salmo gairdneri*). Respir. Physiol. 42, 351-372.

Woodbury, J.W. (1965). Regulation of pH. In: Physiology and Biophysics (Ruch, T.C. and Patton, H.D., eds.), pp. 899-934, Saunders, Philadelphia.

3

Effects of Variations in Water pH on Fish

David J. Randall and Hong Lin

INTRODUCTION

Changes in freshwater pH do not have any direct effect on either oxygen uptake (Ye *et al.*, 1991) or carbon dioxide excretion (Lin and Randall, 1990) in resting fish, unless there is a large production of mucus which increases diffusion resistance, reduces gas transfer and generates a hypoxemia (Packer and Dunston, 1972). Reductions in water pH inhibit proton excretion (Lin and Randall, 1991a) and sodium chloride influx (McDonald *et al.*, 1983; Wright and Wood, 1985). Exposure to acid water results in a decrease in blood NaCl levels and eventually a blood acidosis develops (McDonald *et al.*, 1983; Ye *et al.*, 1991). In acid water, low calcium levels exacerbate ion loss across the gills, whereas high calcium levels in water result in an acidosis because impairment of ion regulatory mechanisms are ameliorated (McDonald, 1983).

There is no immediate increase in ventilation in fish exposed to water acidified by the addition of hydrochloric acid (Lin and Randall, 1990). A decrease in water pH due to a rise in carbon dioxide, however, will cause a marked rise in ventilation (Janssen and Randall, 1975). In this case the decrease in water pH is associated with an immediate and marked blood acidosis due to an increase in $Paco_2$. This acidosis will decrease blood oxygen content via the Root effect and be associated with an increase in circulating catecholamine levels, both of which can cause an increase in gill ventilation. There may also be a direct stimulation of ventilation by increased levels of $Paco_2$.

Exposure of fish to alkaline water is less well studied. Short term exposure of trout to alkaline water caused a marked blood alkalosis and a reduction in sodium influx and ammonia excretion (Wright and Wood, 1985).

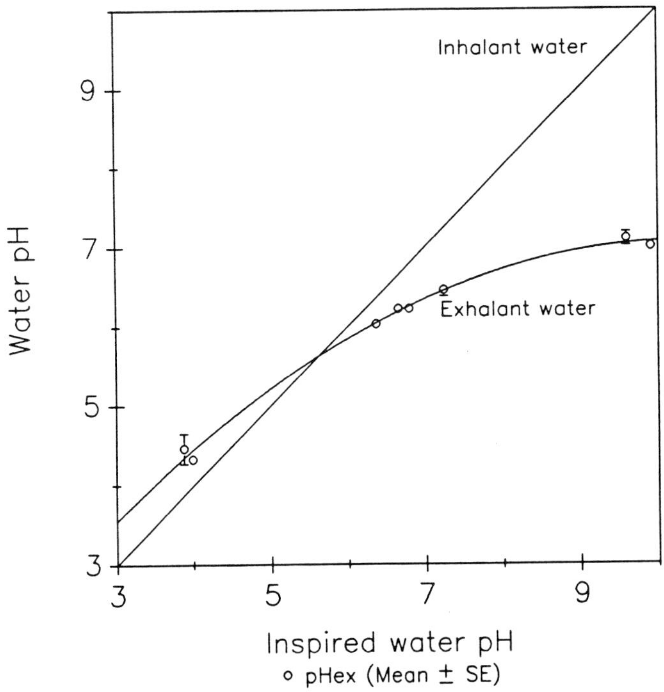

Fig.1. The relationship between pH of exhalant water (o) and inhalant water of rainbow trout (from Lin and Randall, 1991b).

Water passing over the gills of fish is acidified if the pH is above about 5.5 but is alkalinized at lower pH values (Fig.1). The alkalinization at low pH can be accounted for by assuming all ammonia is excreted as NH_3 and then binds a proton raising water pH. The degree of acidification at high pH, however, is too large to be accounted for by CO_2 hydration, even if it is assumed that all carbon dioxide excretion occurs as molecular CO_2. Thus at high pH there must be acid excretion in excess of that due to carbon dioxide hydration. In fact, acidification is not due to carbon dioxide hydration because bicarbonate levels decrease as water passes over the gills (Fig.2), indicating that

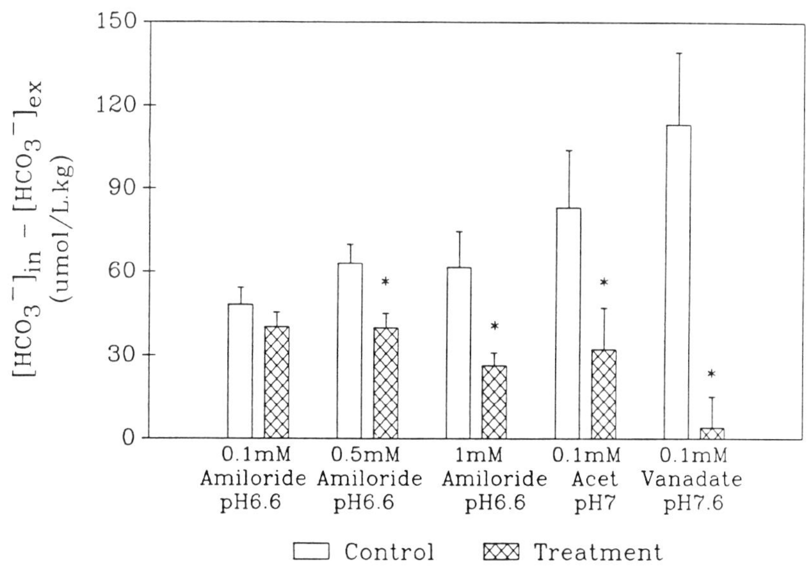

Fig.2. Bicarbonate concentration differences between inspired and expired water of rainbow trout under control, amiloride, acetazolamide and vanadate treatments. * indicates a significant difference between the control and treatment values ($P<0.05$). (from Lin and Randall, 1991a).

proton excretion causes bicarbonate dehydration even though there is simultaneous carbon dioxide excretion. That is carbon dioxide levels in water passing over the gills increase due to both carbon dioxide excretion and bicarbonate dehydration as a result of proton excretion by the fish.

PROTON EXCRETION

It has long been hypothesised that a $Na^+/H^+(NH_4^+)$ antiport occurring in the apical membrane of the fish gill epithelia is the major pathway for sodium uptake and proton excretion and an optional excretory pathway for ammonia. The degree of coupling between sodium absorption and ammonia excretion, however, was loose in goldfish, carp and trout in vivo (Payan, 1978) and trout *in vitro* (Avella and Bornancin, 1989). As a result Avella and Bornancin proposed an indirect coupling of an electrogenic proton pump and a passive sodium entry, instead of a $Na^+/H^+(NH_4^+)$ exchanger, as the mechanism of sodium and proton transport across the fish gill epithelium, similar to that hypothesized for frog skin and turtle bladder (Ehrenfeld *et al*, 1985; Steinmetz & Andersen, 1982).

In vivo studies of proton excretion in freshwater rainbow trout have provided evidence for the existence of an active proton pump (proton ATPase) in the gill epithelium (Lin & Randall, 1991a). Proton excretion rate was determined in trout from measurements of inspired and expired water pH and the buffering capacity of the water, plus ammonia excretion. This estimate of proton excretion is not equivalent to net acid excretion (McDonald and Wood, 1981) because it includes effects of carbon dioxide hydration/dehydration on gill water acidification.

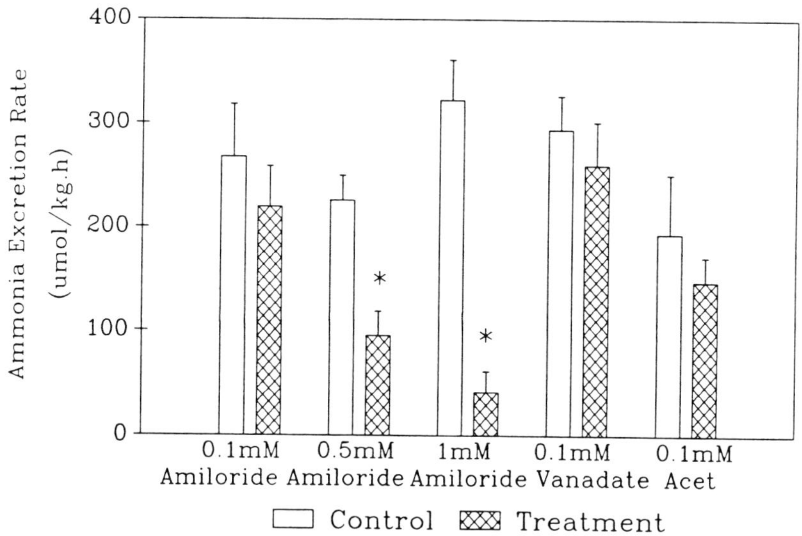

Fig.3. Ammonia excretion rates of rainbow trout under control, amiloride, acetazolamide and vanadate treatments. * indicates a significant difference between control and treatment values ($P<0.05$). (from Lin and Randall, 1991a).

Ammonia excretion across the gills was found to be unaffected by 0.1mM amiloride in external water (Fig.3), a treatment reported to have caused a 94% reduction of gill sodium uptake in the same species under similar experimental conditions (Wright & Wood, 1985). This indicates that sodium uptake and ammonia excretion are not coupled by a $Na^+/H^+(NH_4^+)$ exchanger. Higher concentrations of amiloride reduced ammonia excretion (Fig.3), probably because application of amiloride above 0.1 mM from the mucosal side of the epithelium can interfere with the $Na^+/H^+(NH_4^+)$-ATPase in the basolateral membrane (Knauf *et al*, 1976; Kleyman & Cragoe, 1988), which transfers potassium, and perhaps NH_4^+ from the blood into the epithelial cell. Alternatively, amiloride may have had a direct inhibitory effect on gill epithelial proton ATPase.

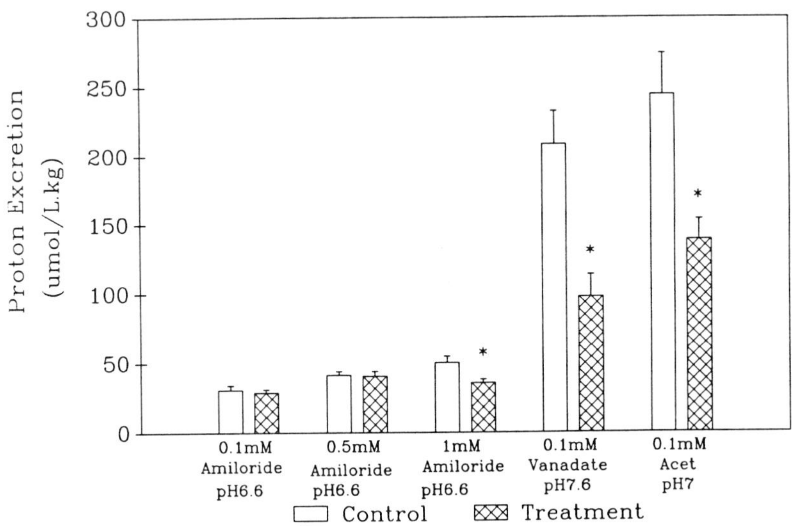

Fig.4. Proton excretion across the gill epithelium of rainbow trout in control, amiloride, acetazolamide and vanadate treatments. * indicates a significant difference between control and treatment values ($P<0.05$).

0.1 mM vanadate added to the water resulted in a reduction of proton excretion across the gills by more the 50%

(Fig.4) whereas ammonia excretion was unaffected (Fig.3). A similar inhibitory effect of vanadate on proton transport mediated by an electrogenic proton pump was demonstrated in frog skin and turtle urinary bladder (Ehrenfeld *et al*, 1985; and Arruda *et al*, 1981). Treatment with 0.1mM acetazolamide, another proton transport inhibitor in frog skin and turtle bladder (Ehrenfeld & Garcia-Romeu, 1977; Steinmetz & Andersen, 1982), also caused a 48% reduction in proton excretion (Fig.4). Proton excretion in trout increased with increasing carbon dioxide levels in a neutral external environment (Lin and Randall, 1991a). Elevated ambient CO_2 levels also enhance proton transport in frog skin and turtle bladder (Ehrenfeld & Garcia-Romeu, 1977; Steinmetz & Andersen, 1982). Given all these similarities between fish gills and frog skin, we suggest that proton excretion across the gills is mediated by a active proton pump in trout.

Although sodium and proton transport is not linked by $Na^+/H^+(NH_4^+)$ antiport, they are electrically coupled. The negative potential in the inner side of the apical membrane generated by the primary proton pump is the driving force for sodium uptake. This is why freshwater fish can maintain sodium uptake in a very dilute media. Proton excretion was reduced when external sodium was replaced by choline and restored by the addition of sodium to sodium-free water. Theoretically, when sodium transport is inhibited, either by using sodium-free external media or by blocking the sodium channel with 0.1mM amiloride, the negative potential in the inner side of the apical membrane should build up and eventually stop the operation of the active proton pump. Since removal of sodium from water or the addition of amiloride did not completely block proton excretion, another counter ion must also be involved. The most probable candidate would seem to be chloride through a chloride channel.

In conclusion, the results of these studies indicate that proton excretion across the gills of freshwater trout is mainly mediated by an active proton pump in the gill epithelium, which

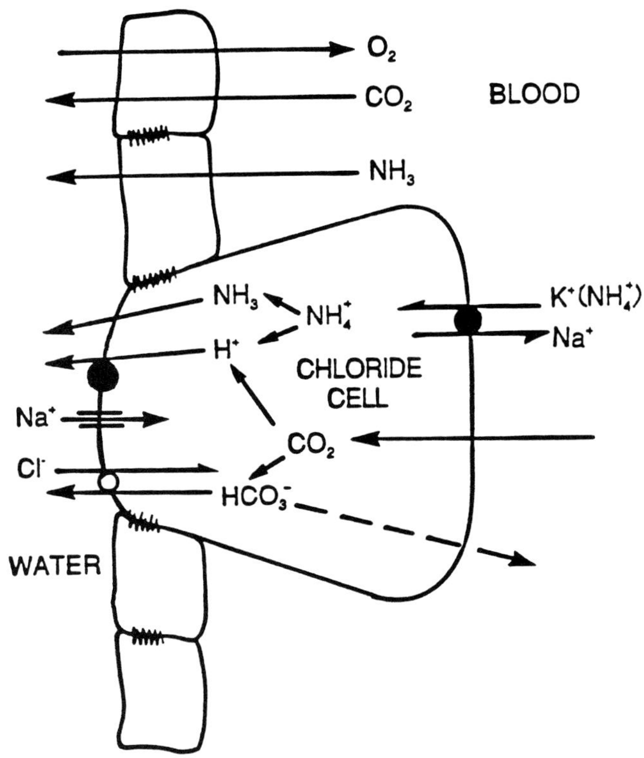

Fig.5. Schematic representation of gas and ion transport across the gill epithelium of rainbow trout. ATP-driven pumps are denoted by filled circles, passive diffusion by arrows.

resembles the proton pump in turtle bladder and frog skin in several major characteristics. Figure 5 illustrates an ion transport model for the freshwater trout gill epithelium which summarises the relationships between the electrogenic proton pump and other ion transport pathways.

EFFECT OF WATER pH ON ION TRANSFER

When external water pH was below 5.5, proton excretion was completely inhibited, presumably due to the large hydrogen ion electrochemical gradient operating against the proton pump (Fig.6). Sodium uptake was also found to be inhibited by low

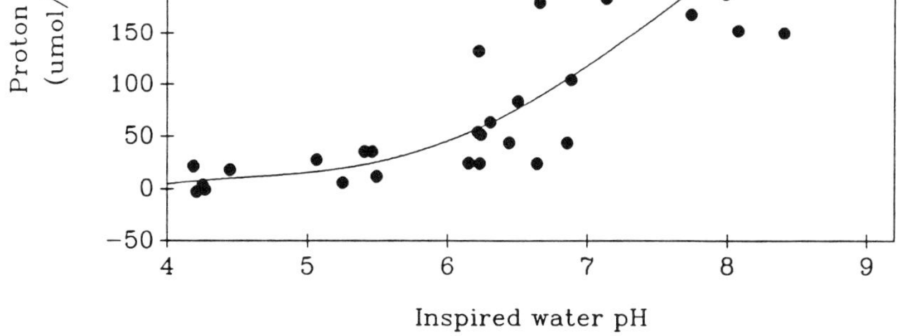

Fig.6. The relationship between the proton excretion across the gills of freshwater rainbow trout and the external water pH, expressed by a five-degree regression curve (R^2 = 0.896).

water pH (Wright and Wood, 1985). Above pH 5.5, acid excretion increased linearly with pH. A linear relationship between proton transport and mucosal pH was demonstrated in both frog skin and turtle bladder (Ehrenfeld *et al*, 1985; Steinmetz & Andersen, 1982).

Sodium uptake by fish is blocked by amiloride (Perry and Randall, 1981; Wright and Wood, 1985) and exposure to acid or alkaline water (Wright and Wood, 1985). Exposure of trout to 0.1 mM amiloride in the water resulted in an 84% (Perry and Randall, 1981) and 94% (Wright and Wood,1985) inhibition of sodium uptake, indicating that most if not all sodium is taken up through an amiloride sensitive sodium channel. Acid conditions inhibit proton excretion and therefore sodium uptake. The effects of high water pH on sodium uptake, however, cannot be simply related to a membrane potential generated by activity of the proton pump, as seen in frog skin, because membrane potential is increased under alkaline conditions, and this should enhance, not inhibit, sodium uptake.

Acid conditions cause a marked increase in chloride efflux but a decrease in chloride influx across trout gills (McDonald *et al.*, 1983). This is associated with a decrease in the chloride electrochemical gradient, so the increased efflux must be due to a large increase in epithelial permeability (Ye *et al.*, 1991). Chloride flux is considered to be associated with bicarbonate transfer, because it is inhibited by SITS (4-acetamido-4'-isothiocyanatostilbene-2-2'-disulphonic acid; Perry and Randall, 1981). SITS resulted in a reduction of proton excretion in fish pretreated with amiloride. The electrochemical gradient for chloride across the apical membrane favours chloride efflux rather than influx and so under some conditions Cl^-/HCO_3^- exchange may be involved in chloride efflux. Thus the effect of SITS on proton excretion could be due to an inhibition of bicarbonate uptake (and chloride efflux). The addition to the water of the drug 9-anthroic acid (9-AA), which is thought to inhibit chloride channels in turtle bladder (Stetson *et al.*, 1985), had no effect on proton excretion in fish in sodium free water (Randall and Lin, 1992). This indicates an absence of chloride channels in the apical membrane of the gill epithelium of fish.

PROTON, CO_2 AND AMMONIA INTERACTIONS

The substantial proton excretion across the gill epithelium of freshwater rainbow trout in neutral and alkaline environments caused an acidification of water flowing over the gills (Lin & Randall, 1990) and bicarbonate in expired water was dehydrated despite considerable carbon dioxide excretion. Although proton excretion might change the CO_2/HCO_3^- level in the boundary water layer, the CO_2 excretion rate was unaffected by the proton pump inhibitors except acetazolamide (Lin & Randall, 1991a) which caused an increase in carbon dioxide excretion. Acidification of the boundary layer, whether by carbon dioxide hydration or proton excretion, will augment ammonia excretion by trapping ammonium ions and enhancing

NH_3 gradients (Wright *et al.*, 1989; Randall *et al.*, 1991).

Ammonia excretion is reduced during the initial stages of exposure to both acid and alkaline water (Wright and Wood, 1985) but then returns to normal, with elevated plasma ammonia levels (Randall and Wright, 1989). It is not clear why acid exposure should inhibit ammonia excretion as one might expect enhanced excretion due to increased ammonium ion trapping in the acid water. It is possible that as the fish becomes acidotic, ammonia stores in the fish increase and ammonia excretion rates are reduced but eventually return to normal when ammonia gradients are re-established.

Under alkaline conditions NH_3 levels in the water are elevated, inhibiting ammonia excretion and increasing body total ammonia content of trout in soft water to levels that can be toxic. Lake Magadi tilapia, however, survive in water of pH 9-10.5, at least in part, because they can detoxify ammonia by producing urea via the ornithine urea cycle (Randall *et al.*, 1989). Exposure of the tilapia, *Oreochromis mossambica*, to water of high bicarbonate concentrations results in a marked increase in urea excretion (McKenzie, unpublished observations), indicating that the induction of urea production may be more related to elevated bicarbonate than pH, as suggested by Atkinson and Bourke (1987).

REFERENCES

Arruda,J.A.L., Sabatini,S. and Westenfelder,C. (1981). Vanadate inhibits urinary acidification by the turtle bladder. *Kidney International*, 20: 772-779.

Atkinson, D.E. and Bourke, E. (1987). Metabolic aspects of the regulation of systemic pH. *Am. J. Physiol.* 252: F947-F956.

Avella,M. and Bornancin,M. (1989). A new analysis of ammonia and sodium transport through the gills of the freshwater rainbow trout (*Salmo gairdneri*). *J.exp.Biol.* 142: 155-175.

Ehrenfeld,J. and Garcia-Romeu,F. (1977). Active hydrogen excretion and sodium absorption through isolated frog skin. *Am.J.Physiol.* 233: F46-F54.

Ehrenfeld,J., Garcia-Romeu,F. and Harvey,B.J. (1985). Electrogenic active proton pump in *Rana esculenta* skin and its role in sodium ion transport. *J.Physiol.,Lond.* 359: 331-355.

Janssen, R.G. and Randall, D.J. (1975). The effects of changes in pH and Pco_2 in blood and water on breathing in rainbow trout, *Salmo gairdneri*. *Respir. Physiol.* 25: 235-245.

Kleyman, T.R. and Cragoe, E.J.Jr. (1988). Amiloride and its analogs as tools in the study of ion transport. *J.Membr.Biol.* 105: 1-21.

Knauf,H., Simon,B. and Wais,U. (1976). Non-specific inhibition ofmembrane-ATPase byamiloride. *Naunyn-Schmiedeberg's Arch. Pharmac.* 292: 189-192.

Lin,H. and Randall,D.J. (1990). The effect of varying water pH on the acidification of expired water in rainbow trout. *J.exp.Biol.* 149: 149-160.

Lin, H. and Randall, D.J. (1991a). Evidence for the presence of an electrogenic proton pump on the trout gill epithelium. *J.exp.Biol.* In press.

Lin, H. and Randall, D.J. (1991b). Effects of varying water pH on gill function. *Proceeding of an internation symposium on fish physiology, toxicology and water quality managment.* In press.

McDonald, D.G. (1983). The interaction of environmental calcium and low pH on the physiology of the rainbow tro *Salmo gairdneri.* I. Branchial and renal net ion and H^+ fluxes. *J.exp.Biol.* 102: 123-140.

McDonald, D.G. Wood, C.M. (1981). Brancial and renal acid and ion fluxes in rainbow trout, *Salmo gairdneri* at low environmental pH. *J.exp.Biol.* 93: 101-118.

McDonald, D.G.; Walker, R.L. and Wilkes, P.R.H. (1983). The interaction of environmental calcium and low pH on the physiology of the rainbow trout, *Salmo gairdneri.* II. Branchial ionoregulatory mechanisms. *J.exp.Biol.* 102: 141-155.

Packer, R.K. and Dunson, W.A, (1972) Anoxia and sodium loss associated with the death of brook trout at low pH. *Comp. Biochem. Physiol.* 41A: 17-26.

Payan, P. (1978). A studies of the Na^+/NH_4^+ exchange across the gill of the perfused head of the trout (*Salmo gairdneri*). *J.comp.Physiol.* 124: 181-188.

Perry,S.F. and Randall,D.J. (1981). Effects of amiloride and SITS on branchial ion fluxes in rainbow trout, (*Salmo gairdneri*). *J.exp.Zool.* 215: 225-228.

Randall, D.J. and Lin H. (1992). Effects of water pH on gas and ion transfer across fish gills. In: *Fish ecophysiology*. Ed. Rankin J.C. and Jensen F.B. Chapman and Hall, London U.K. In press.

Randall, D.J.; Lin, H. and Wright, P.A. (1991). Gill water flow and the chemistry of the boundary layer. *Physiol.Zool.* 64(1): 26-38.

Randall, D.J.; Wood, C.M.; Perry, S.F.; Bergman, H.; Maloiy, G.M.O.; Mommsen, T.P. and Wright, P.A. (1989). Ureotelism in a completely aquatic teleost fish: a strategy for survival in an extremely alkaline environment. *Nature*, 337(12): 165-166.

Randall, D.J. and Wright, P.A. (1989). The interaction between carbon dioxide and ammonia excretion and water pH in fish. *Can.J.Zool.* 67: 2936-2942.

Steinmetz, P.R. and Andersen, O.S. (1982). Electrogenic proton transport in epithelial membrane. *J.Membrane Biol.* 65: 155-174.

Stetson, D.L.; Beauwens, R.; Palmisano, J.; Mitchell, P.P. and Steinmetz, P.R. (1985). A double-membrane model for urinary bicarbonate secretion. *Am.J.Physiol.* 249: F546-F552.

Wright, P.A.; Randall, D.J. and Perry, S.F. (1989). Fish gill water boundary layer: a site of linkage between carbon dioxide and ammonia excretion. *J.Comp.Physiol.B*, 158: 627-635.

Wright,P.A. and Wood,C.W. (1985). An analysis of branchial ammonia excretion in the freshwater rainbow trout: effects of environmental pH change and sodium uptake blockage. *J.exp.Biol.* 114:329-353.

Ye, X., Randall, D.J. and He, X. (1991). The effect of acid water on oxygen consumption, circulating catecholamines and blood ionic and acid-base status in rainbow trout (*Salmo gairdneri*, Richardson). *Fish Physiol.Biochem.* 9(1):23-30.

4

Calcium Regulation in Fish

S.E. Wendelaar Bonga and G. Flik

Calcium is important for growth, reproduction, and many other physiological processes such as muscle contraction, nerve signal transduction, and control of membrane permeability and cellular metabolism. These processes are dependent on an effective control of the ionic calcium levels of the extracellular fluid (ECF). Although fishes seem to be slightly more resistant to changes of ionic calcium of the ECF than the terrestrial vertebrates, deviations from the normal physiological concentration ("setpoint") have essentially the same effects in all vertebrates: hypocalcemia leads to increased excitability of cellular membranes and may result in tetanic seizures, whereas hypercalcemia reduces membrane excitability with lethargy and cardiovascular dysfunction as consequences. Precise control of the setpoint of ionic calcium is the rule in fish as well as in other vertebrates.

The maintenance of constant calcium levels in the ECF is effected in fish by the continuous exchange of calcium between the animals and the environment, mainly via the gills. In full-grown animals influx and efflux of calcium are in equilibrium, whereas during periods of growth and reproduction net calcium accumulation occurs. During such periods large amounts of calcium are transported from the sites of entry through the blood and other compartments of the ECF to bones or ovaria. This represents a special challenge for the mechanisms that control the calcium concentration of the ECF.

There are several important differences between fish and the terrestrial vertebrates with respect to the organs involved in the maintenance of the calcium balance of the body, and in the endocrine mechanisms controlling these organs. Firstly, whereas in terrestrial vertebrates calcium is obtained exclusively from the food via the gut, in fish the gills are in general the main route of uptake of calcium with the gut in a secondary role. The ambient water seems sufficient as a calcium source in fish. Our unpublished studies on the African cichlid *Oreochromis mossambicus* have shown that calcium-deficient food does not have any effect on the calcium balance of tilapia, which confirmed similar observations on goldfish by Ichii and Mugiya (1983). The uptake of calcium in fish then is an essentially continuous process, in contrast to the calcium uptake in terrestrial vertebrates, which is episodic and restricted by the calcium content of the food. The second important difference between fish and terrestrial vertebrates is the role played by the skeleton. In the absence of a continuous access to external calcium sources, terrestrial animals have developed the bone tissue as a calcium store which is used for the minute-to-minute control of the calcium level of the body fluids. We will discuss below that a calcium exchange capacity of the skeleton in fish is not required because the control of calcium homeostasis in fish is mainly effected at the level of the gills, in particular by regulation of the calcium influx from the water. Thirdly, and not surprisingly, given the above-mentioned characteristics of calcium metabolism in fish, there are essential differences in the hormonal mechanisms regulating the calcium balance between fish and the terrestrial vertebrates. In this paper we will concentrate on these regulatory aspects.

A. Terrestrial Vertebrates

In the terrestrial vertebrates two hormones are specifically involved in the control of calcium metabolism: parathyroid hormone (PTH) and calcitonin. PTH is the dominant hormonal factor. It stimulates the intestinal uptake and reduces the renal excretion of calcium, processes mediated by a vitamin-D3-metabolite 1,25-dihydroxycholecalciferol. PTH further mobilizes calcium from the skeleton. When the ionic calcium concentration of the ECF falls below the physiological setpoint (about 1.24 mM Ca^{2+}) the

PTH-cells respond within seconds with the enhanced secretion of PTH (Brown *et al.* 1987). The hormone stimulates the osteoclasts, and this leads to a rapid rise of the extracellular Ca^{2+} level. Calcium concentrations above the setpoint inhibit PTH secretion. The rapid response of the PTH cells is mediated by the special Ca^{2+}-sensing capacity of these cells (Jones and Fitzpatrick 1990).

Calcitonin has long been considered the hormonal antagonist of PTH, because of its reportedly hypocalcemic action. Although hypocalcemic effects have been shown after injection of the hormone in young and growing mammals, birds and reptiles, consistent hypocalcemia could not be demonstrated in adult animals (Copp and Kline 1989). Neither chronically elevated calcitonin levels, nor removal of the calcitonin-producing C-cells by thyroidectomy (in mammals) or ultimobranchialectomy (in birds and amphibians) are inducing persistent imbalance of the extracellular Ca^{2+} levels (see Wendelaar Bonga and Pang 1991). This sharply contrasts with the effects of parathyroidectomy, which include hypocalcemia, tetany and death. One role of calcitonin is the reduction of hypercalcemia that occurs after intestinal absorption of calcium following a meal (VanderWiel and Talmage 1981). It promotes the deposition of ingested calcium into the bone. It further protects the skeleton against excessive demineralization during periods of increased calcium demand, such as pregnancy and lactation in mammals (Stevenson *et al.* 1979). In conclusion, in the terrestrial vertebrates the exchange of calcium with the environment is mainly regulated by PTH, that also controls the Ca^{2+} concentration of the ECF. The bone plays an essential role as a calcium buffer. Compared to PTH, calcitonin only has a minor function in the homeostatic control of extracellular Ca^{2+}.

B. Fish

Our knowledge of the calcium metabolism and the endocrine control in fish is mainly based on freshwater teleosts and therefore we will mainly deal with this group. Teleost fish maintain plasma total and ionic calcium within narrow limits, but these limits are to some extent dependent on the growth rate and on the cal-

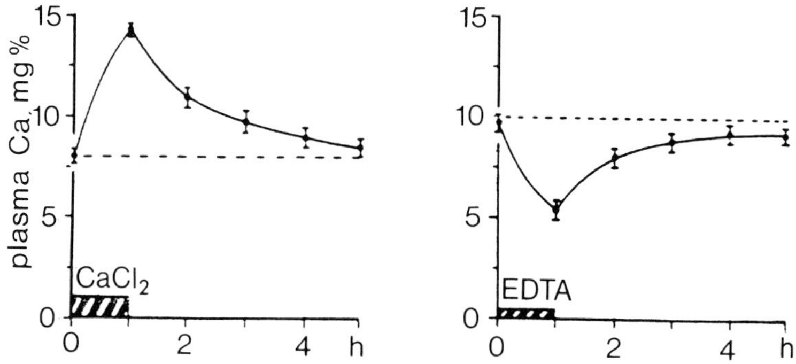

Fig. 1. Recovery from induced hyper- or hypocalcemia in rainbow trout (Copp *et al.* 1985).

cium concentration of the ambient water (Urasa and Wendelaar Bonga 1987). When plasma calcium levels of teleost fish are changed experimentally, normal levels are rapidly restored. In rainbow trout (Copp *et al.* 1985) and American eels (Fenwick and Brasseur 1991) it took 3 to 6 h before a twofold increase of plasma total calcium had decreased to control values (Fig. 1). Restoration of a 50% drop in plasma calcium induced by EGTA infusion in rainbow trout lasted a similar period (Copp *et al.* 1985; Fig. 1). These data show that teleost fish can successfully restore changes in the calcium concentration of the extracellular fluid. The response may be similar to or even more rapid than in the higher vertebrates. The changes in calcium levels in the terrestrial animals are compensated by modulating the exchange of calcium between the bone and the ECF. How is this compensation effected in fish?

B. 1. Bone

In amphibians and fishes, the calcium-exchange capacity of the bones has been frequently considered as limited. This does certainly not imply that calcium mobilization from bone does not occur. There are several reports for teleost fish with cellular as well as acellular bone indicating that the calcium concentration of bone varies in response to environmental or physiological changes (Fleming *et al.* 1973). In tilapia, a species with acellular bone, reduction of the water calcium concentration or of the water pH - both stimulate calcium efflux across the gills - causes a transient decrease of the calcium concentration of the bone (Wendelaar Bonga and Dederen 1985). Induction of scale regeneration by removal of mature scales is accompanied by a transient reduction of the calcium concentration of the remaining scales (Weiss and Watabe 1978). Forced gonadal maturation by injection of estrogens may lead to calcium mobilization from the bone and/or the scales (Mugiya 1982). Thus, calcium can be mobilized from skeleton and scales in fish. The contribution of calcium mobilization from the bone to the calcium homeostasis of the ECF is unclear, however. Except for estrogens no hormone is known that stimulates bone calcium release in fish. The function of bone as a readily accessible and hormonally controlled calcium store seems to have been evolved after the water-to-land transition of the vertebrates, in conjunction with the appearance of PTH and a system of bone cells with calcium-releasing activity that rapidly responds to this hormone (Pang *et al.* 1980; Wendelaar Bonga and Pang 1991).

B.2. Gills

If the bone does not function as a fast-acting calcium buffer, the most likely mechanism for the minute-to-minute control of the extracellular calcium concentration is the regulation of the calcium-exchange between the fish and its environment. The exchange of calcium is much higher in fish than in terrestrial vertebrates. This is mainly caused by the calcium fluxes across the branchial epithelium. Although this epithelium is very tight, in particular in fresh water, there is still a substantial passive efflux of calcium ions. This loss of calcium ions is compensated by the active uptake of calcium ions through the gills and, to a lesser extent, the intestine. Whereas the branchial calcium efflux most likely follows a paracellular route, the calcium influx is an active, transcellular process. We have investigated this process in detail, and these studies have resulted in a

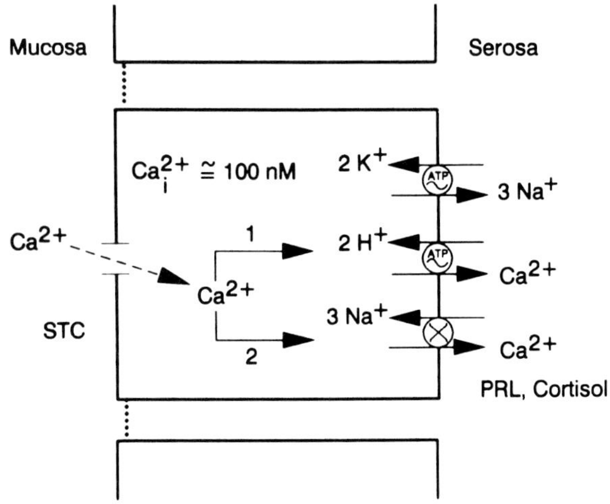

Fig. 2. Transcellular transport of Ca^{2+} in fish gills. Ca^{2+} from the water may enter the cell passively, down an electrochemical gradient. Stanniocalcin (STC) regulates calcium channels in the apical membrane; STC signal transduction requires second (cAMP, DAG, IP_3) and third (Ca^{2+}) messengers. Extrusion of Ca^{2+} from the cell could occur via a Ca^{2+}-ATPase (route 1) or via a Na^+/Ca^{2+} exchanger (route 2). Both extrusion mechanisms are present in branchial epithelium. In gill cells route 1 probably dominates. Hormones with hypercalcemic effects such as prolactin (PRL) in several species and cortisol (in trout) determine the relative abundance of the ATPases and by doing so may determine the calcium transporting capacity of the epithelium (Flik *et al.* 1992).

model that has been verified for a number of teleost species, including trout, tilapia and two eel species (Fig. 2). The calcium-influx is effected by the chloride cells, the cells specialized for ion transport in fish gills. Calcium ions from the water are entering the cytoplasm of the chloride cells via the apical membrane of these cells, probably through voltage-independent calcium channels, and following the electrochemical gradient (Verbost *et al.* 1989). The cytoplasmic ionic calcium concentration is two to three orders of magnitude lower than in the ambient water or in the ECF. This gradient favours the entrance of calcium into the cell. However, special mechanisms are required for the extrusion of calcium ions from the cytoplasm across the basolateral membrane into the ECF. We have identified these mechanisms as ATP-driven Ca^{2+}-pumps and Na^+/Ca^{2+}-exchangers. Studies on the activity of both mechanisms as a function of the Ca^{2+} concentration have shown that the Ca^{2+}-pump is more important than the Na^+/Ca^{2+}-exchanger for the extrusion of calcium from the chloride cells. The reverse is indicated for the intestine (for a review of our studies see Flik *et al.* 1992).

B. 3. Endocrine Control of Calcium Metabolism.

a. Prolactin

The absence of parathyroid glands from purely aquatic amphibians and fish has now been well established. This absence has stimulated the search for a hormone with a similar hypercalcemic action. Several studies have indicated that the pituitary gland is the source of such a hormone, with prolactin as the most likely candidate. In the terrestrial vertebrates prolactin has some effects on calcium metabolism, in particular during the female reproductive period: it stimulates intestinal calcium uptake and the transfer of calcium to the offspring via the placenta or the mammary glands (see Wendelaar Bonga and Pang 1991).

In fish the role of prolactin in calcium metabolism seems more important than in the terrestrial animals. Removal of the pituitary gland in freshwater fish usually leads to mild hypocalcemia although data in the literature are not always consistent (Wendelaar Bonga and Pang 1989). Pang *et al.* (1978) have shown that the hypocalcemia of hypophysectomized killifish from low-calcium freshwater can be corrected by injections of pituitary homogenates or mammalian prolactin preparations. The latter produced hypercalcemia in e.g. intact sticklebacks, tilapia, trout, salmon or eel (Wendelaar Bonga and Pang 1989). Recently, using recombinant hormone preparations, we have shown that both forms of tilapia prolactin induce hypercalcemia in intact freshwater tilapia (Swennen *et al.* 1991). The most likely mechanism of action of prolactin is the stimulation of the activity of the ATP-dependent Ca^{2+}-pumps of the chloride cells: prolactins of mammalian or teleostean origin, including the recombinant tilapia preparations, increase the Ca^{2+}-ATPase activity and stimulate the Ca^{2+}-influx of the gills (Flik *et al.* 1986, 1989a). It takes several days for these effects to develop, and this contrasts sharply with the immediate responses to PTH in terrestrial animals. It is therefore unlikely that in teleosts prolactin accounts for the rapid restoration of experimentally induced hypocalcemia such as shown in fig. 1, or for the minute-to-minute control of the calcium concentration of the ECF in general. The primary function of prolactin in fish most likely is the control of integumental permeability to water and ions (Hirano 1986; Wendelaar Bonga and Pang 1989). Hypercalcemic actions have been ascribed to some other pituitary hormones of teleosts, but these have not been confirmed (see Wendelaar Bonga and Pang 1989). In trout cortisol was shown to produce hypercalcemia by increasing total body calcium uptake (Flik and Perry 1989). This hormone had no effect on tilapia (our unpublished observations) and it is unlikely to be a hormone with a specific hypercalcemic function.

b. Calcitonin

Calcitonin has been demonstrated in all major groups of fishes with exception of the agnathans. Its function in fish is still in dispute. Administration of mammalian or teleost calcitonin has only occasionally resulted in hypocalcemia. The doses required were high. In many experiments the hormone had no effect on plasma calcium (see Wendelaar Bonga and Pang 1991). An antihypercalcemic action of the hormone was demonstrated by Fenwick and Lam (1988) in mudskippers that were made hypercalcemic by taking them out of the water. On the other hand, removal of the Stannius bodies leads to a marked hypercalcemia, due to elevated total and ionic calcium levels (Hanssen *et al.* 1989) and develop in the presence of the ultimobranchial bodies. These bodies even show signs of hypersecretion of calcitonin in stanniectomized fish (Wendelaar Bonga and Greven 1978). Thus, the antihypercalcemic potency of the hormone is limited, in particular when compared with stanniocalcin, the hormone of the Stannius bodies present in teleostean and holostean fish. Enhanced plasma calcitonin levels were reported for salmonids during gonadal maturation (Björnsson *et al.* 1989), and enlarged ultimobranchial glands were reported for several female fish during the reproductive period. This is consistent with the elevated plasma calcitonin levels that occur during the female reproductive period of some terrestrial vertebrates, and that have led to the hypothesis that calcitonin protects the skeleton during reproduction (see under A). Calcitonin may have a similar function in fish.

c. Stanniocalcin

Stanniocalcin is a glycoprotein that has recently been isolated from the Stannius bodies of some teleost species. It is most likely identical with the proteinaceous hypocalcemic factor called hypocalcin by Pang *et al.* (1974), and teleocalcin, purified and partially sequenced by Wagner *et al.* (1986). Stanniocalcin is different from the small glycoprotein isolated from salmon that was originally called teleocalcin (Ma and Copp, 1978), and also different from the glycoprotein called parathyrin of the corpuscles of Stannius isolated from European eels by Milet *et al.* (1980). With the identification (Wagner *et al.* 1986; Lafeber *et al.* 1988) and sequencing (Butkus *et al.* 1987) of stanniocalcin, the confusion about the nature of the major calcium-regulating factor of the Stannius bodies has been solved, and stanniocalcin has now been established as a hormone with unique structural and functional characteristics. It is a dimeric glycoprotein with monomeric constituents with a molecular radius of about 30 kDa (Wagner *et al.* 1986; Flik *et al.* 1989b). The 30 kDa form is dominating in the circulation (Wagner *et al.* 1991).

Injection of stanniocalcin effectively reduces the hypercalcemia that occurs after operative removal of the Stannius bodies. It acts by reducing the branchial influx of calcium, and its effects become noticeable within 15 min after injection (Lafeber *et al.* 1988b). Because it leaves the passive efflux of calcium unaffected, the result is a rapid decrease of the calcium concentration of the blood plasma (Wagner *et al.* 1986; Lafeber *et al.* 1988b). Verbost *et al.* (1989) have shown that stanniocalcin inhibits $^{45}Ca^{2+}$ uptake from the water into the branchial epithelial cells, and this indicates that it acts by inhibiting the entrance of Ca^{2+} across the apical membranes of the chloride cells. The authors therefore suggested that stanniocalcin is a specific blocker of voltage-independent Ca^{2+}-channels in these membranes. The hormone did not influence the Ca^{2+}-extruding mechanisms in the basolateral membranes of the chloride cells (Flik *et al.*, unpublished). Because stanniocalcin is unlikely to influence the Ca^{2+}-channels of the apical membranes directly, we have postulated that the receptors for stanniocalcin are located in the basolateral membrane, and that the signal is mediated to the channels by second messenger. The hormone indeed modifies the second messenger systems in these cells: it reduces the branchial cAMP content and inhibits adenylate cyclase activity and stimulates the production of 1,2-sn-diacylglycerol in tilapia branchial cells (Flik *et al.* 1992). The latter effect indicates that stanniocalcin acts via the phosphoinositol pathway.

Our present concept of the endocrine control of calcium influx in teleost fish is illustrated in Fig. 2. As discussed above (see under B 2) Ca^{2+} can enter the fish - at least its chloride cells - passively, as long as the Ca^{2+}-channels are open and the electrochemical gradient is favourable. The latter is likely to be the case in most fresh waters, even with very low Ca^{2+} levels. This means that under normal freshwater conditions teleost fish may face hypercalcemia rather than hypocalcemia, even when the water Ca^{2+}-concentration is well below that of the ECF, unless the Ca^{2+} channels are closed by stanniocalcin. This is the essence of our model. It explains why the Ca^{2+} concentration of the blood plasma is rising to extremely high levels when the Stannius bodies are removed. This effect has been consistently observed in freshwater and seawater species and is caused by increased Ca^{2+} influx (So and Fenwick 1979). The total and ionic Ca^{2+} levels produced as a result of the absence of stanniocalcin are much higher than those that can be obtained by injection of hormones with hypercalcemic actions such as prolactin or cortisol. Apparently, in freshwater as well as seawater, the total body uptake of calcium and the control of the Ca^{2+} concentration of the ECF are under inhibitory control by stanniocalcin.

In conclusion then, the endocrine control of the calcium concentration of the ECF is most likely dominated by one hormone only in terrestrial vertebrates and in teleost fish. This contrasts with the earlier concept of two antagonistically acting hormones. In the terrestrial vertebrates it is PTH, which increases the Ca^{2+}-concentration of the ECF, mainly by stimulating the intestinal uptake of calcium and by mobilizing calcium from the skeleton. In teleost fish it is stanniocalcin, which lowers the Ca^{2+}-concentration of the ECF by inhibiting the passive influx of calcium from the ambient water into the branchial chloride cells and, possibly, the epithelial cells of the intestine.

REFERENCES

Brown EM, Leboff MS, Oetting M, Posillico JF, Chen C (1987) Secretory control in normal and abnormal parathyroid tissue. Recent Prog Horm Res 43: 337-382

Björnsson BT, Haux C, Bern HA, Deftos LJ (1989) 17-Beta-estradiol increases plasma calcitonin levels in salmonid fish. Endocrinology (Baltimore) 125: 1754-1760

Butkus A, Roche PJ, Fernley RT, Haralambidis J, Penschow JD, Ryan JB, Trahair JF, Tregéar GW, Coghlan JP (1987) Purification and cloning of a corpuscle of Stannius protein from *Anguilla australis*. Mol Cell Endocrinol 54: 123-133

Copp DH, Kline LW (1989) Calcitonin. In: Pang PKT, Schreibman MP (eds), Vertebrate endocrinology: fundamentals and biomedical applications, Vol 3. Academic Press, San Diego, pp 79-103

Copp DH, Ma SWY, Qiu MC, Sim M, Szeto K (1985) Role of the pituitary and the corpuscles of Stannius in calcium regulation in teleosts. In: Lofts B, Holmes WN (eds), Current trends in comparative endocrinology. Hong Kong University Press, Hong Kong, pp 819-821

Fenwick JC, Lam TJ (1988) Effects of calcitonin on plasma calcium and phosphate in the mudskipper, *Periophthalmodon schlosseri* (Teleostei), in water and during exposure to air. Gen Comp Endocrinol 70: 224-230

Fenwick JC, Brasseur JG (1991) Effects of stanniectomy and experimental hypercalcemia on plasma calcium levels and calcium influx in American eels, *Anguilla rostrata*, Lesueur. Gen Comp Endocrinol 82: 459-465

Fleming WR, Brehe JE, Hanson R (1973) Some complicating factors in the study of the Ca metabolism of teleosts. Amer Zool 13: 793-797

Flik G, Perry SF (1989) Cortisol stimulates whole body calcium uptake and the branchial calcium pump in freshwater rainbow trout. J Endocrinol 120: 75-82

Flik G, Fenwick JC, Kolar ZI, Mayer-Gostan N, Wendelaar Bonga SE (1986) Effects of ovine prolactin on calcium uptake and distribution in *Oreochromis mossambicus* Am J Physiol 250: R161-

R166
Flik G, Fenwick JC, Wendelaar Bonga SE (1989a) Calcitropic actions of prolactin in freshwater North American eel (*Anguilla rostrata* LeSueur). Am J Physiol 257: R74-R79
Flik G, Labedz T, Lafeber FPJG, Wendelaar Bonga SE, Pang PKT (1989b) Studies on teleost corpuscles of Stannius: physiological and biochemical aspects of synthesis and release of hypocalcin in trout, goldfish and eel. Fish Physiol Biochem 7: 343-349
Flik G, van der Velden JA, Dechering KJ, Verbost PM, Schoenmakers TJM, Kolar ZI, Wendelaar Bonga SE (1992) Ca^{2+} and Mg^{2+} transport in gills and gut of tilapia, *Oreochromis mossambicus*: a review. J Exp Zool (in press)
Hanssen RGJM, Lafeber FPJG, Flik G, Wendelaar Bonga SE (1989) Ionic and total calcium levels in the blood of the European eel (*Anguilla anguilla*): effects of stanniectomy and hypocalcin replacement therapy. J Exp Biol 141: 177-186
Hirano T (1986) The spectrum of prolactin actions in teleosts. In: Ralph CL (ed), Comparative endocrinology: developments and directions. AR Liss, New York, pp 53-74
Ichii T, Mugiya Y (1983) Effects of dietary deficiency in calcium on growth and calcium uptake from the aquatic environment in the goldfish, *Carassius auratus*. Comp Biochem Physiol 74A: 259-262
Jones JI, Fitzpatrick LA (1990) Binding of [^{125}I]iodipine to parathyroid cell membranes: evidence of a dihydropyridine-sensitive calcium channel. Endocrinology (Baltimore) 126: 2015-2020
Lafeber FPJG, Flik G, Wendelaar Bonga SE, Perry SF (1988a) Hypocalcin from Stannius corpuscles inhibits gill calcium uptake in trout. Am J Physiol 254: R891-R896
Lafeber FPJG, Hanssen RGJM, Choy YM, Flik G, Herrmann-Erlee MPM, Pang PKT, Wendelaar Bonga SE (1988b) Identification of hypocalcin (teleocalcin) isolated from trout Stannius corpuscles. Gen Comp Endocrinol 69: 19-30
Ma WY, Copp DH (1978) Purification, properties and action of a glycopeptide from the corpuscles of Stannius which affect calcium metabolism in the teleost. In: Gaillard PH, Boer HH (eds), Comparative endocrinology. Elsevier Biomedical Press, Amsterdam, pp 283-286
Milet C, Hillyard CJ, Martelly E, Girgis S, MacIntyre L, Lopez E (1980) Similitudes structurales entre l'hormone hypocalcémiante des corpuscules de Stannius de l'anguille et l'hormone parathyroidienne mammalienne. C R Acad Sci (Paris) D291: 977-980
Mugiya y (1982) Calcium metabolism in estrogenized fish. in: Oguro C, Pang PKT (eds), Comparative endocrinology of calcium regulation. Japan Scientific Societies Press, Tokyo, pp 109-114
Pang PKT, Pang RK, Sawyer WH (1974) Environmental calcium and the sensitivity of killifish (*Fundulus heteroclitus*) in bioassays for the hypocalcemic response to Stannius corpuscles from killifish and cod (*Gadus morhua*) Endocrinology (Baltimore) 94: 548-555
Pang PKT, Schreibman MP, Balbontin F, Pang RK (1978) Prolactin

and pituitary control of calcium regulation in the killifish, *Fundulus heteroclitus*. Gen Comp Endocrinol: 306-316

Pang PKT, Kenny AD, Oguro C (1980) The evolution of the endocrine control of calcium regulation. In: Pang PKT, Epple A (eds), Evolution of vertebrate endocrine systems. Texas Tech University Press, Lubbock, pp 323-356

So YP, Fenwick JC (1977) Relationship between net Ca^{2+} influx across a perfused eel gill and the development of poststanniectomy hypercalcemia. J Exp Zool 200: 259-264

Stevenson JC, Hillyard CJ, MacIntyre I, Cooper H, Whitehead MI (1979) A physiological role for calcitonin: protection of the maternal skeleton. Lancet 2: 769-770

Swennen D, Rentier-Delrue F, Auperin B, Prunet P, Flik G, Wendelaar Bonga SE, Lion M, Martial JA (1991) Production and purification of biologically active recombinant tilapia (*Oreochromis niloticus*) prolactins. J Endocrinol 131: 219-227

Urasa FM, Wendelaar Bonga SE (1987) Effects of calcium and phosphate on the corpuscles of Stannius of the teleost fish *Oreochromis mossambicus*. Cell Tiss Res 249: 681-690

VanderWiel CJ, Talmage RV (1981) Ultrastructural and physiological evidence for calcitonin-induced postprandial calcium storage in bones of rats. Calcif Tiss Int 33: 417-424

Verbost PM, van Rooij J, Flik G, Lock RAC, Wendelaar Bonga SE (1989) The movement of cadmium through freshwater trout branchial epithelium and its interference with calcium transport. J Exp Biol 145: 185-197

Wagner GF, Hampong M, Park CM, Copp DH (1986) Purification, characterisation, and bioassay of teleocalcin, glycoprotein from salmon corpuscles of Stannius. Gen Comp Endocrinol 63: 481-491

Wagner GF, Milliken C, Friesen HG, Copp DH (1991) Studies on the regulation and characterization of plasma stanniocalcin in rainbow trout. Mol Cell Endocrinol 79: 129-138

Weiss RE, Watabe N (1978) Studies on the biology of fish bone - I. Bone resorption after scale removal. Comp Biochem Physiol 60A: 207-211

Wendelaar Bonga SE, Greven JAA (1978) The relationship between prolactin cell activity, environmental calcium and plasma calcium in the teleost *Gasterosteus aculeatus*. Observations on stanniectomized fish. Gen Comp Endocrinol 36: 90-101

Wendelaar Bonga SE, Dederen LHT (1986) Effects of acidified water on fish. Endeavour 10: 198-201

Wendelaar Bonga SE, Pang PKT (1989) Pituitary hormones. In: Pang PKT, Schreibman MP (eds), Vertebrate endocrinology: fundamental and biomedical implications, vol. 1. Academic Press, San Diego, pp 105-137

Wendelaar Bonga SE, Pang PKT (1991) Control of calcium regulating hormones in the vertebrates: parathyroid hormone, calcitonin, prolactin, and stanniocalcin. Int Rev Cytol 128: 139-213

5

Seawater Adaptation Strategies in Salmonids

Gilles Boeuf

INTRODUCTION

As exposed in a recent paper (Boeuf 1992), salmonids are a small teleosts family with species only able to live in freshwater (FW) and others which migrate actively from FW to seawater (SW) and can naturally adapt to this new environment. Migrating species already develop in FW many deep changes preparing them to the life in hyperosmotic surrounding, and called "parr-smolt transformation" or "smoltification" (Hoar 1988, Boeuf 1992).

In aquaculture, both species are used, naturally migrating and adapting to SW and others, naturally spending their entire life in FW. We can easily understand the FW rearing of these last salmonids but it is more difficult to imagine how it is possible to rear these same fish into SW. However a few countries produce rainbow (*Oncorhynchus mykiss*) and brown trouts (*Salmo trutta*) in SW (Denmark, Norway, Canada, Chile, France...), on an industrial scale, even sometimes at full salinity (> 34 ‰) with good results in terms of survival and growth. Although they are not migrating species (sedentary strains), these fish can adapt to SW and grow under specific conditions that we propose to develop here, comparing their adaptation strategies and osmotic capacities to naturally migrating salmonids.

Anyway, in marine aquaculture, in all cases the transfer from FW to SW will represent a very critical phase in the rearing cycle, using either smoltifying or naturally FW species (see table 1). Naturally FW sedentary species (or strains) find the physiological resources to adapt after direct or progressive exposure to SW because they have a "good base of euryhalinity", but we have to bear in mind every time, that it is a possible response to an "agression" and it is under the human constraint that the fish reacts. For naturally migrating species (or strains), the situation is totally different and, at the right moment, the fish will look for the salinity and will prefer to live in SW. The most difficult problem for the grower will be the precise determination of this narrow time "window" and to choose the good fish in the population, all the salmon of the same age and broodstock not being able to adapt and to grow in SW. A few difficulties can also appear when the transfer to SW is direct to high salinity, even with true smolts, the fish progressively adapting in the wild and staying a few moments in the estuary (Zaugg *et al* 1985). Parr-smolt transformation is ended by a natural will in the fish to leave the river, to migrate to the estuary and to adapt and grow in SW and finally SW transfer does not represent more than the completion of a natural pre-adaption, already occurring in FW (Hoar 1988, Boeuf 1992). However, deep disturbances can appear if the transfers from FW to SW are made with non-totally smoltified fish (mortalities or "stunting" phenomenon).

On the other hand we know cases where "normally" migratory species (or strains) can also remain their whole life in FW, grow and reproduce (naturally "landlocked" Atlantic salmon, or introduced coho salmon in Great Lakes). All these facts demonstrate the remarkable osmotic "plasticity" of salmonids and allow us to think, that according to the FW physiological status prior to the tranfer, the SW adaptation capacities and mechanisms will be different.

Table 1 : Main species (or strains) produced in SW

Species naturally spending their entire life in FW		Active migrators
Rainbow trout	*Oncorhynchus mykiss*	Steelhead trout
Brown trout	*Salmo trutta*	Sea trout
		Atlantic salmon (*Salmo salar*)
		All Pacific salmons (g. *Oncorhynchus*)
In mariculture, possible adaptation to SW		Natural parr-smolt transformation and growth in SW

Critical phase : transfer to SW

GENERAL PHYSIOLOGICAL BASES

Teleosts, having developed deep mechanisms of control of the hydromineral balance, require more energy than osmoconforming in the marine environment. The anatomical adaptations and physiological costs demanded for complete iono and osmoregulation became energetically feasible in ancestors of the actinopterygians as the abundant oxygen would support greater rates of aerobic metabolism (Ballantyne *et al* 1987). Salmonids are well-known for controlling the *milieu intérieur* and corresponding to fish having an "anisosmotic extracellular regulation" and maintaining the osmotic pressure of the organism to a range from 280 to 360 $mOsm.l^{-1}$ (Holmes and Donaldson 1969). They are particularly interesting for the physiologist in their capacities to change from FW to SW and *vice versa* and represent very adapted models in both osmoregulation basic studies and applied aquaculture experiments.

We can consider three compartments in fish with different biochemical and ionic composition (see table

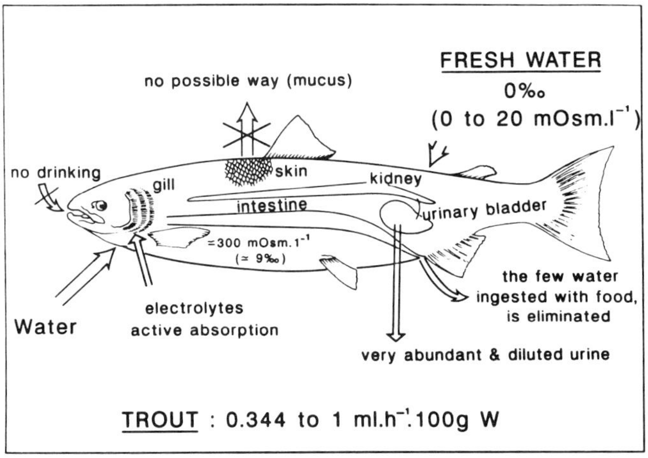

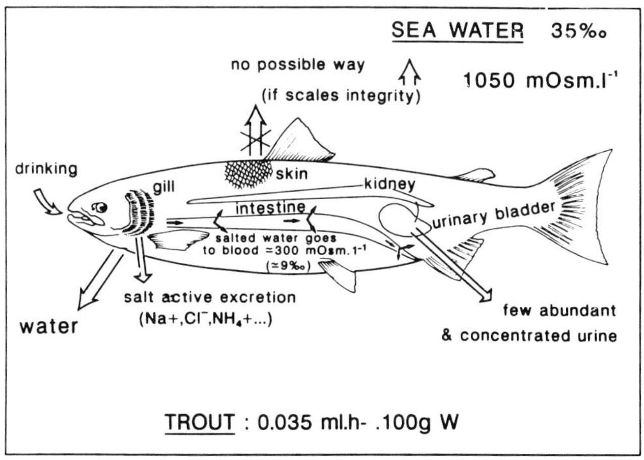

Figures 1 and 2 : The rainbow trout in freshwater (FW), or in seawater (SW), main patterns of the osmoregulation. The figures below the drawings are the level of urine production.

2) but with a good osmotic homeostasy. Figures 1 and 2 briefly summarize the situations in salmonids according to the external salinity.

The same fish will be able to change and to adapt to full salinity SW in precise and specific conditions we will develop in the following parts. Apart from these conditions, the salmonid will not be able to survive, adapt and grow in this new surrounding, then behaving like a stenohaline species. The direct transfer from FW to full salinity SW then poses many problems which can lead to deep water losses, ionic hyperconcentrations, then death.

The main tissues assuring a role in osmotic and ionic regulation in salmonids are the skin, the intestine, the kidney-urinary bladder and the gill. The skin is of the highest importance, representing a "shield" avoiding the water and electrolytes fluxes from the body towards the environment or *vice versa*. There are not chloride cells in salmonids in the skin apart from the opercular membrane in a few species. In FW fish, the skin is very thick and the mucus very abundant. In SW, salmonids are silvery, due to purines deposition (Markert and Vanstone 1966).

In salmonids, the intestine is the most important part of the digestive tract involved in osmoregulation. The fish drinks a lot in SW and the increase of water gut absorption is very high in rainbow trout after SW transfer (Shehadeh and Gordon 1969, Lahlou *et al* 1975). Nonnotte *et al* (1987) demonstrated the transient changes of the intestinal water, sodium and chloride absorption occurring after SW contact : Na^+ and Cl^- transports follow in two phases, first a permeabilization of the epithelium for one day, then a transient impermeabilization and increase of the mucosa protein content and a progressive increase of both unidirectional Na^+, Cl^- fluxes and ATPase activity. These patterns are in accordance with the important ultrastructure intestine mucosa changes occurring just after abrupt transfer into SW (Nonnotte *et al* 1986).

In FW, teleosts produce a very abundant and diluted urine (see Figure 1), whereas in SW, it is rare and relatively concentrated. For example, in rainbow trout, it changes from 0.344 to 1 in FW to 0.035 $ml.h^{-1}.100\ g^{-1}$ in SW (Holmes 1961, Lahlou *et al* 1975). In FW, kidney and urinary bladder are of considerable importance in osmoregulation, their action consisting in eliminating the excess of water without losing electrolytes. The glomerular filtration rate is very high and tubular reabsorption low.

Table 2 : Osmotic and ionic values in different compartments in rainbow trout (these data are valid for fish < 15 °C in SW).

External medium		*Milieu intérieur*	Intracellular medium
Freshwater			
OP	0 to 20 $mOsm.l^{-1}$	280 to 310	280 to 310
Na^+	< 1 $mEq.l^{-1}$	140 to 155	20
K^+	0.1	1 to 4	110 to 130
Cl^-	< 1	125 to 135	8 to 10
Seawater (35 ‰)			
OP	1050	300 to 330	300 to 330
Na^+	450	150 to 165	26
K^+	11	1 to 5	140 to 160
Cl^-	560	130 to 145	8 to 12

The gill is mainly active in SW in osmoregulation enabling the fish to actively eliminate the excess of electrolytes (Payan *et al* 1984, McCormick *et al* 1989, Avella and Bornancin 1990) through the action of mitochondria-rich cells (chloride cells). These specific cells play a major role and considerably increase in number and activity in SW, compared to FW fish (Conte and Lin 1967, Payan *et al* 1984). The ultrastructure is quite different between a FW fish and a SW one (Pisam *et al* 1990). Avella and Bornancin (1990) recently reviewed the role of (Na^+-

K^+)-ATPase activity in the basolateral plasma membranes and tubular system of chloride cells in salmonids. The gill microsomes ATPase activity is directly involved in ionic extrusion in SW and this enzymatic potential activity, revealed in vitro, is often used as a good criterion of osmotic capacities in FW migrating salmonids (reviews in Hoar 1988, Boeuf 1992).

SEAWATER ADAPTATION IN NON SMOLTIFYING SALMONIDS

A lot of attempts in Norway, then in North America, Japan, Denmark, France, South America... have revealed the SW adaptation capacities in both rainbow and brown trouts. However the strains used for rearing are not migrating steelhead or sea trout. How do they react ?

These species naturally only live in FW and spend their whole life in river, this fact not excluding local migrations in streams, but never towards open ocean. Directly immersed in full salinity SW, they rapidly suffer deep osmotic and ionic disturbances and die at small sizes. Larger fish can react and adapt, by beginning to decrease the gill integument water permeability (Gordon 1963) to prevent too massive water losses, this capacity consisting in an important part of the salinity adaption, probably according to immediate endocrine changes (prolactin is acutely involved in this phenomenon). Thereafter the fish begins to drink 9 to 12 h after the SW contact (Evans 1979, Bath and Eddy 1979b) to try to compensate the water losses, mainly through the gills. On the other hand, in juveniles, the increase of drinking rate is much slower after abrupt transfer from FW to dilute SW (Tytler *et al* 1990), but they don't survive in full salinity SW. Branchial chloride efflux increases immediately on transfer and Cl^- urine excretion begins to develop after 14 h, following 30 ‰ SW transfer (Bath and Eddy 1979a). Na^+ efflux only begins after 4-5 h, following the SW contact, but anyway after 24 h in SW, the effluxes for both sodium and chloride are much lower than expected for a fish in marine environment (Bath and Eddy 1979a). It was suggested (Bath and Eddy 1979b) that during SW transfer, Cl^- influx is greater than Na^+ influx due to steeper chemical and electrical gradient across the gills and that this may be a possible stimulus for the immediate increase in Cl^- efflux. Intracellular ionic content increases, mainly chloride (Bath and Eddy 1979b). (Na^+-K^+)-ATPase activity progressively increases in the gill microsomes (see Figure 3), from the third day in SW to reach a maximum after 2-3 weeks in 35 ‰ SW (Boeuf 1987, Madsen and Naamansen 1989)).

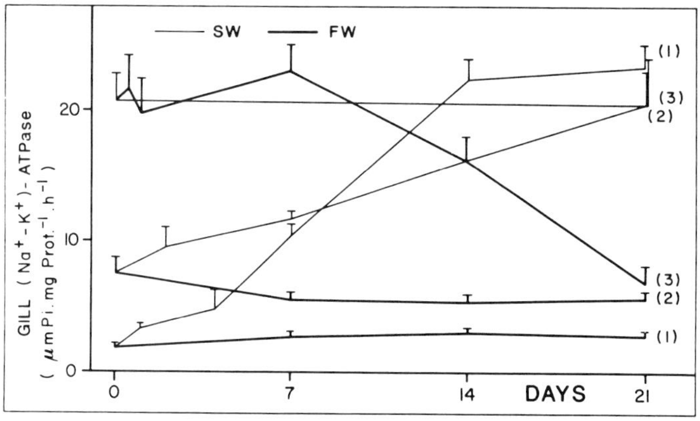

Figure 3 : Gill ATPase activity in rainbow trout after direct exposure from FW to full salinity SW (1 and 2), and from SW to FW (3). After 2-3 weeks, the fish reaches the "normal" enzyme activity level (original data). For all figures, each point represents the mean ± the standard error of the mean for 10 fish.

The "classical" response of the trouts to a direct SW transfer is represented in Figures 4 and 5. Due to delayed reactions to drinking mechanisms and ions extrusion, the fish blood concentration increases after the SW transfer : described for the first time by Houston (1957), then by Jackson (1981), two phases follow one after the other. Firstly, there is an osmotic adjustment phase, essentially the fact of dehydration, ionic diffusion and intestinal ionic absorption (Shehadeh and Gordon 1969) and secondly a regulation phase, after 3-4 days consisting in a much lower diffusion through te gill and mainly an active ionic excretion through the action of the gill ATPase activity (Boeuf 1987, Avella and Bornancin 1990). Progressive adaptation to SW seriously improves the situation (Figure 4a), giving the fish the time to react and to adapt. The larger the fish is the better the response (Figure 4c). From Figure 5, we can see the physiological response of rainbow trouts to direct SW transfer in relationships with both SW temperature and fish size.

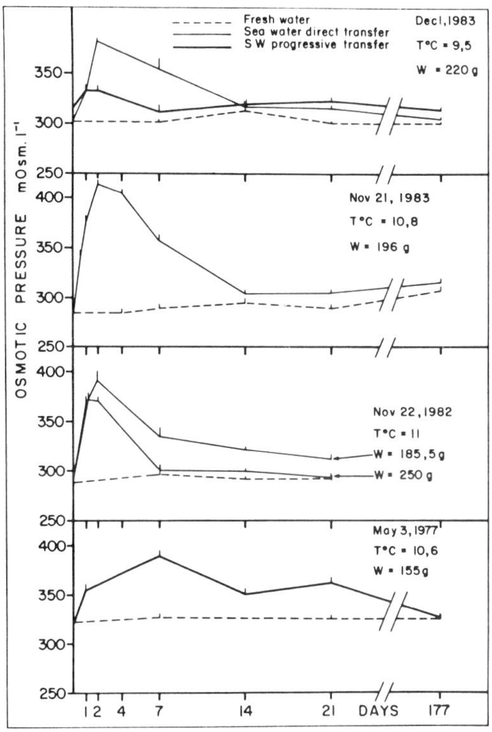

Figure 4 : Blood plasma osmotic pressure changes after direct (1,2 3 and 4) or progressive (on 10 days, 1) transfer from FW to full salinity SW (35 ‰) in rainbow trout. Date, SW temperature and fish average weights are indicated (original data).

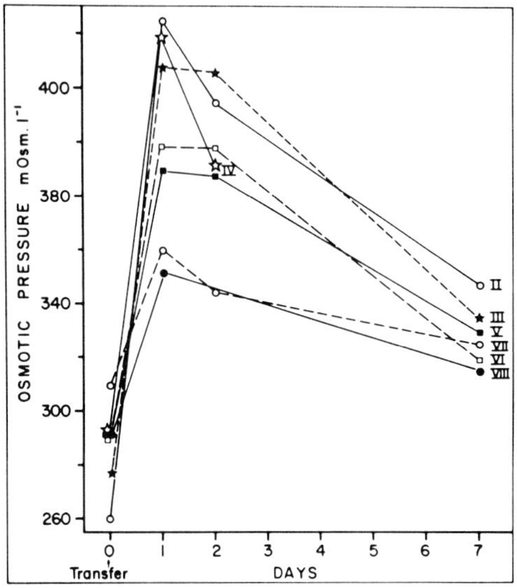

Figure 5 : Plasma osmotic pressure changes after direct exposure from FW to 35 ‰ SW in rainbow trout. Mortality appears well correlated to osmotic disequilibrium intensity (original data).

Code	Date of transfer	SW temperature	weight	7 days mortality
I	Aug. 3	17.5°C	20.3 g	100 %
II	Aug. 17	17.5	30.9	90
III	Aug. 31	16.8	36.4	78
IV	Sept. 14	17.0	55.5	90
V	Sept. 27	16.5	62.2	25
VI	Oct. 12	14.6	112.2	18
VII	Oct. 26	13.0	127.5	5
VIII	Nov. 9	12.5	122.0	5

About this last point, Figures 4 and 5 well illustrate that larger trouts are physiologically more capable of dealing with SW : this might simply be a consequence of the volume, surface area ratio being inversely proportional to weight (Bath and Eddy 1979a). These same authors showed that in the first hours into SW, the oxygen consumption decreased in big fish to go back to the FW fish values after 6 h : both beginning of Na^+ extrusion and O_2 uptake seem linked. When the trout enters SW there is a complete reversal of the osmotic gradient across the gill which must necessitate water loss and shrinkage of the branchial epithelium so reducing the exchange surface (Bath and Eddy 1979a). On the other hand, contrarily to these authors, we are capable to directly adapt rainbow trouts to full salinity SW (35 ‰) in France without problems in terms of survival and subsequent growth capacity (Boeuf and Harache 1984a, Boeuf 1987). Maybe the initial FW "quality" can represent the explanation for this fact : high Ca^{2+} and mainly very low Na^{+} and Cl^{-} FW levels can lead to higher difficulties to adapt to SW for salmonids. After 2-3 weeks, the trouts (Figures 4, 5 and 6) are perfectly adapted and plasma electrolyte concentrations vary only sligthly between FW and SW fish (Lahlou *et al* 1975).

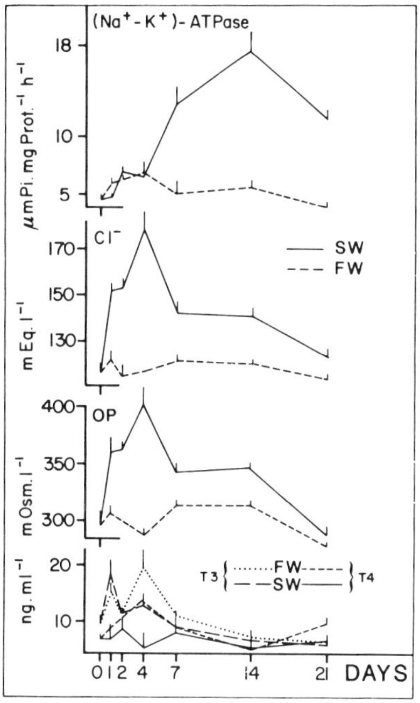

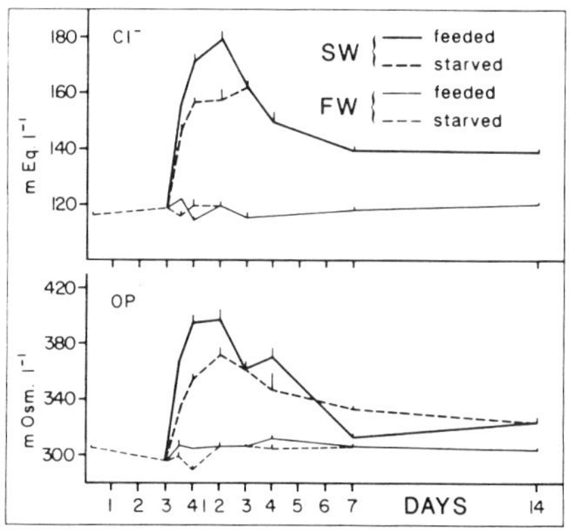

Figure 6 : Gill ATPase activity, plasma T_3, T_4, chloride and osmotic pressure levels in brown trout (*Salmo trutta*) after direct exposure to 35 ‰ SW (original data).

Figure 7 : Influence of starvation on SW adaptation capacities in rainbow trout ; plasma chloride and osmotic pressure after direct exposure to SW (original data).

The drinking rate in adapted rainbow trout can reach 5.4 $ml.h^{-1}.kg^{-1}$ body weight in full salinity

SW, the intestinal absorption of water, sodium and chloride increasing deeply after complete transfer (Shehadeh and Gordon 1969) : during passage through the intestine, ingested SW and dilutions thereof are gradually depleted of monovalent ions and calcium and enriched in magnesium and sulfate. SW adaptation changes Na^+ intestinal absorption principally in its posterior part, under the ATPase activity action (Colin *et al* 1985). The fish, just after direct exposure to SW, can remain a few days without urine production (Holmes 1961). Values are indicated in Figures 1 and 2.

Amino acids of white muscle increase in SW fish (Jurss *et al* 1983) and ionic compartments arrangements are of the highest importance in SW adaptation, the transfers occurring shortly after exposure and associated with energy expenditure for regulation (Kostecki 1984) and the interstitial fluids representing almost seven times plasma volume.

Increasing salinity and decreasing weights appear adversely to affect the trout ability to adapt to SW and high plasma osmotic (> 360 $mOsm.l^{-1}$) and ionic concentrations are shown to coincide with loss of appetite and eventually death (Jackson 1981, Boeuf 1987 ; see Figure 5). Anyway, domestic trouts are able to adapt to SW, to survive (> 95 %) and grow under specific conditions which are :

- a minimum size : for a direct transfer to 35 ‰, we found in France > 110 g for domestic rainbow trout of 8-9 months in the first autumn (Figure 5) and > 40 g for brown trouts of the same age, > 60 g for yearling in the second spring (Quillet et al 1986),

- non sexual maturation : both male and female cannot survive if transferred to full salinity,

- SW temperature not higher than 15 °C (Figure 5, Lahlou *et al* 1975) ; to take care also of the difference between FW and SW temperatures, to transfer the fish,

- the fish has to be starved before transfer (see Figure 7 and Nance *et al* 1987) : this status enables the fish to better control the osmotic and ionic disequilibrium in clearly decreasing the gill ions fluxes,

- the fish needs a good "general health status" and a perfect integrity of the skin ("descaled" salmonids cannot survive in SW) and the gill integument, decisive in breathing, osmoregulation and excretion ; no veterinary treatments, no ammonia exposure, no high loads in FW just prior to SW transfer,

- to avoid, if possible, long and difficult transportation.

In these conditions, trouts are capable to adapt, but every time it has to be after a transitory disturbances phase (2 weeks), ineluctable at high salinity. The fish is perfectly adapted to SW and can be considered as a marine fish after 3 weeks. However, rainbow trout will not be able to survive in SW during the summer in France when both salinity and temperature are high (Boeuf and Harache 1984a) while brown trout can adapt much more easily and survive (Boeuf and Harache 1984b).

SEAWATER ADAPTATION IN MIGRATING SPECIES

These salmonids naturally begin their life in FW and spend from six weeks to six years in river, depending on the species and the strain. At a very definite moment (end of the parr-smolt transformation), they leave the river and migrate actively to the estuary and the sea. In these conditions they have a very high SW adaptation potential, smolting occurring in FW, but corresponding to a SW pre-adaptation phase (Hoar 1988, Boeuf 1992). Of course, if they are easily able to adapt at that moment, they will face many difficulties in other periods in the year. The main problem in aquaculture will be to select the good fish and the good time in the year to realize the transfer from FW to SW (see Boeuf 1992).

Figure 8 shows the osmotic and ionic changes occuring in Atlantic salmon of the same strain, reared in the same conditions in FW, after a direct exposure to 35 ‰ SW (Boeuf *et al* 1989) at four times in the year. According to the moment, the size of the fish and mainly the physiological status, the young salmon will be capable or no to adapt, and SW adaptation mechanisms will be totally different. Only the smolt is capable to survive, adapt and grow in SW.

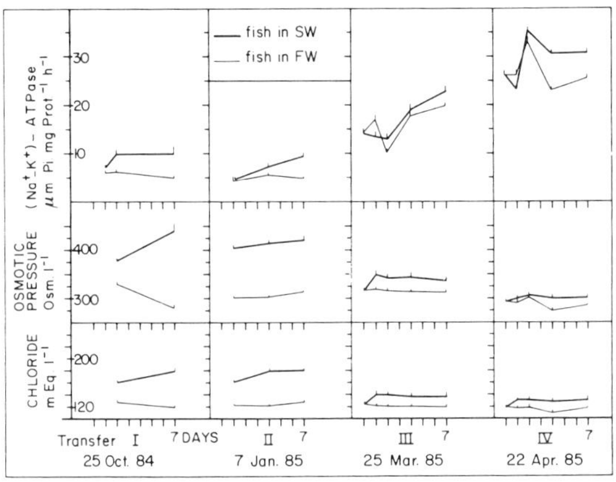

Figure 8 : Gill ATPase activity, plasma chloride and osmotic pressure in Atlantic salmon, following a direct transfer from FW to SW at four times in the year ; for more explanation see Boeuf *et al* 1989 ; presented with the kind permission of Elsevier, reproduced from *Aquaculture*.

In smolt, drinking commences 4-6 h after abrupt transfer from FW to SW at a rate of 2.4 $ml.h^{-1}.kg^{-1}$ to reach 7.94 in feeding fish after 10 days (Usher *et al* 1988). This delay is longer than in eel where Hirano (1974) showed an immediate triggering of drinking, just after SW contact, this probably mediated by peripheral chloride receptors. In fully adapted salmon, the quantity of food in the gut does not appear to affect water ingestion (Usher *et al* 1988), like in trout (Dabrowski *et al* 1986). To explain this delay to begin to drink in smolt and however the absence of osmotic disequilibrium after direct exposure to SW (Figure 8), Nance *et al* (1990) checked the impact of SW contact on both capacity to impermeabilize the gill and to actively excrete ions : in fact, the terminal Atlantic salmon smolt seems perfectly capable to totally impermeabilize the gill and begin to eliminate Na^+ in a few minutes, the excretion mainly effected through the filaments. Gill Na^+-K^+-ATPase activity is already high in FW smolt (Zaugg and McLain 1972, Saunders and Henderson 1978, Boeuf *et al* 1985, Figure 9) and enables the fish to immediately react in SW. Apart from this enzymatic system, citrate synthase and cytochrome C oxydase specific ativities do not really change after transfer from FW to SW (McCormick *et al* 1989), indicating no substantial increases in metabolic capacity of the gill.

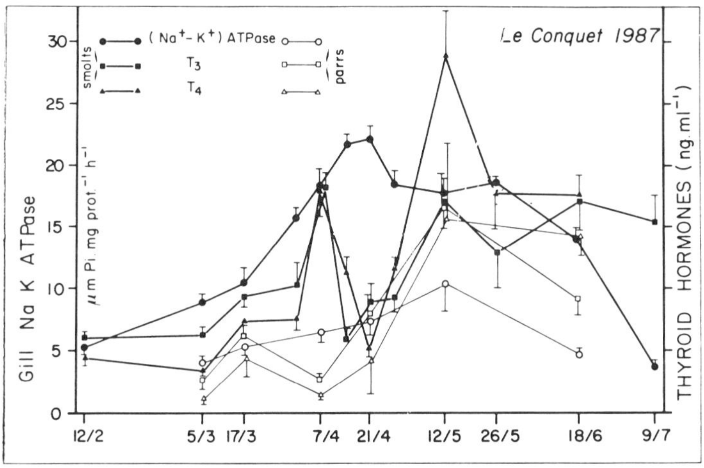

Figure 9 : Plasma T_3, T_4 and gill (Na^+-K^+)-ATPase activity in Atlantic salmon in FW in upper mode fish ("smolts") and lower mode fish ("parr"), original data.

A seawater challenge test is "classically" used in many countries, using the salmon aptitude to control blood Na^+ (Blackburn and Clarke 1987) or Cl^- after direct exposure to high salinity SW. Before or after smolting, the fish can react as a small trout, not at all capable to survive and to adapt (Parry 1960, Farmer *et al* 1978, Boeuf *et al* 1978, 1989) or as a large trout, able to survive and adapt after a transitory osmotic and ionic disequilibrium . But very often, intermediery responses appear with fish surviving, but without growth (stunting, see Hoar 1988 and Boeuf 1992). Parr-smolt transformation criteria are very useful to forecast the adaptation capacities of the fish. A good criterion of the success of the SW transfer for the farmer is the feeding capacity of the SW fish, even if very often, appetite and growth are transitorily suppressed following direct exposure to full salinity (Usher *et al* 1991), this fact highly depending on the temperature. Progressive adaptation from FW to SW, closer the natural behaviour, can highly improve the "quality" of the SW transfer (Bergheim *et al* 1990).

In fact the smolt, still living in FW, is perfectly preadapted to SW at the level of the skin, intestine, kidney-urinary bladder and gill systems (Hoar 1988, Boeuf 1992), this allowing it to react very fast and to adapt, survive and grow in good conditions in SW.

Salinity tolerance is affected by environmental factors, mainly temperature (Figure 5, Usher *et al* 1991) and photoperiod. Both very high (> 18°C) and very low (< 6°C) temperatures impair SW adaptation in smolts. Several studies have demonstrated that artificially extended day length is an effective way to stimulate growth and smolting in salmonids (McCormick *et al* 1987, Saunders and Harmon 1990).

Hormones are of course involved with nervous system in the mechanisms of SW adaptability and several authors have tried to trigger or to improve SW tolerance, using hormonal treatments :these experiments have been preferentially carried out in smoltifying species because they react much more clearly than non migratory trouts, this treatment in fact triggering earlier natural processes the fish would have developed later in the cycle.

Among a few hormones, cortisol and growth hormone appear as effective in smoltification triggering and SW tolerance improvement (Collie *et al* 1989, Boeuf *et al* 1990, Madsen 1990a). These hormones can also improve the development of hypoosmoregulatory mechanisms in rainbow trout (Madsen 1990b).

CONCLUSIONS

Salmonids are capable to adapt to full salinity SW, maintaining a good osmotic homeostasy of the *milieu intérieur*. To reach this aim, patterns will be very different according to the species or the strain :

- FW "sedentary" species or strains are sometimes able to adapt, survive and grow but every time following a transitory osmotic and ionic disequilibrium, depending on the water temperature and salinity, the size of the fish and its maturity status,
- migratory, smoltifying species or strains will be able to adapt without transitory osmotic disequilibrium, but at very precise moments in the year.

SW adaptation difficulties are directly linked to the external salinity, the higher it is and the more difficult the transfer to SW. A "salinity threshold" seems to appear beyond 31-32 ‰, salmonids clearly preferring to live at lower salinities. Progressive adaptation is much better born by all the fish.

Inadequate rearing or transfer conditions can severely affect the fish and involve mortalities or stunting responses. For non migratory species, there is not of course a preadaptation phase in FW but however some of them, beyond a certain size, can bear high salinity and grow : if they are able to survive, they will be able to grow, stunting phenomenon mainly affecting salmons. It can be interesting to rear them in SW in winter, temperature generally being higher in SW than in FW and stay in SW improving the quality. For smoltifying species, the transfer is possible without any problem provided the end of the smolting time is respected.

In fact, the grower has the possibility to choose between two situations, the physiology of the fish involving biotechnical aspects :

- to use non migratory trouts and to transfer them to SW by respecting the conditions we precedently exposed, almost all around the year, but only at large size,
- to use smoltifying species and to transfer them at low size, but in very specific moments in the year. Anyway, the grower has to think that a direct exposure to SW is not easy to bear for the fish and to take a maximum of precautions, mainly at high salinity.

Acknowledgements

I want to thank Annick Le Roux, Armelle Sévère and Bruno Petton for their excellent technical assistance having enabled us to produce these results, Sylvie Gros for her help in drawing the figures, Marie Moal in putting the text in correct English and Patrick Payan in revising the original manuscript.

REFERENCES

Avella, M. and Bornancin, M. (1990) Ion fluxes in the gills of freshwater and seawater salmonid fish. In Comparative Physiology, (eds R.K.H. Kinne, E. Kinne–Saffran, K.W. Beyenbach), Basel, Karger, 6, pp. 1–13.

Ballantyne, J.S., Moyes, C.D. and Moon, T.W. (1987) Compatible and counteracting solutes and the evolution of ion and osmoregulation in fishes. Can. J. Zool., 65(8), 1883–1888.

Bath, R.N. and Eddy, F.B. (1979a) Salt and water balance in rainbow trout *Salmo gairdneri* rapidity transferred from fresh to seawater. J. Exp. Biol., 83, 193–202.

Bath, R.N. and Eddy, F.B. (1979b) Ionic and respiratory regulation in rainbow trout during rapid transfer to seawater. J. Comp. Physiol., 134, 351–357.

Bergheim, A., Kroglund, F., Vatne, D.F. and Rosseland, B.O. (1990) Blood plasma parameters in farmed Atlantic salmon (*Salmo salar* L.) transferred to sea cages at age eight to ten months. Aquaculture, 84, 159–165.

Blackburn, J. and Clarke, W.C. (1987) Revised procedure for the 24 hours seawater challenge test to measure seawater adaptability of juvenile salmonids. Can. Tech. Rep. Fish. Aquat. Sci., 1515, 35 pp.

Boeuf, G. (1987) Contribution à l'étude de l'adaptation à l'eau de mer chez les Poissons Salmonidés. Détermination de critères de smoltification par mesures de l'activité (Na^+–K^+)–ATPasique des microsomes de la branchie et des hormones thyroïdiennes plasmatiques. Thèse de Doctorat d'Etat, Université de Bretagne Occidentale, pp. 369.

Boeuf, G. (1992) Salmonid smolting : a pre–adaptation to the oceanic environment. In Fish Ecophysiology, J.C.Rankin and J.B.Jensen eds, Chapman and Hall, London, in press.

Boeuf, G. and Harache, Y. (1984a) Salmonid marine rearing in France. J. World Mar. Soc., 15, 246–253.

Boeuf, G. and Harache, Y. (1984b) Adaptation osmotique à l'eau de mer de différentes espèces (*Salmo trutta, Salmo gairdneri, Salvelinus fontinalis*) et hybride (*Salmo trutta* femelle x, *Salvelinus fontinalis* mâle) de salmonidés. Aquaculture, 40(4), 343–358.
Boeuf, G., Lasserre, P. and Harache, Y. (1978) Osmotic adaptation of *Oncorhynchus kisutch* Walbaum. II. Plasma osmotic and ionic variations and gill Na+–K+–Adenosine Triphosphatase activity of yearling coho salmon transferred to seawater. Aquaculture, 15(1), 35–52.
Boeuf, G., Le Roux, A., Gaignon, J.L. and Harache, Y. (1985) Gill (Na+–K+)–ATPase activity and smolting in Atlantic salmon *Salmo salar* L. in France. Aquaculture, 45(1–4), 73–81.
Boeuf, G., Le Bail, P.Y. and Prunet, P. (1989) Growth hormone and thyroid hormones during Atlantic salmon, *Salmo salar* L, smolting and after transfer to seawater. Aquaculture, 82(1–4), 257–268.
Boeuf, G., Prunet, P. and Le Bail, P.Y. (1990) Un traitement à l'hormone de croissance peut–il stimuler la smoltification du saumon atlantique ? C. R. Acad. Sci. Paris, 310(III), 75–80.
Colin, D.A., Nonnotte, G., Leray, C. and Nonnotte, L. (1985) Na transport and enzyme activities in the intestine of the freshwater and sea–water adapted trout (*Salmo gairdneri* R.). Comp. Biochem. Physiol., 81A(3), 695–698.
Collie, N.L., Bolton, J.P., Kawauchi, H. and Hirano, T. (1989) Survival of salmonids in seawater and the time–frame of growth hormone. Fish Physiol. Biochem., 7(1–4), 315–321.
Conte, F.P. and Lin, D.H. (1967) Kinetics of cellular morphogenesis in gill epithelium during salt water adaptation of *Oncorhynchus walbaum*. Comp. Biochem. Physiol., 23, 945–957.
Dabrowski, K., Leray, C., Nonnotte, G. and Colin, D.A. (1986) Protein digestion and ion concentrations in rainbow trout (*Salmo gairdneri* Rich.) digestive tract in sea– and fresh water. Comp. Biochem. Physiol., 83A(1), 27–39.
Evans, R.M. (1979) Onset and rate of drinking in rainbow tout *Salmo gairdneri* following transfer to dilute seawater. Can. J. Zool., 57, 1863–1865.
Farmer, G.J., Ritter, J.A. and Ashfield, D. (1978) Seawater adaptation and parr–smolt transformation of juvenile Atlantic salmon, *Salmo salar*. J. Fish.Research Bd Can., 35(1), 93–100.
Gordon, M.S. (1963) Chloride exchanges in rainbow trout *Salmo gairdneri* adapted to different salinities. Biol. Bull., 124, 45–54.
Hirano, T. (1974) Some factors regulating water intake by the eel, *Anguilla japonica*. J. Exp. Biol., 61, 737–747.
Hoar, W.S. (1988) The physiology of smolting salmonids, in Fish Physiology (eds W.S.Hoar and D.J.Randall), Academic Press, XIB, 275–343.
Holmes, R.M. (1961) Kidney function in migrating salmonids. Rep. Challenger Soc., 3, 23–28.
Holmes, R.M. and Donaldson, E.M. (1969) The body compartments and the distribution of electrolytes. In : Fish Physiology, vol 1 "Excretion, Ionic regulation and Metabolism", Hoar, W.S and Randall, D.J. (Eds), 1–79.
Houston, A.H. (1957) Responses of juvenile chum, pink and coho salmon to sharp seawater gradients. Can. J. Zool., 35, 371–383.
Jackson, A.J. (1981) Osmotic regulation in rainbow trout *Salmo gairdneri* following transfer to seawater. Aquaculture, 24(1–2), 143–151.
Jurss, K., Bittorf, T., Vokler, T. and Wacke, R., 1983. Influence of nutrition on biochemical seawater adaptation of the rainbow trout *Salmo gairdneri* Richardson. Comp. Biochem. Physiol., 75B(4), 713–717.
Kostecki, P.T. (1984) The effect of osmotic and ion osmotic stresses on the blood and urine composition and urine flow of rainbow torut *Salmo gairdneri*. Comp. Biochem. Physiol., 79A(2), 215–221.
Lahlou, B., Crenesse, D., Bensahla–Talet, A. and Porthe–Nibelle, J. (1975) Adaptation de la truite d'élevage à l'eau de mer. Effets sur les concentrations plasmatiques, les échanges branchiaux et le transport intestinal du sodium. J. Physiol., Paris, 70, 593–603.
Madsen, S.S. and Naamansen E.T. (1989) Plasma ionic regulation and gill Na+/K+–ATPase changes during rapid transfer to sea water of yearling rainbow trout, *Salmo gairdneri* : time course and seasonal variation. J. Fish Physiol., 34, 829–840.

Madsen, S.S. (1990a) The role of cortisol and growth hormone in seawater adaptation and development of hypoosmoregulatory mechanisms in sea trout parr (*Salmo trutta* trutta). Gen. Comp. Endocrinol., 79, 1–11.
Madsen, S.S. (1990b) Effect of repetitive cortisol and thyroxine injections on chloride cells number and Na^+–K^+–ATPase activity in gills of freshwater acclimated rainbow trout, *Salmo gairdneri*. Comp. Biochem. Physiol., 95A, 171–175
Markert, J.R. and Vanstone, W.E. (1966) Pigments in the belly skin of coho salmon *Oncorhynchus kisutch*. J. Fisheries Research Bd Can., 23(7), 1095–1098.
McCormick, S.D., Moyes, C.D. and Ballantyne, J.S. (1989) Influence of salinity on the energetics of gill and kidney of Atlantic salmon (*Salmo salar*). Fish Physiol. Biochem., 6(4), 243–254.
McCormick, S.D., Saunders R.L., Henderson, E.B. and Harmon, P.R. (1987) Photoperiod control of parr–smolt transformation in Atlantic salmon (*Salmo salar*) : changes in salinity tolerance, gill Na+, K+–ATPase activity and plasma thyroid hormones. Can. J. Fisheries Aquat. Sci., 44, 1462–1469.
Nance, J.M., Bornancin, M., Sola, F., Boeuf, G. and Dutil, J.D. (1990) Study of transbranchial Na^+ exchange in *Salmo salar* smolts and post–smolts directly transferred to sea water. Comp. Biochem. Physiol., 96A(2), 303–308.
Nance, J.M., Masoni, A., Sola, F.and Bornancin, M. (1987) The effects of starvation and sexual maturation on Na+ transbranchial fluxes following direct transfer from freshwater to seawater in rainbow trout (*Salmo gairdneri*). Comp. Biochem. Physiol., 87A, 613–622.
Nonnotte, L., Nonnotte, G. and Leray, C. (1986) Morphological changes in the middle intestine of the rainbow trout, *Salmo gairdneri*, induced by a hyperosmotic environment. Cell Tissue Res., 243, 619–628.
Nonnotte, G., Aubree–Pelletier, A. and Colin, D.A. (1987) Transient changes of the intestinal absorption of sodium and chloride in the rainbow trout after abrupt transfer into sea–water. Comp. Biochem. Physiol., 88A(3), 437–441.
Parry, G. (1960) The development of salinity tolerance in the salmon *Salmo salar* and some related species. J. Exp. Biol., 37, 425–434.
Payan, P., Girard, J.P. and Mayer–Gostan, N. (1984) Branchial ion movements in teleost : the roles of respiratory and chloride cells, in Fish Physiology (eds W.S. Hoar and D.J. Randall), Academic Press, XB, 39–63.
Pisam, M., Boeuf, G., Prunet P. and Rambourg, A. (1990) Ultrastructural features of mitochondria–rich cells in stenohaline freshwater and seawater fishes. Am. J. Anat., 187, 21–31.
Quillet, E., Krieg, F., Happe, A. et Chevassus, B. (1986) Etudes des possibilités de transfert automnal en mer des juvéniles de truite fario (*Salmo trutta*). Bull. Français Pêche Pisc., 303, 125–133.
Saunders, R.L. and Henderson, E.B. (1978) Changes in gill ATPase activity and smolt status of Atlantic salmon *Salmo salar*. J. Fish. Research Bd Can., 35(12), 55–66.
Saunders, R.L. and Harmon, P.R. (1990) Influence of photoperiod on growth of juvenile Atlantic salmon and development of salinity tolerance during winter–spring. Trans. Amer. Fish. Soc., 119, 689–697.
Shehadeh, Z.H. and Gordon, M.S. (1969) The role of the intestine in salinity adaptation of the rainbow trout *Salmo gairdneri*. Comp. Biochem.Physiol., 30, 397–418.
Tytler, P., Tatner, M. and Findlay, C. (1990) The ontogeny of drinking in the rainbow trout, *Oncorhynchus mykiss* (Walbaum). J. Fish Biol., 36, 867–875.
Usher, M.L., Talbot, C. and Eddy, F.B. (1988) Drinking in Atlantic salmon smolts transferred to seawater and the relationship between drinking and feeding. Aquaculture, 73, 237–246.
Usher, M.L., Talbot, C. and Eddy, F.B. (1990) Effects of transfer to seawater on digestion and gut function in Atlantic salmon smolts (*Salmo salar* L.). Aquaculture, 90, 85–96.
Zaugg, W.S. and McLain, L.R. (1972) Changes in gill ATPase activity associated with parr smolt transformation in steelhead trout *Salmo gairdneri*, coho *Oncorhynchus kisutch*, and spring chinook salmon *O.tshawytscha*. J. Fish. Research Bd Can., 29, 167–171.
Zaugg, W.S., Prentice, E.F. and Waknitz, F.N. (1985) Importance of river migration to the development of seawater tolerance in Columbia river anadromous salmonids. Aquaculture, 51, 33–47.

6

Photoperiodism and the Control of Reproduction and Development in Farmed Fish

N. Bromage, C. Randall, B. Davies, M. Thrush, J. Duston, M. Carillo, and S. Zanuy

Introduction

For most temperate and longer latitude organisms including fish, life history is dominated by the seasonal changes in climate and food supplies. As a consequence control systems have evolved which enable developmental events such as reproduction and smoltification to be timed by seasonal changes in the external environment. This ensures that maturation and the production of young fish occur at appropriate times of the year for optimum survival of the species.

Amongst the large number of environmental factors, implicated as possible cues or "zeitgebers" for the synchronisation of reproduction and development with season in fish, the seasonal waxing and waning of photoperiod or daylength has been shown to be of prime importance in the salmonids (Whitehead *et al* 1978; Bromage & Duston, 1986; Bromage & Cumaranatunga, 1988) and also the flatfish, bass, breams and mullet (Girin &

[1] Instituto de Acuicultura, Torre de la Sal, Ribera de Cabanes, 12595 Castellon, Spain.

Devauchelle, 1978; Bye, 1987; Carrillo *et al* 1989, 1991). Even the reproduction and spawning of many cyprinids and some tropical and sub-tropical species which existing literature (de Vlaming, 1972; Sundararaj & Vasal, 1976; Lam & Munro, 1987) would suggest are cued by temperature or rainfall, almost certainly depend on photoperiod as a measure of daily and seasonal time (Brook & Bromage, 1988; Bromage *et al* 1990).

Although seasonal reproduction and developmental change are of considerable adaptive significance for wild stocks of fish, this seasonality is a major disadvantage in intensive fish farming where supplies of eggs and fry and table fish production need to be maintained throughout the year (Bromage & Duston, 1986). The precision and reliability of the reproductive and developmental responses of fish to specific photoperiod cues, however, provides the means by which the timing of maturation and spawning can be adjusted so that eggs and fry can be produced on demand for commercial farms (Bromage & Cumaranatunga, 1988).

In the present review we consider the ways in which fish respond to altered seasonal and constant light regimes, how seasonal and daily time is measured and how these processes can be manipulated by aquaculturists to optimise the all-year-round production of farmed fish. Data will be included for a range of intensively-reared fish whilst concentrating on results for the rainbow and brown trout (*Oncorhynchus mykiss* and *Salmo trutta*) and the sea bass (*Dicentrarchus labrax*).

Seasonal Light Cycles

One of the first demonstrations that daylength exerts an

influence on the timing of reproduction in fish was the work of Hoover (1937) and Hoover & Hubbard (1938) on the brook trout (*Salvelinus fontinalis*); these authors showed that fish exposed to seasonal light cycles compressed into periods of time shorter than one year spawned 3 - 4 months earlier than control fish under ambient conditions. The response of brook trout to advanced seasonal photoperiods has been confirmed on a number of occasions (Hazard & Eddy, 1951; Corson, 1955; Carlson & Hale, 1973). Modified seasonal photoperiods have also been shown to alter the timing of spawning in other species including the rainbow trout (Whitehead *et al* 1978. 1983; Bromage *et al* 1982a; Bromage & Duston, 1986) and other Oncorhynchid species (MacQuarrie, 1978, 1979), the brown trout (Bromage *et al* 1990), the sea bass, gilthead bream (*Sparus aurata*), sole (*Solea solea*) and turbot (*Scophthalmus maximus*: Girin & Devauchelle, 1978). Spawning has also been shown to be delayed if the seasonal photoperiod is extended into periods of time greater than one year (MacQuarrie *et al* 1978; Bromage *et al* 1984; Bromage & Duston, 1986). An example of the advancing and delaying effects of compressed and extended seasonal photoperiod regimes respectively is shown in Fig 1, overleaf, (from Bromage & Duston, 1986).

Constant Photoperiod Regimes

Although modifications in the rate of change of the annual seasonal light cycle induce profound alterations in the timing of maturation and spawning, it is evident from a number of studies that the increasing and decreasing daylength experienced during the earlier and later parts of the annual cycle can be replaced by constant long and short daylengths respectively. Thus, rainbow trout, exposed to LD 18:6 from January to May

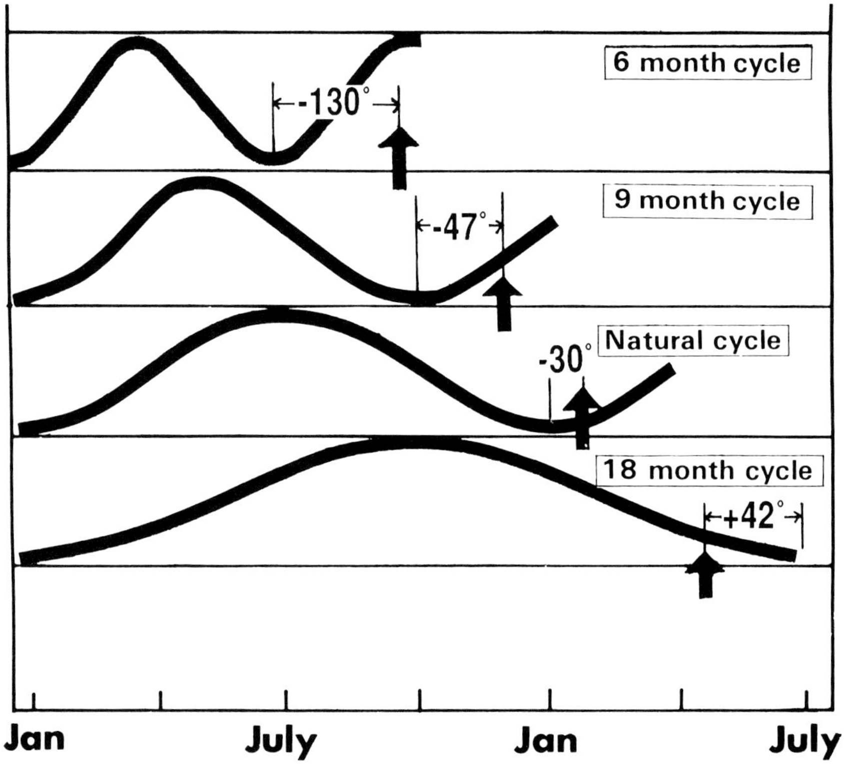

Fig 1 Effects of altered seasonal photoperiods on rainbow trout spawning (arrow is time of first spawning in each group)

followed by a sudden reduction to LD 6:18, spawn 3 - 4 months earlier than controls (Whitehead & Bromage, 1980; Bromage *et al*, 1984). Delaying the timing of the sudden increase to long days until February in this species results in a less pronounced advancement of spawning indicating that the timing of the increase is a significant determinant of maturation rate (Bromage *et al*, 1984; Duston & Bromage, 1988). The subsequent decrease to short days and its timing is also important, for fish which remain on long days throughout the year, although spawning earlier than controls, are less advanced than those which receive the "long-to-short" switch (Duston & Bromage 1987).

Advancements of spawning following exposure to constant long and then short days have also been reported for the masu salmon (*Oncorhynchus masou* : Takashima & Yamada, 1984), the brook trout (Henderson, 1963) the dace (*Leuciscus leuciscus*: Brook & Bromage, 1988), the brown trout (Bromage *et al* 1990, see Fig 2) and the sea bass

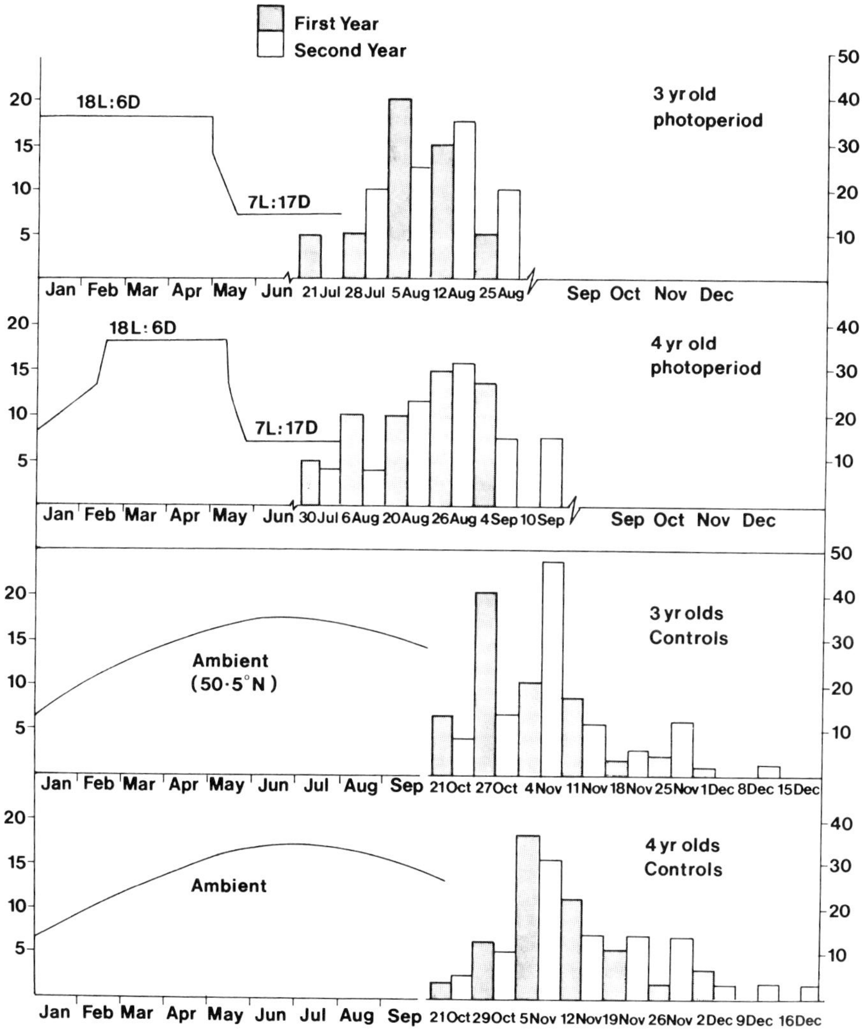

Fig 2 Effects of long-to-short constant photoperiods on spawning of stocks of 3 and 4 year old brown trout. Left and right vertical axes are hours of light/day and % fish spawning resp.

(*Dicentrarchus labrax*: Fig 3 after Carrillo *et al*, 1989, 1991). Furthermore, as little as one month of long days

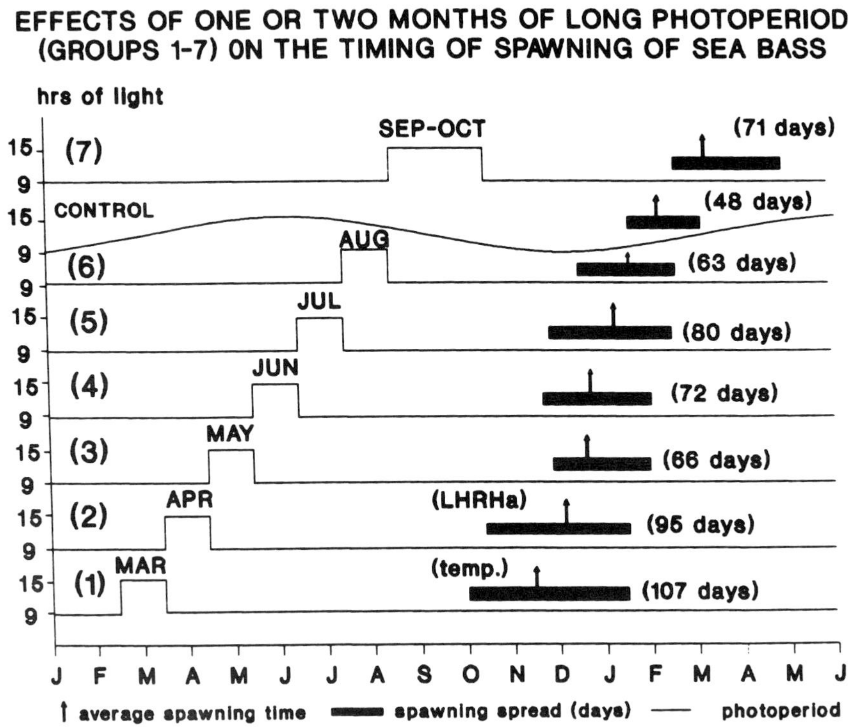

Fig 3 Effects of the timing of one or two months of long days in an otherwise short day regime on sea bass spawning.

during the early part of the cycle is all that is necessary to advance the timing of spawning (Bromage *et al* 1984; Takashima & Yamada, 1984; Carrillo *et al* 1989, 1991).

Constant daylengths have also been shown to delay spawning. Thus, constant short days either throughout the year or during the first 7 or 8 months of the year delay spawning as does exposure of fish to periods of constant long days after the summer solstice (Allison, 1951; Henderson 1963; Eriksson & Lundqvist 1980; Whitehead & Bromage, 1980; Bromage *et al*, 1982b, 1990, 1991).

Most of the experiments with salmonids have been conducted with constant water temperatures, thus raising the question of the relative importance of temperature and photoperiod in the timing of reproduction. Recently, Davies & Bromage (1991) confirmed, at least as far as the rainbow trout is concerned, that photoperiod is the principal determinant of the timing of maturation, with water temperature providing only a secondary influence. These authors showed that long days followed by short days produced similar 3 - 4 month advancements of spawning in fish maintained in river water, which had seasonally-varying temperatures (2 - 21°C), to those maintained in constant 9°C borehole water; there was, however, a modifying influence of temperature with the controls and photoperiodically-advanced fish in river water in each case spawning 3 - 4 weeks after their counterparts in borehole water (Fig. 4, overleaf, from Davies & Bromage, 1991). It should also be mentioned that very low temperatures may also delay maturation probably as a result of direct effects on oocyte growth (Bromage & Cumaranatunga 1988).

There are, also, other complications with temperature relating to its effect on final maturation of the oocytes, ovulation and oviposition. Davies and Bromage (unpublished) found in trout, maintained in river water and photoperiodically-induced to spawn in mid-summer during periods of very high ambient temperatures, that the eggs undergoing final maturation in the ovary appeared to overripen whilst still within the gland and

Spawning times for fish on borehole water exposed to ambient (T3) or advancing (T4) photoperiod.

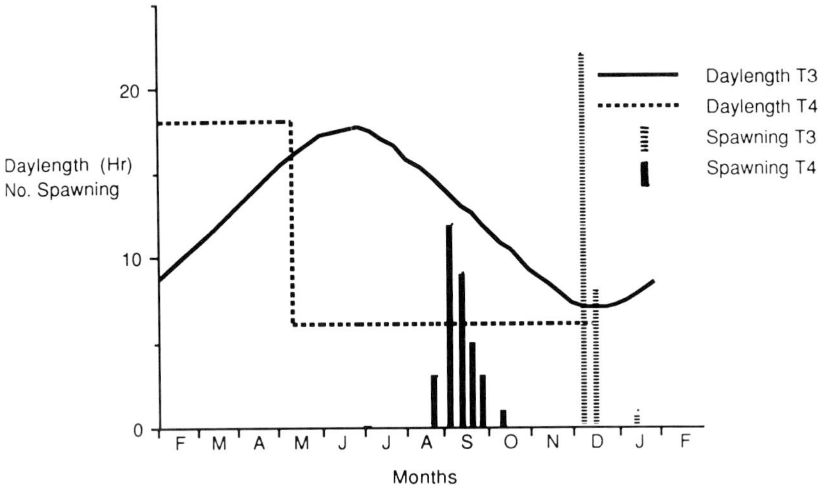

Spawning times for fish held on river water exposed to ambient (T1) or advancing (T2) photoperiod.

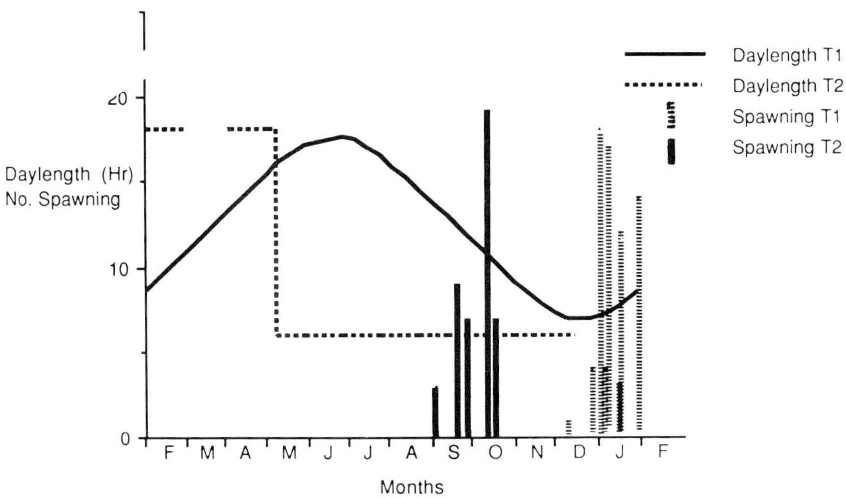

Fig 4 Spawning times for rainbow trout maintained on seasonally-varying river water and constant temeprature borehole water and exposed to ambient light or a stimulatory long-to-short regime.

before ovulation had occurred, resulting in no eggs being released. In a similar way, Carrillo *et al* (1991) reported that sea bass, which had been programmed to spawn in September using a modified photoperiod, failed to spawn naturally unless the water temperatures were reduced below 16°C or the fish received LHRHa injections (Fig 3). By contrast, fish in which final maturation was adjusted so that it occurred a little later in the Autumn or Winter, when ambient water temperatures had cooled, were able to spawn naturally without temperature or hormonal manipulation. Clearly, in both these species of fish the high temperatures were in some way interfering with the normal processes of final maturation, ovulation and/or oviposition.

Critical Daylength and Photoperiodic History

At one time it was considered that the absolute length of the photoperiod was of critical importance in fish, as in some other photoperiodic organisms, and that once this threshold daylength was reached in a seasonal cycle, then maturation would be induced. Certainly, in many birds and some mammals photoperiod serves to drive maturation in this way with gonadal recrudescence following a short time after exposure to a specific critical daylength (Follett, 1984). However, we find in trout exposed to modified seasonal or constant photoperiods that the induction of spawning does not bear a direct or constant relationship to the prevailing daylength. Thus, under altered seasonal cycles, spawning occurs at a different phase of the light cycle (i.e. a different daylength) depending on the degree of compression or extension of the modification. Under compressed cycles, spawning is phase-delayed with regards to the daylength at which spawning occurs in control fish under ambient conditions, whereas under extended 18-month seasonal cycles it is advanced. (Fig 1 shows these phase differences). Similar differences in phase were also reported by Devauchelle and Girin (1978) in their study of the effects of seasonal cycles in the turbot, sea bass and gilthead bream.

There are also variations in the timing of spawning of trout exposed to long and then short photoperiods at different times of the year with spawning time not relating in a consistent way to the onset of either the long or short daylengths (Bromage *et al* 1991). Many fish also continue to spawn under a range of year-round constant daylengths from LD 6:18 to LD 22:4 and including continuous light (LL) and dark (DD) (Bromage & Duston 1986; Bromage *et al* 1991). It would, thus, appear that hypotheses relating the timing of maturation to critical daylength are probably not applicable to fish.

This contention is further supported by a series of experiments in trout in which the effects on spawning of different lengths of long and short days have been investigated (Randall *et al* 1987, 1991). These show that any daylength can be considered long or short providing that the previous photoperiod to which the fish have been exposed is shorter or longer respectively. Fig 5, overleaf, based on data from Randall *et al* (1991) shows that daylengths as "long" as LD 14:10 can be considered short if fish have previously been exposed to longer daylengths. Even a daylength of LD 6:18 which would generally be agreed to be a short daylength can produce an advancement of spawning if followed by exposure to an even shorter daylength of LD 2:22 (Randall *et al* 1987, 1991). Clearly, photoperiodic history and the direction of change of photoperiod is of far greater importance in the timing of maturation and

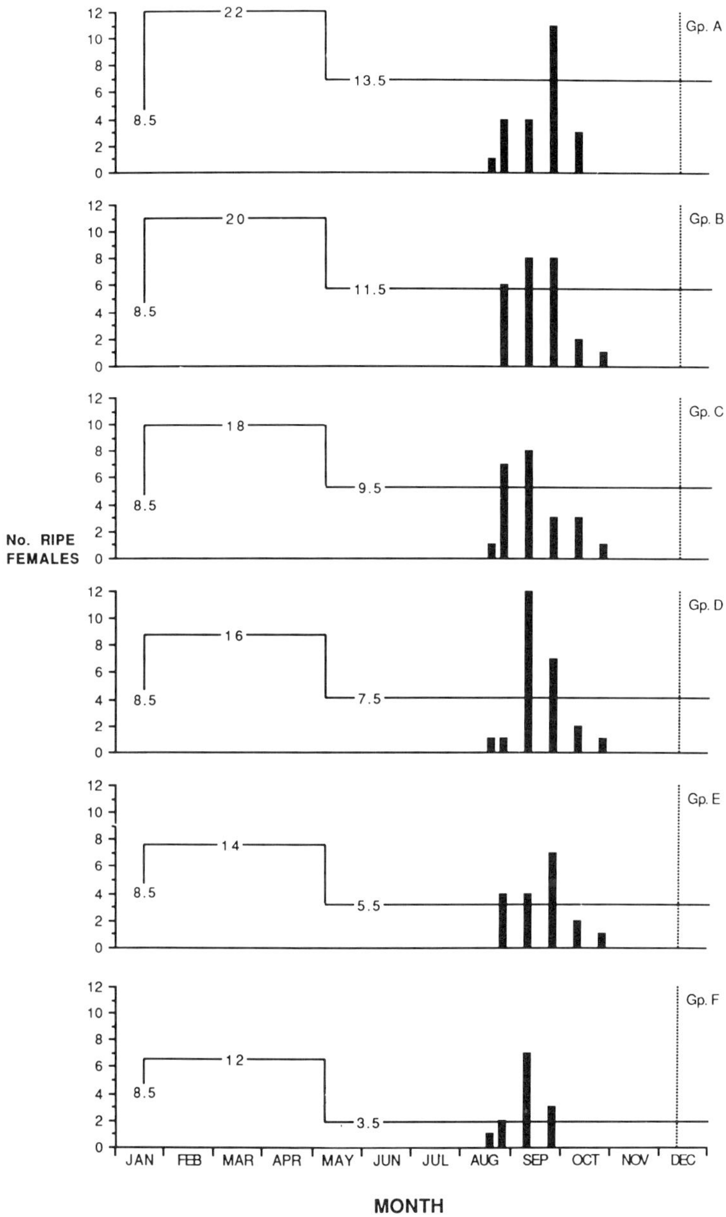

Fig 5 Effects of different lengths of "long" or "short" daylengths on spawning of rainbow trout.

spawning in fish than any concept of a rigid critical daylength.

This dependence of spawning time on photoperiodic history may also lead to complications in the interpretation of results of experiments where the timing of introduction of fish into photoperiod may have been carried out at different times of the year under quite different daylengths. In each case the response achieved will be dependent on the photoperiod from which the animals will have been transferred at the beginning of the experiment.

Endogenous Timing Mechanisms

From the foregoing discussion of the relative importance of photoperiodic history and critical daylength in the timing of fish reproduction, it is clear that the response to seasonal changes in daylength, exhibited by many fish, can not be accommodated within classical concepts of photoperiodism as exemplified by many birds (Follett, 1984). A particular difficulty is the ability of fish to continue to spawn at approximately annual intervals under constant conditions of light, temperature, food, etc (ie. in the absence of any obvious environmental cue). This has lead to suggestions that endogenous timing mechanisms may be in operation (Whitehead *et al* 1978; Sundararaj *et al* 1982; Bromage & Duston 1986) and, because they relate to annual rhythms, they have been described as circannual (see Gwinner, 1986).

Full acceptance of a rhythm as circannual depends on a number of criteria being met: firstly, the rhythm must be represented over more than one and preferably several cycles; secondly, under constant conditions it should "free-run" with a periodicity which approximates to, but is significantly different from a year; thirdly, it should be entrainable by environmental factors or cues; and lastly, it should not be unduly influenced by variations in temperature. Many of the earlier studies on fish do not satisfy all these criteria generally because they had only considered data from a single reproductive cycle or that all conditions had not been maintained at constant levels throughout the experiments (see Bromage & Duston 1986; Duston & Bromage 1986, 1987, 1988 for further discussion). Amongst the earlier studies, only Sundararaj *et al* (1982) working with the Indian catfish (*Heteropneustes fossilis*) provided good evidence of a self-sustaining circannual rhythm, although such a rhythm has also been proposed for the stickleback (*Gasterosteus aculeutus* : Baggerman, 1980).

More recently Bromage and co-workers working with trout, and Carrillo, Zanuy and Bromage working with the sea bass, have followed spawnings of fish maintained under constant condition for up to 5 years. Trout maintained under constant LD 6:18 exhibited circannual rhythm of gonadal maturation and spawning (Duston & Bromage 1986, 1991), whose periodicity varied between fish, and in successive cycles in the same fish, ranging in length from 11-15 months (Fig 6, overleaf). Under LD 18:6 or LL the periodicity of the rhythm is much shorter (Bromage *et al* 1984; Duston & Bromage 1986). Similar free-running circannual periodicities in spawning have also been recorded for sea bass maintained under constant LD 15:9 and LD 9:15 although with this species spawnings approximated to a year under both photoperiod regimes (Fig 7, overleaf, Carrillo, Bromage & Zanuy unpublished).

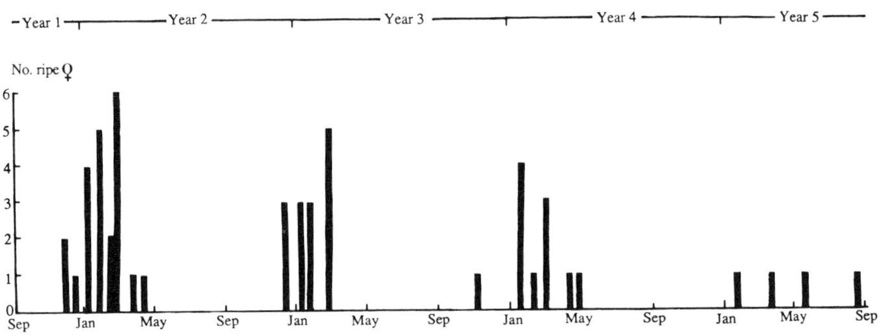

Fig 6 Circannual rhythms of spawning in the rainbow trout maintained under constant LD 6:18, water temperature, water quality and feed rate.

If one accepts that endogenous mechanisms underlie timing mechanisms in fish, then it is easier to understand how the altered seasonal and constant photoperiod regimes exert their effects, why photoperiodic history and the direction of change of daylength are important and why the same photoperiod can

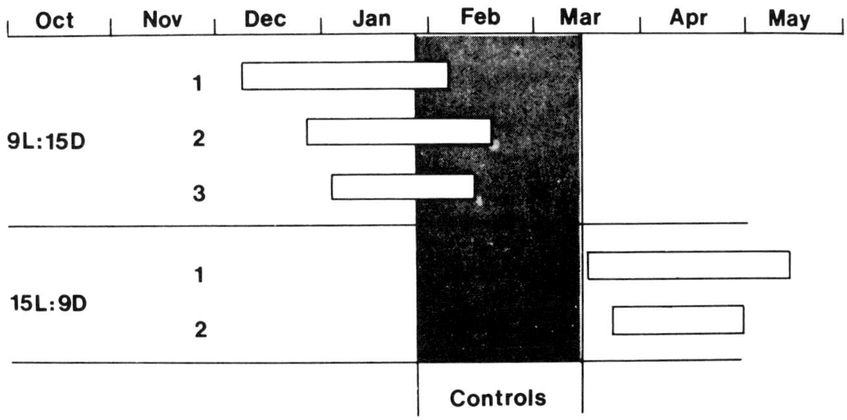

Fig 7 Circannual rhythms of spawning in the sea bass maintained under constant LD 15:9 and LD 9:15.

induce different effects if administered at different times of the year. Under natural conditions, the circannual rhythm or clock would be fully in phase with the ambient light cycle as a result of ongoing re-entrainment by the seasonally-changing daylength. Long days earlier in the year, particularly if followed by short days, would be expected to lead to corrective phase advances of the rhythm because it will have been perceived as running behind real time. Conversely, short days early in the year or long days after the Summer Solstice would both be perceived as the clock being ahead of real time and would be followed by phase delays in the rhythm. In a similar fashion compressed and seasonal light cycles would lead respectively to phase advances and delays of the endogenous rhythm. In many ways, the entrainment behaviour of circannual rhythms appears to parallel that seen with circadian systems. With circadian rhythms, on which much more work has been conducted, characteristically one observes an initial lag between the modifying photoperiod cue and the perceived response, which is more pronounced the more out-of-phase the zeitgeber is to the underlying rhythm. Such differences in phase can be readily seen with the alterations in spawning times of trout following exposure to modified seasonal light cycles (Fig. 1).

Melatonin and the Transmission of Information on Daily and Seasonal Time

Despite widespread recognition of the importance of photoperiod in the timing of reproduction and extensive knowledge of the hypothalamic - pituitary mechanisms involved in its control, it is only recently that we have begun to understand how animals obtain information on the daily and seasonal changes in daylength. It is now clear that this information is provided by the changing patterns of secretion of melatonin from the pineal gland (Bartness & Goldman, 1989). In all vertebrates so far examined, including fish, levels of circulating melatonin are raised during the night and fall to basal during the day. Fig 8 shows the diel profile of change of plasma melatonin in the rainbow trout in response to changes in light intensity during an ambient light cycle (approximately LD 12:12).

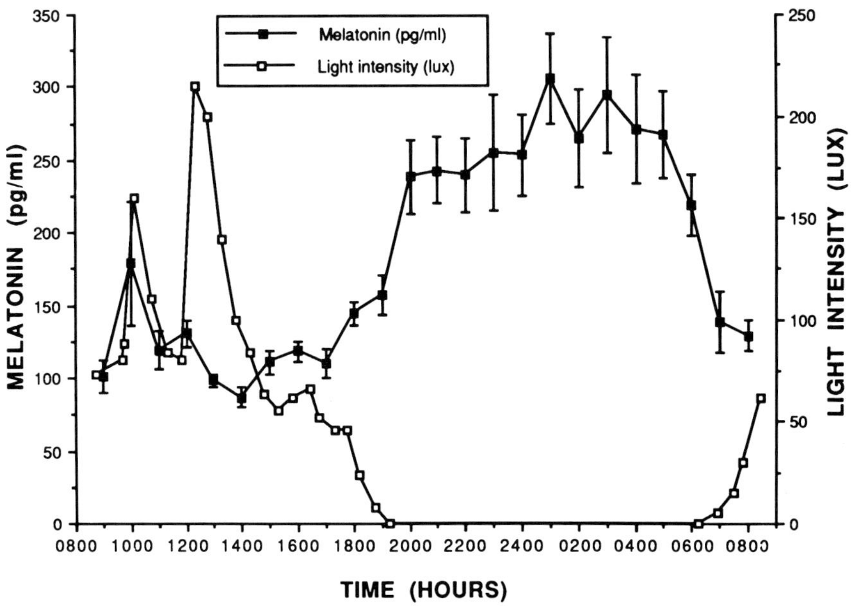

Fig 8 Diel profile of change of melatonin and light intensity under natural daylength (approximately LD 12:12).

Exposure of fish to either constant long or short days leads to a curtailment or extension of the duration of raised melatonin levels respectively (Randall *et al* 1991). Furthermore, measurements of diel patterns of melatonin secretion at the times of the Summer and Winter solstices and the two Equinoxes, have shown that on each occasion the duration of raised melatonin levels faithfully represents the length of the night and hence, the daylength (Randall, Bromage, Thorpe & Miles, unpublished). In this way all organ systems would be provided with information regarding daily and seasonal time.

There is also evidence, at least as far as higher vertebrates are concerned, that infusions or implants of melatonin, designed to simulate melatonin secretion at different seasonal daylengths, are able to modify the time of breeding (Bartness & Goldman, 1989). However, as yet no such experiments have been conducted with fish. It is also not known how the patterns of melatonin secretion are able to entrain or modify the timing of the endogenous clock(s) and how this in turn, activates the neuroendocrine cascade which brings about seasonal breeding.

Photoperiodism and its Importance to Aquaculture

The principal objective of all commercial hatcheries is to provide adequate numbers of good quality eggs and fry at the times of the year required by the farmers who grow the juveniles up to table-size fish. Aquaculture is at its most effective when farms are able to provide product of consistent size, quality and specification throughout the year, an objective which can only realised if there is full control of the reproduction of the species under cultivation.

Hypophysation has been used for several decades to induce spawning, although significant changes in spawning time as yet, have not proved possible with this technique. Only photoperiod manipulation has been shown to achieve the degree of control and simplicity of operation required by commercial farmers. Such techniques are effective because they act at the highest centres in the brain-pituitary-gonadal axis and as a consequence use the fish's own neuroendocrine system and natural levels of hormones to induce changes in timing of all stages of gonadal development up to and including spawning. Methods are now in use which enable supplies of eggs and fry of salmonids, bass, breams and flatfish to be made available as required by the industry.

From the practical standpoint our knowledge of the mechanism involved in the control of fish reproduction has provided a number of important technical advances. Firstly, the reliance of fish on the amount of light received each day rather than its seasonal rate of change has enabled farmers to use constant rather than seasonal light cycles to modify spawning time, thus avoiding possible problems with frequent adjustments of timeclocks and difficulties in introducing new stocks of fish into photoperiod control.

Possibly, of more significance is the realisation that photoperiod does not have a direct driving influence on maturation, rather that it serves to entrain an endogenous circannual rhythm. This will enable farmers to predict more accurately the response of fish to photoperiodic change.

Acknowledgements: N.B. is grateful for grants from BP Nutrition (UK) Ltd and Scottrout Ltd which helped to support this work.

References

Allison, L. (1985) Delay of spawning in eastern brook trout by means of artificially prolonged light intervals. Prog. Fish. Cult. 13:111-116.

Baggerman, B. (1980) Photoperiodic and endogenous control of the annual reproductive cycle in teleost fishes. In Ali M A (ed): "Environmental Physiology of Fishes", Plenum Press, pp 553-568.

Bartness, T. and Goldman, B. (1989) Mammalian pineal melatonin : a clock for all seasons. *Experient. 45:*939-45.

Bromage, N. Cumaranatunga, R. (1988) Egg production in the rainbow trout. Rec. Adv. Aquacult 3:63-138.

Bromage, N. and Duston, J. (1986) The control of spawning in the rainbow trout using photoperiod techniques. *Rep. Inst. Fres. Res. Drottn., 63:*26-35.

Bromage, N.R., Whitehead, C. and Breton, B. (1982a) Relationships between serum levels of gonadotropin, oestradiol-17β and vitellogenin in the control of ovarian development in the rainbow trout. II. Effects of alterations in environmental photoperiod. *Gen. Comp. Endocrinol.*, 47:366-76.

Bromage, N.R., Whitehead, C., Elliot, J., Breton, B. and Matty, A. (1982b) Investigations into the importance of daylength on the photoperiodic control of reproduction in the female rainbow trout. In C. Richter and H. Th. Goos (ed). *Reproductive physiology of fish,* Pudoc Press, Wageningen, pp 233-6.

Bromage, N. R., Elliott, J. A., Springate, J. R. C. and Whitehead, C. (1984) The effects of constant photoperiods on the timing of spawning in the rainbow trout. *Aquaculture, 43,* 213-23.

Bromage, N., Duston, J., Randall, C., Brook, A., Thrush, M., Carrillo, M. and Zanuy, s. (1990) Photoperiodic control of teleost reproduction. Prog. in Comp. Endocr. 342, 620-626.

Bromage, N., Jones, J., Randall, C., Thrush, M., Davies, B., Springate, J., Duston, J., and Barker, G. (1991) Broodstock management, fecundity, egg quality and the timing of egg production in the rainbow trout (*Oncorhynchus mykiss*) Aquaculture, 100 (in press).

Brook, A., and Bromage, N. (1988) Photoperiod: the principal environmental cue for reproduction in the dace. J. Interdisc Cycle Res 19:165-166.

Bye, V. (1987) Environmental management of marine fish reproduction. Proc III Internat. Symp. Repro. Physiol. Fish St John's Canada, pp 289-298.

Carlson, A.R. and Hale, J.G. (1973) Early maturation of brook trout in the laboratory. *Prog. Fish Cult.,* 35:15

Baggerman, B. (1980) Photoperiodic and endogenous control of the annual reproductive cycle in teleost fishes. In Ali M A (ed): "Environmental Physiology of Fishes", Plenum Press, pp 553-568.

Bartness, T. and Goldman, B. (1989) Mammalian pineal melatonin : a clock for all seasons. *Experient. 45*:939-45.

Bromage, N. Cumaranatunga, R. (1988) Egg production in the rainbow trout. Rec. Adv. Aquacult 3:63-138.

Bromage, N. and Duston, J. (1986) The control of spawning in the rainbow trout using photoperiod techniques. *Rep. Inst. Fres. Res. Drottn., 63*:26-35.

Bromage, N.R., Whitehead, C. and Breton, B. (1982a) Relationships between serum levels of gonadotropin, oestradiol-17β and vitellogenin in the control of ovarian development in the rainbow trout. II. Effects of alterations in environmental photoperiod. *Gen. Comp. Endocrinol.*, 47:366-76.

Bromage, N.R., Whitehead, C., Elliot, J., Breton, B. and Matty, A. (1982b) Investigations into the importance of daylength on the photoperiodic control of reproduction in the female rainbow trout. In C. Richter and H. Th. Goos (ed). *Reproductive physiology of fish,* Pudoc Press, Wageningen, pp 233-6.

Bromage, N. R., Elliott, J. A., Springate, J. R. C. and Whitehead, C. (1984) The effects of constant photoperiods on the timing of spawning in the rainbow trout. *Aquaculture, 43,* 213-23.

Bromage, N., Duston, J., Randall, C., Brook, A., Thrush, M., Carrillo, M. and Zanuy, s. (1990) Photoperiodic control of teleost reproduction. Prog. in Comp. Endocr. 342, 620-626.

Bromage, N., Jones, J., Randall, C., Thrush, M., Davies, B., Springate, J., Duston, J., and Barker, G. (1991) Broodstock management, fecundity, egg quality and the timing of egg production in the rainbow trout (*Oncorhynchus mykiss*) Aquaculture, 100 (in press).

Brook, A., and Bromage, N. (1988) Photoperiod: the principal environmental cue for reproduction in the dace. J. Interdisc Cycle Res 19:165-166.

Bye, V. (1987) Environmental management of marine fish reproduction. Proc III Internat. Symp. Repro. Physiol. Fish St John's Canada, pp 289-298.

Carlson, A.R. and Hale, J.G. (1973) Early maturation of brook trout in the laboratory. *Prog. Fish Cult.,* 35:150-3.

Carrillo, M., Bromage, N., Zanuy, S., Prat, F., Serrano, R. (1989) The effects of modifications in photoperiod on spawning time, ovarian development, and egg quality in the sea bass. Aquaculture 81:351-365.

Carrillo, M., Bromage, N., Zanuy, S., Serrano, R. and Ramos, J. (1991) Egg quality and fecundity in sea bass (*Dicentraxchus labrax*) and the effects of photoperiodically-induced advances and delays of spawning time. *Proc. IVth Internat. Symp. Repro. Physiol. Fish.* Norwich, UK:259-261.

Corson, B. (1955) The use of artificially controlled light to induce early spawning of brook trout. Prog. Fish Cult. 17:99-102.

Davies, B. and Bromage, N. (1991) The effects of fluctuating advancement of reproduction in female rainbow trout. *Proc. IVth Internat. Symp. Repro. Physiol. Fish. Norwich, UK*:154-156.

Duston, J. and Bromage, N.R. (1986) Photoperiodic mechanisms and rhythms of reproduction in the female rainbow trout. *Fish Physiol. Biochem., 2:*35-51.

Duston, J. and Bromage, N.R. (1987) Constant photoperiod regimes and the entrainment of reproduction in the female rainbow trout (*Salmo gairdneri*). *Gen. Comp. Endocrinol., 65:*373-84.

Duston, J. and Bromage, N. (1988) The entrainment and gating of the endogenous circannual rhythm of reproduction in the female rainbow trout. J. Comp. Physiol. A. 164:259-268.

Duston, J. and Bromage, N. (1991) Circannual rhythms of gonadal maturation in female rainbow trout (*Oncorhynchus mykiss*). J. Biol. Rhythms 6:49-53.

Eriksson, L, and Lundqvist, H. (1980) Photoperiod entrains ripening by its different effect in salmon. *Naturwissen.* 67:202-3.

Follett, B. (1984) Birds. In Lamming, G. (ed): "Marshall's Physiology of Reproduction", 4th ed Vol. 1, Churchill Livingstone, pp 283-350.

Girin, M., and Devauchelle, N. (1978) Decalage de la periode de reproduction par raccourcissement des cycles photoperiodique et thermique chez des poissons marin. Ann. Biol. Anim. Bioch. Biophys. 18:1059-1066.

Gwinner, E. (1986) Circannual Rhythms. Springer-Verlag, Berlin. 154pp.

Hazard, T.P. and Eddy, R.E. (1951) Modification of the sexual cycle in brook trout (*Salvelinus fontinalis*) by control of light. *Trans Am. Fish Soc. 80:*158-62.

Henderson, N. (1963) Influence of light and temperature on the reproductive cycle of the eastern brook trout. J. Fish. Res. Bd. Can. 20:859-897.

Hoover, E.E. (1937) Experimental modification of the sexual cycle in trout by control of light. *Science, 86:*425-6.

Hoover, E., and Hubbard, H. (1937) Modification of the sexual cycle in trout by control of light. Copeia 1937 4:206-210.

Lam, T. and Munro, A. (1987) Environmental control of reproduction in teleosts : an overview. *Proc. Internat. Symp. Repro. Physiol. Fish. St Johns Canada*:279-288.

MacQuarrie, D.W., Markert, J. and Vanstone, W.E. (1978) Photoperiod induced off-season spawning of coho salmon (*Oncorhynchus kisutch*). *Ann. Biol. Anim. Biochem. Biophys., 18:*1051-8.

MacQuarrie, D., Vanstone, W. and Markert, J. (1979) Photoperiod off-season spawning of pink salmon. *Aquaculture, 18*:289-302.

Randall, C., Duston, J., Bromage, N. (1987) Photoperiodic history and the entrainment of the annual cycle of reproduction in the female rainbow trout. Proc. IIIrd. Internat. Symp. Repro. Physiol. Fish St. John's Canada: 310.

Randall, C., Bromage, N., Thrush, M. and Davies, B. (1991) Photoperiodism and melatonin rhythms in salmonid fish. *Proc. IVth Internat. Symp. Repro. Physiol. Fish. Norwich UK:*136-138.

Sundararaj, B., and Vasal, S. (1976) Photoperiod and temperature control in the regulation of reproduction in the female catfish. J. Fish. Res. Bd. Can. 33:959-973.

Sundararaj, B. Vasal, S., Halberg, F. (1982) Circannual rhythmic ovarian recrudescence in the catfish. Adv. Biosci. 41:319-337.

Takashima, F., Yamada, Y. (1984) Control of maturation in masu salmon by manipulation of photoperiod. Aquaculture 43:243-257.

de Vlaming, V.L. (1972) Environmental control of teleost reproductive cycles; a brief review. *J. Fish Biol., 4*:131-40.

Whitehead, C., Bromage, N., Forster, J. and Matty, A. (1978) The effects of alterations in photoperiod on ovarian development and spawning time in the rainbow trout. *Ann. Biol. Anim. Biochem. Biophys., 18:*1035-53.

Whitehead, C. and Bromage, N. (1980) Effects of constant long and short day photoperiods on the reproductive physiology and spawning of the rainbow trout. *J. Endocrinol., 87:*6-7.

Whitehead, C., Bromage, N. and Breton, B. (1983) Changes in serum levels of gonadotropin, oestradiol-17β and vitellogenin during the first and subsequent reproductive cycles of female rainbow trout. *Aquaculture, 34:*317-26.

7

The Metabolism of Phospholipids and Polyunsaturated Fatty Acids in Fish

J.R. Sargent, J.G. Bell, M.V. Bell, R.J. Henderson, and D.R. Tocher

I. INTRODUCTION.

Of all the nutrients required by fish lipids remain the least well understood and, arguably, the major remaining practical and theoretical problems in fish nutrition centre on lipids. This situation is not unique to fish in that a strong parallel occurs in human nutrition where major controversies continue to exist over what constitutes an "optimal" or even "desirable" dietary intake of lipid, particularly in relation to the balance between saturated and unsaturated fat in the diet and, above all, the balance between the (n-6) and (n-3) polyunsaturated fatty acids (PUFA), i.e. the (n-6) and (n-3) essential fatty acids (EFA). There is growing awareness that some Western societies have an over-high dietary ratio of (n-6)/(n-3)PUFA which is related to their high incidences of human stress-related disorders. This is of direct relevance to fish nutrition because fish, with their natural abundance of (n-3)PUFA, constitute a major source of these nutrients for human diets. It is advantageous, therefore, to maintain high levels of (n-3)PUFA in fish farmed for human consumption, so that it is necessary that fish nutritionists solve the technical and theoretical problems posed by (n-3)PUFA in fish production. Simultaneously, fish nutritionists have unusual opportunities to exploit fish as model species to investigate the basic roles of (n-3)PUFA in the nutrition of animals generally including man.

Lipids presented externally to animals through their diets probably have a greater direct influence on the metabolism of their intracellular counterparts than is the case for any other class of nutrient. This is very likely a direct consequence of the specificities of enzyme-substrate interactions in the metabolic pathways through which the major nutrients are metabolised. Thus, enzyme - substrate interactions in amino acid and protein metabolism depend for the most part on strong ionic and hydrogen bond interactions, whereas enzyme - substrate interactions in lipid and fatty acid metabolism are much more dependent on weaker "hydrophobic" interactions such as van der Waals and dispersion forces. Thus, the exact and invariant amino acid compositions of proteins reflect the high specificity of coupling of amino acids with transfer RNAs whereas the relatively variable fatty acid compositions of glycerides reflect the relatively low specificity of incorporation of fatty acyl groups into these lipids. A noteable example of relatively low substrate specificity in fatty acid metabolism occurs in the fatty acid desaturases where, for example, 16:1(n-7), 18:1(n-9), 18:2(n-6) and 18:3(n-3) are all substrates for the enzyme delta-6 fatty acid desaturase, and a variety of C20 and C22PUFA are all effective inhibitors of the enzyme. It is inevitable, therefore, that the metabolic conversions of PUFA within cells and, therefore, their final concentrations and cellular functions are strongly influenced by variations in the fatty acid compositions of dietary lipids. This contrasts with the situation for the essential amino acids whose cellular functions are to a large extent independent of the amino acid compositions of dietary proteins.

The foregoing has an important corollary which is that the introduction of polar groups into lipids may be expected to enhance their specificity in enzyme-substrate interactions. Thus, the introduction of oxygen, usually by peroxidative reactions, into the EFA to form the eicosanoids greatly enhances the molecular specificity of the products as reflected in their highly specific cellular and metabolic functions.

These considerations may account, at least partly, for ongoing difficulties in lipid nutritional biochemistry and, in particular, for our inability after many years of research to comprehend fully the mechanisms of action of and the precise dietary requirements for lipids, especially the EFA, in any animal species. Such difficulties stem fundamentally from the relative ease with which hydrophobic interactions within cells can be influenced by hydrophobic materials assimilated from the environment.

II. LINKS BETWEEN TRIACYLGLYCEROLS, PHOSPHOLIPIDS AND ESSENTIAL FATTY ACIDS.

Dietary lipid is required by animals including fish both for the provision of metabolic energy, generated as ATP through the oxidative metabolism of fatty acids, and for the production of polar lipids, i.e. phospholipids and sphingolipids, required for the formation of cell membranes. This dual role is reflected in the compartmentalisation of body lipid into adipose tissue, comprised mainly of triacylglycerols, and cell membrane lipid composed mainly of polar lipids and cholesterol. The former compartment is variable and labile and is utilised principally as a source of fatty acids destined for oxidation to produce ATP, although it can be exploited as a source of fatty acids for polar lipid biosynthesis. The cell membrane compartment can *in extremis* be exploited as a source of fatty acids for oxidation, but it is generally much more stable than adipose stores reflecting its critical roles in maintaining cell structures and cell functions, with EFA being involved in both of these roles.

The implications of this duality of lipid function are considerable, especially in species such as fish where lipid rather than carbohydrate is a major source of energy for growth, reflected in the frequently large adipose stores in fish. Moreover, animal tissues generally have not less than 10% of their dry weight as lipid, largely polar lipid in cell membranes, with the brain, a key organ, having circa 50% of its dry weight as polar lipid. Therefore lipid, which includes the EFA present in both triacylglycerols and polar lipids, can form a large part of animal biomass, i.e. it is a quantitatively major nutrient for the growth and well being of all animals including fish.

The situation is illustrated by the nutrient composition of the eggs of marine fish and salmonids (Tocher and Sargent, 1984; Cowey et al, 1985) which contain approximately 50-70% of their dry weight as protein and 10-30% of their dry weight as lipid. In terms of essential nutrients, fish eggs contain approximately 5 times as much essential amino acids (EAA) as essential fatty acids (EFA) on a weight basis. Some of the EFA will be combusted to provide metabolic energy during embryonic development, but so also will some of the EAA, and the remaining EAA and EFA are destined for tissue growth. Such simple considerations do not define quantitative nutritional requirements but they do reveal that EFA, like EAA, are quantitatively major nutrients for fish growth. Moreover, the fish's quantitative requirements for lipid as an energy source cannot be separated from its quantitative requirements for lipid as a constituent of cell membranes, in that both requirements have to be met simultaneously from the same diet which in nature invariably contains both triacylglycerols and phospholipids, albeit in variable amounts.

III. PHOSPHOLIPID METABOLISM

1. Digestion and Assimilation. Few if any detailed studies have been carried out on mechanisms whereby ingested phospholipids are hydrolysed and assimilated by fish, and our present understanding of this area in fish rests heavily on extrapolations from the situation in mammals (see Gurr and Harwood, 1991), which is itself imperfectly understood. Dietary lipid in mammals, generally a mixture of triacylglycerols and phospholipids, is emulsified with bile salts secreted from the liver to form mixed micelles whose glycerides are cleaved by pancreatic lipases, principally the triacylglycerol lipase that removes esterified fatty acids from the sn1 and sn3 positions of the triacylglycerols and the phospholipase A_2 that removes esterified fatty acids from the sn2 position of the phospholipids. The mixed micelles, now containing 2-monoacylglycerols, 1-acyl lysophosphatides, free fatty acids and bile salts, are absorbed into the intestinal mucosal cells where the triacylglycerols and phospholipids are reconstituted by ATP- and coenzyme A-dependent re-esterification of the free fatty acids into the partial glycerides. The reconstituted triacylglycerols and phospholipids are then assembled in the intestinal mucosal cells into lipoprotein particles that are released into the intestinal lymph from where they migrate to the blood to be delivered to the tissues, principally adipose tissue and the liver. The adipose tissue is fundamentally a store of triacylglycerols whereas the liver plays a major role in the further processing of dietary triacylglycerols and phospholipids, much of this being directed towards modifying both the chain lengths and degrees of unsaturation of dietary fatty acids. These modifications are particularly important in relation to dietary PUFA.

Several studies in recent years have demonstrated that phospholipids are superior to triacylglcyerols in promoting growth in larval fish (Kanazawa et al 1985; Teshima et al 1987; Olsen et al 1991; Koven, Tandler and Kissel, personal communication). Both phospholipids and triacylglycerols have PUFA esterified preferentially at the sn2 position of their glycerol backbones, whereas saturated and monounsaturated fatty acids are preferentially esterified at the sn1 and sn3 positions of triacylglycerol glycerol and the sn1 position of phospholipid glycerol. Thus, the maximum % of PUFA in the total fatty acids of triacylglcyerols and phospholipids is generally one third and one half respectively. However, the higher % of PUFA in fatty acids presented as phospholipids as compared to fatty acids presented as triacylglceryols cannot itself account for the superior growth promoting effects of phospholipids in larval fish, since at least one of the phospholipids with growth accelerating properties, soya lecithin (Kanazawa et al 1985; Teshima et al., 1987), is not a rich source of the (n-3)PUFA required in large amounts by fish.

It may be the case that phospholipids are superior to triacylglycerols for larval fish growth because, being more polar, they are more easily emulsified and, therefore, more rapidly hydrolysed and assimilated than triacylglceryols. Thus triacylgycerol assimilation is likely to be much more susceptible to bile salt limitation than phospholipid assimilation, especially in larval fish fed diets rich in triacylglycerols whose digestive capacity may not be fully developed. Definition of the development and control of hepatic and pancreatic functions in larval fish is required to elucidate this area.

Very relevant to the foregoing is the recent finding that the intestines of cod (*Gadus morhua*) contain a single bile - activated lipase that is capable of hydrolysing both triacylglycerols and phospholipids to yield 2-monoacylglycerols and (presumably) 2-acyl-lysophosphatides (Gjellesvik et al, 1989; Gjellesvik and Walther 1991). That fish apparently lack the A_2 phospholipase described earlier for mammals implies that the "essential" components of phospholipids, i.e. their EFA, phosphate and "base" are assimilated by fish as a single entity, obviating the *de novo* biosynthesis of phospholipids required when these entities are assimilated separately.

2. Intracellular Biosynthesis.

(a) De novo and acyl exchange pathways in glyceride biosynthesis. There are no published studies on the pathways of biosynthesis of glycerophospholipids in fish, but there are no reasons to believe that these pathways differ from the well established pathways in mammals (see Gurr and Harwood, 1991). Glycerophospholipids in mammals can be biosynthesised *de novo* by sequential acylation of the sn1 and sn2 positions of glycerol-3-phosphate to yield 1,2-diacylglycerol-3-phosphate, i.e. phosphatidic acid. The latter is cleaved by a phosphomonoesterase to yield diacylglycerol which can be further acylated on the sn3 position to yield triacylglycerol. Alternatively the phosphatidic acid can react with the cytidyl (CMP) derivatives of choline or ethanolamine to generate phosphatidylcholine (PtdCho) and phosphatidylethanolamine (PtdEtn). PtdEtn can be further converted to phosphatidylserine (PtdSer) by direct exchange of serine for ethanolamine. Phosphatidylinositol (PtdIns) and diphosphatidylglycerol (cardiolipin) are formed by direct reaction of the "base", i.e. inositol or glycerol-3-phosphate, with the CMP derivative of phosphatidic acid.

Complex and incompletely understood controls operate in these pathways, particularly in relation to determining relative rates of biosynthesis of triacylglycerols and phosphoacylglycerols where phosphatidic acid phosphoesterase probably plays a pivotal role. Equally important are the mechanisms that incorporate fatty acids selectively into the sn1 and sn2 positions of the phosphoglycerides during their *de novo* biosynthesis (Sprecher, 1991). Such selectivity involves fatty acyl CoA synthetases that have not been fully characterised and whose specificities have not been defined beyond the generalisation that saturated and unsaturated fatty acids are preferentially esterified into positions sn1 and sn2 respectively of the glycerol backbone. In addition, fatty acyl exchange reactions with the same broad specificities operate to exchange fatty acyl groups in existing phospholipids with the intracellular fatty acyl CoA pool. This pool is derived from both exogenous (mainly dietary) fatty acids delivered to the cell and from fatty acids biosynthesised *de novo* within the cell from glucose and amino acid precursors. Thus, the final fatty acid composition of individual phospholipids in individual membranes is a function of (i) the selectivity of uptake of exogenous fatty acids by the cell, (ii) the cell's internal *de novo* fatty acid biosynthetic activity, (iii) the selectivity for fatty acyl incorporation operating in the *de novo* biosynthesis of phospholipids, (iv) selective fatty acyl exchange reactions operating on intact phospholipids, (v) selectivity in incorporating *de novo* biosynthesised phospholipids from their site of synthesis in the endoplasmic reticulum to their final destination in individual cellular membranes, this being an example of selective processes in "membrane trafficing" (Van Meer, 1989).

(b) Dietary PUFA and endogenous PUFA conversions. Interactions between dietary fatty acids and fatty acids generated by intracellular *de novo* biosynthesis are important in determining the composition of membrane phospholipids in several respects. First, an excess of dietary fatty acids, especially PUFA, suppresses *de novo* fatty acid biosynthesis. Second, this holds particularly in the liver which is the major site of *de novo* fatty acid biosynthesis in most animals including fish. Third and notwithstanding the foregoing, modifications of dietary fatty acids can still proceed, principally in the liver, this being particularly important for dietary PUFA.

It is generally held at the present time (reviewed by Henderson and Tohcer, 1987 and Sargent et al 1989), albeit on the basis of detailed studies with few species, that freshwater fish are capable of converting C18 PUFA of both the (n-3) and (n-6) series, i.e. 18:2(n-6) and 18:3(n-3), to their highly unsaturated homologues (HUFA), i.e. 20:4(n-6) together with 22:5(n-6) and 20:5(n-3) together with 22:6(n-3) respectively. Marine fish, however, are not capable of these conversions, i.e. they require preformed HUFA, and there is accumulating evidence that this fundamental difference stems from a deficiency or impairment of delta-5 fatty acid desaturase in marine fish

(Owen et al 1975; Tocher et al 1989; Tocher and McKinlay 1990; Tocher and Sargent 1990).

The situation described has several consequences, the first of which is that 18:3(n-3), 18:2(n-6) and 18:1(n-9) all compete for the delta-6 fatty acid desaturase which rate limits the conversion of C18PUFA to C20 and C22HUFA. The efficiency of competition, at least as established in the rat, is 18:3(n-3)>18:2(n-6)>>18:1(n-9) (Sprecher, 1989). In addition the delta-6 desaturase is subjected to retro-inhibition by HUFA. This is particularly important for freshwater fish including the salmonids because it is the balance of dietary PUFA and HUFA that determines which HUFA are formed in liver and, therefore, the final EFA composition of phospholipids in cell membranes. Because of these complications it has, thus far, proved extremely difficult to predict accurately which EFA will finally be deposited in body tissues from given dietary mixtures of PUFA in any animal, including fish (see Lands, 1991).

A second consequence of the competitive interactions between dietary PUFA and HUFA is that uncertainty exists, particularly in marine fish, whether 20:5(n-3) can be converted to 22:6(n-3) at a rate sufficient to meet the requirements of fish, particularly when diets are rich in 20:5(n-3) but relatively deficient in 22:6(n-3). Recent studies in our laboratory on isolated cultured cells from the turbot Scophthalmus maximus, the most intensively investigated marine species with respect to PUFA metabolism, have established that 20:5(n-3) is indeed converted to 22:6(n-3), albeit at a relatively slow rate, consistent with the presence in this species of delta-4 fatty acid desaturase activity (Owen et al 1975; Tocher et al 1989; Tocher and McKinlay 1990; Tocher and Sargent 1990). It is the case, however, that a quantitatively meaningful conversion of 20:5(n-3) to 22:6(n-3) has been difficult to establish with certainty in a range of cell-free and whole cell systems from various species including mammals and fish. This has led some authors to question the existence of a delta-4 fatty acid desaturase *per se*. In fact it has recently been shown by Voss et al. (1991) that delta-4 desaturation in rats can occur by way of delta-6 fatty acid desaturase operating as follows:

$$20{:}5(n\text{-}3) \xrightarrow{a} 22{:}5(n\text{-}3) \xrightarrow{a} 24{:}5(n\text{-}3) \xrightarrow{b} 24{:}6(n\text{-}3) \xrightarrow{c} 22{:}6(n\text{-}3)$$

where a is a microsomal fatty acid elongase, b is the microsomal delta-6 fatty acid desaturase and c is the peroxisomal beta-oxidation chain shortening system. The same scheme can be invoked for the conversion of 20:4(n-6) to 22:5(n-6).

The above scheme implies that very long chain (n-3) and (n-6)PUFA, as well as 18:3(n-3), 18:2(n-6) and 18:1(n-9) all compete for a single enzyme, delta-6 fatty acid desaturase. It emphasises not only the relatively low substrate specificity of the delta-6 fatty acid desaturase referred to earlier, but also the complex interactions and multitude of potential rate controls that can occur in the conversions of dietary unsaturated fatty acids that determine final HUFA compositions of phospholipids in cellular membranes, in marine as well as in freshwater fish. It also highlights the importance of the balance between chain elongating activity, i.e. conversions of C20 to C22 HUFA, and chain shortening activity, i.e. conversions of C22 to C20 HUFA, in determining cellular HUFA levels, particularly since 22:6(n-3) and 22:5(n-6) themselves can readily be chain shortened by peroxisomal beta oxidation to generate 20:5(n-3) and 20:4(n-6) respectively. There is evidence that, in mammals, the balance of elongation/shortening for the (n-6)HUFA is such that 20:4(n-6) is formed and incorporated into phospholipids in preference to 22:5(n-6), whereas for the (n-3)HUFA 22:6(n-3) is formed and incorporated into phospholipids in preference to 20:5(n-3) (Rosenthal et al 1991).

The extent to which the foregoing applies to fish is clearly of major importance to fish nutrition and practical aquaculture, although elucidation of the complex biochemical conversions and their controls presents formidable challenges. In the absence of definitive information, it is sensible in practical diets for aquaculture not to deviate too far from the balance of monounsaturated fatty acids, (n-3) and (n-6)PUFA, and (n-3) and (n-6)HUFA in natural fish diets. In practice this means a diet with a high ratio of (n-3)/(n-6)HUFA and a high ratio of HUFA to PUFA, i.e. a diet rich in 22:6(n-3) with low levels of 18:2(n-6).

IV. PHOSPHOLIPID FUNCTIONS AND THE ESSENTIAL FATTY ACIDS.

Central to the physiological functions of phospholipids in cell membranes is their constituent essential fatty acids (EFA). Deprivation of dietary EFA in mammals and fish leads to the accumulation of the so-called "Mead acid" 20:3(n-9) which differs from arachidonic acid 20:4(n-6) in lacking a single double bond at the latter's delta14,15 position. The biological potency of arachidonic acid 20:4(n-6) is markedly altered by its containing a single additional double bond at the delta 17,18 position in the form of eicosapentaenoic acid, 20:5(n-3). These seemingly simple chemical differences conceal quite complex conformational and reactivity differences between 20:3(n-9), 20:4(n-6) and 20:5(n-3). Such differences account for the markedly different biological properties of the molecules in ways that are not completely understood at present.

Chief among the conformational differences is that adding increasing numbers of double bonds to long chain monounsaturated fatty acids of a given chain length constrains the structures of the molecules such that they form increasingly compact conformations. This effect reaches its maximum in 22:6(n-3) whose minimum energy conformation is a relatively compact helix or "angle iron" structure with an overall length shorter even than a saturated fatty acid such as 18:0 (Applegate and Glomsett, 1986). This property has important implications for the structure of membrane bilayers whose phospholipids are very rich in 22:6(n-3) and it helps to account for particular tissue-specific, structural requirements for EFA.

Chief among the differences in reactivity between 20:3(n-9), 20:4(n-6) and 20:5(n-3) is the fact that 20:3(n-9) is not a substrate for the cyclooxygenase and the lipoxygenases that convert PUFA to biologically active eicosanoids. Both 20:4(n-6) and 20:5(n-3), however, are substrates for these enzymes, being converted to the 2- and 3-series prostaglandins and the 4- and 5-series leukotrienes respectively. The formation of these oxygenated, relatively polar and highly biologically active metabolites goes far towards accounting for particular tissue-specific functions of EFA.

1. Membrane Structure.

The brain, retina and sperm from mammals including man have long been known to be rich sources of 22:6(n-3) (Tinoco, 1982). All fish tissues are rich in (n-3)HUFA, especially 22:6(n-3), but the brain and retina of fish are particularly rich in 22:6(n-3), even more so than their mammalian counterparts. It has long been assumed, therefore, that 22:6(n-3) has a specific role in neural and testicular tissues and it is now established that deficiency of (n-3)PUFA at critical stages in embryonic development in higher primates, including *Homo sapiens*, leads to visual and mental subnormalities (Neuringer et al 1988; Bazan 1990; Neuringer and Connor, 1989; Bjerve 1991; Salem and Carlson 1991). There is no direct experimental evidence thus far that (n-3)PUFA deprivation in fish causes visual or neural impairments, although studies to confirm this very obvious deduction are currently in progress. The issue is critical in fish nutrition and aquaculture because, in contrast to mammals whose neural development occurs for the most part *in utero*, neural development in fish occurs very substantially in free living, predominantly predatory larvae that are entirely dependent for their well being on their ability to capture (n-3)PUFA from their prey.

We have recently established (Mourente et al 1991; Mourente and Tocher, 1991) that the brains of turbot, reared under commercial conditions initially on a diet of rotifers and subsequently on brine shrimp nauplii supplemented with fish oils in an effort to improve their (n-3)HUFA status, are relatively deficient in 22:6(n-3). When the fish are weaned to particulate feeds rich in 22:6(n-3) there is a rapid accumulation of 22:6(n-3) in brain and normalisation of the brain (n-3)HUFA profile. In subsequent studies with cultured brain cells from turbot (Tocher, Mourente and Sargent, unpublished data) we have shown that uptake of 22:6(n-3) into phospholipids, particularly PtdEtn, is rapid but the specificity of uptake of different PUFA is low. These results illustrate the ease with which (n-3)HUFA in turbot brain phospholipids can be influenced by alterations in dietary PUFA.

Up to 40% of the dry matter of fish brain is lipid, approximately 70% of the total lipid is glycerophospholipid and approximately 40% of the glycerophospholipid is PtdEtn (cephalin) (Tocher and Harvie 1988;

Mourente et al 1991). Moreover, 22:6(n-3) can account for up to 40% of the total fatty acids in fish brain PtdEtn and the single molecular species di-22:6(n-3)PtdEtn accounts for circa 15% of the total molecular species of PtdEtn in trout brain (Bell and Tocher, 1989; Bell and Dick, 1991a). An even more striking situation occurs in fish retina where di-22:6(n-3) accounts for over 40% of all the molecular species of PtdEtn present in trout (Bell and Tocher, 1989) and over 70% in cod (Bell and Dick, 1991a). These findings emphasise the necessity of an adequate dietary supply of 22:6(n-3) for the normal development of the retina and brain of marine fish larvae. Therefore, deprivation of dietary 22:6(n-3) will have serious consequences for the behaviour of early fish larvae, including a reduced efficiency as predators. Research to illuminate these issues is currently in progress in our laboratories.

The foregoing sheds no light on the molecular functions of 22:6(n-3) especially in the form of specific molecules such as di-22:6(n-3)PtdEtn in retinal and brain physiology. However, given the previous comments on the marked helicity of 22:6(n-3), it can be deduced that cell membranes such as rod outer segments that are rich in di-22:6(n-3)PtdEtn have a specialised phospholipid bilayer, very probably with novel liquid crystalline properties. Precisely how this specialisation relates to the function of rhodopsin in the eye is currently not known in detail. It has been considered on the basis of physico-chemical studies of mammalian rod outer segment membranes (Dratz and Deese, 1986) that the abundance of di-22:6(n-3)PtdEtn maintains a membrane bilayer with the required balance between fluidity and rigidity necessary to accomodate very rapid protein conformational changes initiated by the *cis -trans* conversion undergone by the retinal chromophore of rhodopsin. Rabinovich and Ripatti (1991) have recently deduced on theoretical grounds that the minimum energy form of 22:6(n-3) is conformationally stable over a wide temperature range, so that membrane bilayers rich in 22:6(n-3) may be essentially "buffered" against environmental change with obvious advantages to their associated physiological processes. A scenario such as the foregoing can be extended in principle to encompass very fast conformational changes undergone by other signal transducers in cell membranes, e.g. those involved in ion current generation, thus accounting for the abundance of 22:6(n-3) in neural tissues generally.

A noteable example of a tissue-specific, structural role for an EFA is the role of 18:2(n-6) in maintaining the water impermeability of rat skin. This is achieved by means of a very long chain fatty derivative of 18:2(n-6), acyl-linked as a ceramide, acting as an inter-cellular, water-impermeable "cement" for rat epidermal cells (Hansen, 1989). This role of 18:2(n-6) accounts for the very marked loss of water through the skins of rats maintained on an EFA-deficient diet. It also contributes substantially to the quantitative dietary requirements for 18:2(n-6) in the rat. Because marine fish have only very low requirements for (n-6)PUFA, it is highly improbable that 18:2(n-6) has a role in water impermeability in fish skin. Nonetheless, skin impermeability is of major importance in fish and the possibility that EFA have a structural role in fish skin deserves investigation in future research.

These considerations shed some light on the nature of tissue-specific structural roles of EFA in animals including fish, but they also reveal the paucity of our present knowledge in this area. In particular we do not know the extent to which we can extrapolate from one species to another, nor do we know the extent to which such structural functions of PUFA account for the quantitative EFA requirements of a given species. It is safe to assume, however, that the requirements for 22:6(n-3) for visual and neural development in rapidly developing larval fish are substantial and are directly reflected in the abundance of phospholipids rich in this fatty acid both in the larvae's original egg yolk nutrients (Tocher and Sargent, 1984) and in their subsequent natural microzooplanktonic prey (Fraser et al 1989).

2. Membrane Function.

As noted above, a major functional role of the EFA present in the phospholipids of cell membranes is as a source of the free HUFA that are the immediate precursors of the eicosanoids, i.e. the prostaglandins and the leukotrienes (Johnson et al 1983). It is noteworthy that, although the cell membranes of the great majority of fish species, especially marine fish, are dominated by (n-3)HUFA, it has long been known that the major biologically active eicosanoids produced by fish are the 2-series prostanglandins and the 4-series leukotrienes (Henderson and Tocher 1987; Tocher and Sargent 1987; Tocher et al 1991), i.e. they are produced from arachidonic acid, 20:4(n-6), which comprises only a very minor percentage of the total fatty acids in the total membrane phospholipids of fish. In recent years we have noted (see Sargent et al 1989; Bell 1989) the extent to which the fatty acids of PtdIns, a quantitatively minor phospholipid of fish and all animal cell membranes, can contain high percentages of 20:4(n-6) such that 18:0-20:4(n-6) is frequently the major molecular species of PtdIns in fish cell membranes. It is not known with certainty in mammals which particular membrane phospholipid class, all of which can be rich in 20:4(n-6), is the substrate for the phospholipase A_2 that generates the free arachidonic acid precursor of the eicosanoids. There are few experimental data on this issue in fish but the concentration of 20:4(n-6) in the PtdIns of fish cell membranes makes it tempting to suggest that this phospholipid is the immediate source of the 20:4(n-6) precursor of the 2-series prostaglandins and 4-series leukotrienes in fish.

More recent work in our laboratory, however, has shown unexpected variation in the molecular species composition of PtdIns in fish phospholipids which has led us to conclude that the composition can be tissue-specific. Thus, 18:0-20:4(n-6)PtdIns is the dominant species in both egg yolk and liver; 18:0-20:5(n-3)PtdIns is a dominant species in brain; 18:0-22:6(n-3) is the dominant species in white muscle; all three species are abundant in the retina (Bell and Dick, 1991b). These findings can be related to increasing evidence for tissue specific isoforms of phospholipase C, a key enzyme in generating the inositol phosphate and diacylglycerol messengers of the "inositol lipid cycle".

That eicosanoids in fish are formed mainly from 20:4(n-6) very strongly supports, if not establishes, the dietary essentiality of (n-6)PUFA in both marine and freshwater fish. However, that eicosanoids are highly biologically active and transient products produced in extremely small amounts implies that the dietary requirements of fish for (n-6)PUFA are small. This is in keeping with the small amounts of (n-6)PUFA including 20:4(n-6) in the natural diets of fish, especially marine fish. There is as yet, however, no direct nutritional evidence for these deductions in marine fish.

All tissues in mammals, and presumably also in fish, are capable of producing eicosanoids. However, in mammals including man it has been argued (Crawford, 1989) that the vascular endothelium has a particularly prominent role in prostaglandin production since its normal operation requires a balance between the production by platelets of the blood clotting agonist thromboxane and the production by the endothelium of the anti-clotting agonist prostacyclin. It has also been argued that the extensive development of the endothelial system, including the placenta, in humans makes endothelial physiology particularly susceptible to imbalances in dietary lipids that perturb eicosanoid production and hence blood flow. In addition, the production of prostaglandins by macrophages is of major importance in the immune response and the production of leukotrienes by neutrophils is central to the inflammatory response. It is interesting, therefore, that salmon white blood cell phospholipids are substantially enriched in (n-6)PUFA as compared to red blood cell phospholipids and phospholipids in other tissues of the species (Bell et al 1991a; Thompson et al 1991). Thus, the prominence of (n-6)PUFA in eicosanoid production in the vascular endothelial system may hold for fish as well as humans.

Despite the foregoing considerations for PtdIns and white blood cells in fish, it is the case that fish phospholipids in general have a marked excess of (n-3)- over (n-6)PUFA, generally approaching one order of magnitude. This is of special interest because it is now well established that dietary supplements of (n-3)PUFA, usually in the form of fish oil fractions (triacylglycerols) enriched in 20:5(n-3) ("Maxepas"), markedly depress eicosanoid production from 20:4(n-6) in terrestrial mammals including man (Johnston 1985; Terano et al 1986; Weber 1990; Knapp 1990; Kinsella et al 1990). This at least partly accounts for the beneficial effects of dietary supplements of fish oils in a variety of human stress-related disorders including ischaemic heart disease, autoimmune disorders and inflammatory disorders, and these beneficial effects are most pronounced in individuals with a high dietary ratio of (n-6)/(n-3)PUFA.

We have recently established (Bell et al 1991b) that elevating the ratio of (n-6)/(n-3)PUFA in the diets of salmon post-smolts, by the addition of vegetable oils rich in 18:2(n-6) to diets with a sufficiency of (n-3)HUFA, can have marked effects on the fish. These include a marked cardiac myopathy accompanied by an increased incidence of atherosclerotic lesions (Fraser, McVicar and Bell, unpublished data), a decreased resistance to infection with pathogenic bacteria (Thompson et al 1991) and an altered ability of the liver to detoxify xenobiotics (George and Henderson, 1991). Particularly notable is that these pathologies developed over a four month period under commercial farming conditions when no differences in the growth or behaviour of the fish were noted. However, subsequent stressing of the fish by handling and transport immediately revealed the cardiovascular pathology. Moreover cell-free studies on these fish have established that eicosanoid production is markedly altered on fish fed the vegetable oil diet (Bell, Dick, Sargent and MacVicar, unpublished data). We conclude that pathologies can be elicited in fish by elevating their dietary ratio of (n-6)/(n-3)PUFA, i.e. the situation in fish is fundamentally the same as in man. The corollory is that dietary supplementation with (n-3)PUFA can, under certain conditions, be beneficial for fish and this is confirmed by reports that enhancement of salmon diets with (n-3)PUFA can markedly increase the efficacy of vaccination in salmon (Salte et al 1988).

These findings are of more than theoretical interest since it can be the case that commercial fish feeds contain oils that are blends of fish oils rich in (n-3)HUFA and vegetable oils rich in 18:2(n-6). Fish fed such diets may be particularly susceptible to stress elicited through overcrowding, disease or environmental pollutants, and this may be particularly important during early larval development.

V. SUMMARY AND THE FUTURE.

This overview has attempted to illustrate the complex interactions that can occur between metabolic conversions of dietary monounsaturated, and dietary (n-6) and (n-3) polyunsaturated fatty acids that generate intracellular levels of essential fatty acids. It has also attempted to illuminate the complex mechanisms that generate specific molecular species of individual phospholipids in specific cell membranes. Finally it has attempted to illustrate the complex mechanisms, both "structural" and functional" whereby EFA incorporated into specific phospholipids exert their biological functions. It remains our basic tenet that the particular difficulties encountered in the nutritional biochemistry of PUFA stem ultimately from the relatively low specificity encountered in lipid enzymology which reflects the relative weakness of the hydrophobic interactions that generally characterise this area of metabolism. Irrespective, the practical outcome of our present uncertainties is that lipid nutrition is fundamentally a matter of balance between dietary fatty acids, this balance being particularly important for the (n-3) and (n-6) PUFA and HUFA. Because fish have a natural abundance of (n-3)HUFA, they present a specialised case in EFA nutritional biochemistry and offer particular challenges and opportunities to advance knowledge in this field.

There is a clear need for further detailed studies of PUFA metabolic conversions, particularly in relation to interactions between different unsaturated fatty acids in the diet. Equally important are studies of the molecular mechanisms whereby (n-6) and (n-3)PUFA exert their essential cellular functions. Although this review has considered these functions as "structural" and "functional", it may be the case that EFA will ultimately prove to have a pervasive and general role in signal transduction processes in specific cell membranes, whether directly as in their role as eicosanoid precursors or indirectly as in their present putative role in light transduction in the retina. The possibility that they may influence membrane receptor systems involved in signal transduction processes, through specific molecular interactions in the bilayer, will be a particularly important issue in future research.

The importance of mechanistic studies of EFA functions in future research, however, does not lessen the need for ongoing conventional nutritional studies of growth rates and efficiencies. More sophisticated research on fish stress and behaviour is also required, the case for such studies being particularly strong in larval fish. It would be unwise to anticipate an over rapid advance in theoretical understanding in this field because, apart from current uncertainties, areas of major importance such as the genetics of the EFA have yet to be studied, a deficiency that is likely to continue until a molecular biological approach to PUFA nutrition has been developed. In the meantime, the best, albeit mundane and self evident approach to practical problems in aquaculture related to EFA metabolism may be not to deviate too far from diets whose composition mimics natural prey, to avoid stressful conditions such as over crowding and to maintain high quality standards of husbandry.

VI. REFERENCES

Applegate KR and Glomset JA (1986) Computer-based modelling of the conformation and packing properties of docosahexaenoic acid. J Lipid Res 27: 658-680

Bazan NG (1988) The supply of omega-3 polyunsaturated fatty acids to photoreceptors and synapses. In "Dietary omega 3 and omega 6 Fatty Acids. Biological Effects and Nutritional Essentiality", edited by C Galli and AP Simopoulos, pp 227-239. Nato ASI Series. Series A: Life Sciences Vol 171 Plenum Press New York and London

Bell JG, Sargent JR and Raynard RS (1991a) Effects of increasing dietary linoleic acid on phospholipid fatty acid composition and eicosanoid production in leucocytes and gill cells of Atlantic salmon (*Salmo salar*). Prostaglandins Leukotrienes and Essential Fatty Acids, in the press.

Bell JG, McVicar AH, Park MT and Sargent JR (1991b) Effects of high dietary linoleic acid on fatty acid compositions of individual phospholiopids from tissues of Atlantic salmon (*Salmo salar*): association with stress susceptibility and cardiac lesion. J Nutrition 121: 1163-1172

Bell MV (1989) Molecular species analysis of phosphoglycerides from the ripe roes of cod (*Gadus morhua*). Lipids 24: 585-588

Bell MV and Tocher DR (1989) Molecular species composition of the major phospholipids in brain and retina from rainbow trout (*Salmo gairdneri*). Biochem J 264: 909-915

Bell MV and Dick J (1991a) Molecular species composition of the major diacyl glycerophospholipids from muscle, liver, retina and brain of cod (*Gadus morhua*) Lipids 26: 565-573

Bell MV and Dick J (1991b) Molecular species composition of

phosphatidylinositol from the brain, retina, liver and muscle of cod (*Gadus morhua*). Lipids 25: 691-694

Bjerve KS (1991) Requirements of adults and elderly. In "Health Effects of Omega3 Polyunsaturated Fatty acids in Seafoods" edited by AP Simopoulos, RR Kifer, SM Barlow and RE Martin, pp. 26-30. Karger Basel.

Cowey CB, Bell JG, Knox D, Fraser A and Youngson A (1985) Lipids and lipid antioxidant systems in developing eggs of salmon (*Salmo salar)*. Lipids 20: 567-572

Crawford, M (1989) The role of fats and EFAs for energy and cell structures in the growth of fetus and neonate. In "The Role of Fats in Human Nutrition", second edition, edited by AJ Vergroessen and M Crawford, pp Academic Press London

Dratz EA and Deese AJ (1986) The role of docosahexaenoic acid, 22:6(n-3), in biological membranes: examples from photoreceptors and model membrane bilayers. In "Health Effects of Polyunsaturated Fatty Acids in Seafoods", edited by AP Simopoulos, RR Kifer and RE Martin, pp 319-351 Academic Press New York

Fraser AJ, Sargent JR, Gamble JC and Seaton, DD (1989) Formation and transfer of fatty acids in an enclosed marine food chain comprising phytoplankton, zooplankton and herring (*Clupea harengus*) larvae. Mar Chem 27: 1-18

George SG and Henderson RJ (1991) Influences of dietary polyunsaturated fatty acid content on cytochrome P450 and transferase activities in Atlantic salmon (*Salmo salar*) Mar Environ Res, in the press

Gjellesvik DR, Raae AJ and Walther BT (1989) Partial purification and characterisation of a triglyceride lipase from cod (*Gadus morhua*). Aquaculture 79: 177-184

Gjellesvik DR and Walther BT (1991) Digestive lipolysis in the cod (*Gadus morhua*): bile salt - dependent lipase as sole pancreatic enzyme. 13th Conf Europ Soc Comp Physiol Biochem. Research for Aquaculture: Fundamental and Applied Aspects, p. 100 (C2.5)

Gurr MI and Harwood JL (1991) "Lipid Biochemistry. An Introduction" fourth edition, 406pp Chapman and Hall London

Hansen HS (1989) Linoleic acid and epidermal water barrier. In "Dietary omega 3 and omega 6 Fatty Acids. Biological Effects and Nutritional Essentiality", edited by C Galli and AP Simopoulos, pp 333-341. Nato ASI Series. Series A: Life Sciences Vol 171 Plenum Press New York and London

Henderson RJ, and Tocher DR (1987) The lipid composition and biochemistry of freshwater fish. Progr Lipid Res 26: 281-347.

Johnson M, Carey F and McMillan RM (1983) Alternative pathways of arachidonate metabolism: prostaglandins, thromboxane and leukotrienes. In "Essays in Biochemistry", edited by PN Campbell and R D Marshall, pp 40-141. Academic Press London

Johnston PV (1985) Dietary fat, eicosanoids and immunity. Adv Lipid Res

21: 103-141.

Kanazawa A, Teshima SI and Sakamoto M (1985) Effects of dietary bonito egg phospholipids and some phospholipids on growth and survival of the larval Ayu, *Plecoglossus altivelis*. Z angew Ichthol 4: 165-170

Kinsella JE, Lokesh B and Stone RA (1990) Dietary n-3 polyunsaturated fatty acids and amelioration of cardiovascular disease: possible mechanisms. Amer J Clin Nutr. 52: 1-28.

Knapp HR (1990) Polyunsaturates, endogenous eicosanoids and cardiovascular disease. J Amer Coll Nutr 9: 344-351.

Lands WEM (1991) Dose-response relationships for omega3/omega6 effects. In "Health Effects of Omega 3 Polyunsatuated Fatty Acids in Sea Foods", edited by AP Simopoulos, RR Kifer, RE Martin and SM Barlow, pp 177-194 Karger Basel

Mourente D. and Tocher, DR (1991) Low levels of docosahexaenoic acid (22:6n-3) in developing turbot (*Scophthalmus maximus*) brain prior to weaning. Aquaculture, in the press

Mourente D, Tocher DR and Sargent JR (1991) Specific accumulation of docosahexaenoic acid (22:6n-3) in brain lipids during development of juvenile turbot *Scophthalmus maximus* L. Lipids 26: in the press

Neuringer M, Andersdon GJ and Connor WE (1988) The essentiality of omega 3 fatty acids for the development and function of the retina and the brain. Ann Rev Nutr 8: 517-541

Neuringer M and Connor WE (1989) Omega-3 fatty acids in the retina. In "Dietary omega 3 and omega 6 Fatty Acids. Biological Effects and Nutritional Essentiality", edited by C Galli and AP Simopoulos, pp 177-190. Nato ASI Series. Series A: Life Sciences Vol 171 Plenum Press New York and London

Olsen R, Henderson RJ and Pedersen T (1991) The influence of dietary lipid classes on the fatty acid composition of small cod (*Gadus morhua* L.) juveniles reared in an enclosure in northern Norway. J Exp Mar Biol Ecol 148: 59-76.

Owen JM, Adron JW, Middleton C and Cowey CB (1975) Elongation and desaturation of dietary fatty acids in turbot *Scophthalmus maximus* and rainbow trout *Salmo gairdneri*. Lipids 10: 528-531.

Rabinovich AL and Ripatti PO (1991) On the conformational, physical properties and functions of polyunsaturated acyl chains. Biochim Biophys Acta 1085: 53-62

Rosenthal MD, Garcia MC, Jones MR and Sprecher H (1991) Retroconversion and delta-4 desaturation of docosatetraenoate [22:4(n-6)] and docosapentaenoate [22:5(n-3)] by human cells in culture. Biochim Biophys Acta 1983: 29-36

Salem N and Carlson SE (1991) Growth and development in infants. In "Health Effects of Omega 3 Polyunsatuated Fatty Acids in Sea Foods", edited by AP Simopoulos, RR Kifer, RE Martin and SM Barlow, pp 20-25 Karger Basel

Salte R, Thomassen MS and Wold K (1988) Do high levels of dietary polyunsaturated fatty acids (EPA/DHA) prevent diseases associated with membrane degeneration in farmed Atlantic salmon at low water temperatures? Bull Eur Ass Fish Pathol 8: 63-65

Sargent JR, Henderson RJ and Tocher DR (1989) Lipids. In "Fish Nutrition" second edition, edited by JE Halver, pp 153-218 Academic Press San Diego

Sprecher H (1989) (n-3) and (n-6) fatty acid metabolism. In "Dietary omega 3 and omega 6 Fatty Acids. Biological Effects and Nutritional Essentiality", edited by C Galli and AP Simopoulos, pp 69-79. Nato ASI Series. Series A: Life Sciences Vol 171 Plenum Press, New York and London

Sprecher H (1991) Enzyme activities affecting tissue lipid fatty acid composition. In "Health Effects of Omega 3 Polyunsatuated Fatty Acids in Sea Foods", edited by AP Simopoulos, RR Kifer, RE Martin and SM Barlow, pp 166-176 Karger Basel

Terano T, Salmon JS, Higgs GA and Moncada S (1986) Eicosapentaenoic acid as a modulator of inflammation - effect on prostaglandin and leukotriene synthesis. Biochem Pharmacol. 35: 779-785.

Teshima SI, Kanazawa A, Horinouchi K, Yamasaki S and Hirata H (1987) Phospholipids of the rotifer, prawn and larval fish. Bull Jap Soc Sci Fish 53: 609-615

Thompson K, Tatner MF and Henderson RJ (1991) Influence of dietary polyunsaturated fatty acids on the immune response of Atlantic salmon (*Salmo salar*) First Nordic Symposium on Fish Immunology in the press

Tinoco J (1982) Dietary requirements and functions of alpha-linolenic acid in animals. Progr Lipid Res 21: 1-46

Tocher DR and Sargent JR (1984) Analyses of lipids and fatty acids in ripe roes of some northwest European marine fish. Lipids 19: 492-499

Tocher DR and Sargent JR (1987) The effect of calcium ionophore A23187 on the metabolism of arachidonic acid and eicosapentaenoic acid in neutrophils from a marine teleost fish rich in (n-3)polyunsaturated fatty acids. Comp Biochem Physiol 87B: 733-740

Tocher DR and Harvie DG (1988) Fatty acid compositions of the major phosphoglycerides from fish neural tissues; (n-3) and (n-6) polyunsaturated fatty acids in rainbow trout (*Salmo gairdneri*) and cod (*Gadus morhua*) brains and retinas. Fish Physiol Biochem 5: 229-239

Tocher DR, Carr J and Sargent JR (1989) Polyunsaturated fatty acid metabolism in fish cells: differential metabolism of (n-3) and (n-6) series acids by cultured cells originating from a freshwater teleost fish and from a marine teleost fish. Comp Biochem Physiol 94B: 367-374.

Tocher DR and Mackinlay EE (1990) Incorporation and metabolism of (n-3) and (n-6) polyunsaturated fatty acids in phospholipid classes in cultured turbot (*Scophthalmus maximus*) cells. Fish Physiol Biochem

8: 251-260.

Tocher DR and Sargent JR (1990) Effect of temperature on the incorporation into phospholipid classes and metabolism via desaturation and elongation of n-3 and n-6 polyunsaturated fatty acids in fish cells in culture. Lipids 25: 435-442

Tocher DR, Bell JG and Sargent JR (1991) The incorporation of [^{3}H]arachidonic acid and [^{14}C]eicosapentaenoic acid into glycerophospholipids and their metabolism via lipoxygenases in isolated brain cells from rainbow trout *Oncorhynchus mykiss*. J Neurochem in the press.

Van Meer G (1989) Lipid traffic in animal cells. Ann Rev Cell Biol 5: 247-275

Voss, A, Reinhart, M, Sankarappa, S and Sprecher, H (1991) The metabolism of 7,10,13,16,19-docosapentaenoic acid to 4,7,10,13,16,19-docosahexaenoic acid in rat liver is independent of a 4-desaturase. J Biol Chem 266: 19995-20000

Weber PC (1990) The modification of the arachidonic acid cascade by n-3 fatty acids. Adv Prostagland Thromboxane Leukotriene Res 20: 232-240

8

Protein Metabolism in Fish

C.B. Cowey

Protein metabolism has been studied much less intensively in fish than in other farm animals, consequently it is not easy to present an integrated account of the subject. This is unfortunate as many practical aspects of fish production such as growth rate, carcass composition, feed conversion efficiency and so on depend on an understanding of the basic biochemistry involved. Growth is in large part a matter of protein deposition and this, in turn, is mainly determined by protein or amino acid supply. The aim, in normal practice, is to provide sufficient dietary protein to meet the needs of the fish for maximal production. In another, more general sense, protein supply might be seen as controlling production and a level of supply might be chosen which would maximise some other criterion - possibly an economic function.

Boorman (1980), referring to data for pigs, noted that if rates of N retention in animals of different weight or age were compared then rate of retention was highest in the young animal, decreased with increasing body weight and tended to zero in the mature animal. Data covering an appropriate size range are not available for fish. However, it seems probable, given that the growth rate of fish decreases with size, that the same sort of relationship holds for fish. With this proviso, net retention of dietary protein by fish appears to be of a similar magnitude to that of terrestrial, omnivorous, warm blooded animals. Bowen (1987) tabulated values for net protein retention of a number of fish species, they were in the range 30-45% with a median value of 31%. The median value for terrestrial vertebrates reported by Bowen was 29%. In both cases much of the dietary protein ingested is used for purposes other than protein deposition.

Many experiments have been carried out on various species of fish with the aim of "sparing" dietary protein, by increasing the concentrations of fat, carbohydrate or both in the diet. These have been generally successful although usually at the expense of an increasingly fatty carcass. Some values of Lee and Putnam (1973) for rainbow trout, shown in Table 1, typify the results obtained. Recently, diets containing very high (30%) concentrations of lipid have been used in the cultivation of Atlantic salmon (Johnsen and Wandsvik, 1991) and these appear to have given net protein retention values greater than 50% with little alteration to carcass composition. Even so, and in common with warm blooded animals, this indicates a very large and apparently inevitable loss of dietary amino acids.

Table 1. Response of rainbow trout to variation in protein:energy ratio in diet

	Water (%)	Protein (%)	Lipid (%)	Starch (%)	Energy (MJ/kg)
Diet 1	-	44	8	18	12.4
Carcass 1	72.6	16.4	9.1	-	
Diet 2	-	44	24	18	18.4
Carcass 2	69.6	15.8	13.0	-	
Initial carcass	75.1	16.7	6.3		

Data of Lee and Putnam (1973)

The use of diets containing different proportions of protein and non-protein energy has brought out the controversy concerning whether "metabolizable" or "digestible" energy values should be used to define their energy content. Digestible energy values of major dietary components have the advantage that they are additive. For rainbow trout, Cho and Kaushik (1990) recommend a protein:energy ratio of 22-25 g digestible protein/MJ digestible energy in diets with an energy density of 13-16 MJ/kg feed.

AMINO ACID SUPPLY

Many measurements have been made on the protein requirement of fish. These have shown that young individuals of most species require 30-50% protein in the diet for maximal weight gain. Several authors (Smith, 1987; Kim et al., 1991) have pointed out that much less protein should be needed for protein synthesis. Recently, Kim et al. (1991) re-evaluated the dietary protein requirement of rainbow trout; they showed that it was not more than 30% of the diet (previous estimates had indicated a requirement of 40%). In addition it was shown that the protein requirement could be met with a diet containing 25% casein and 10% of a mixture of dispensable amino acids. This was interpreted as meaning that the protein requirement of rainbow trout is not more than 25% when appropriate energy sources (i.e. dispensable amino acids) are supplied. It implies a particularly fine distinction between protein required for tissue protein synthesis and that used for other purposes. It is noteworthy that N retention in these experiments was approximately 40% - very much in the range of previous estimates.

Apparent protein digestibility has been measured by an indirect, ratio method in several species of fish. Except for proteins that contain large numbers of disulphide bonds (e.g. feather meal), or those that have suffered heat damage during preparation, values for apparent digestibility coefficient are high - 90% or more (Cho and Kaushik, 1990). Few attempts have been made to measure amino acid absorption by, for example, measurement of the disappearance of amino acids between the stomach and the rectum. Data of Dabrowski and Dabrowska (1981), obtained by this means for rainbow trout showed that availability of indispensable amino acids approached 90% in most cases but tryptophan availability was appreciably lower.

LOSSES OF AMINO ACID NITROGEN

Oxidation of some amino acids occurs during the "first-pass" metabolism in the mucosal cells; glutamine is a preferred substrate of mammalian enterocytes but few studies have been made on mucosal cells from fish.

The time course of N excretion by young sockeye salmon given a single daily meal (3% dry body weight) indicates rapid and extensive deamination of dietary amino acids. A peak of ammonia excretion occurred approximately 4.5 hours following the meal and amounted to 27% of the nitrogenous intake (Brett and Zala, 1975). Whether such losses are typical of fish fed to satiation several times each day is not clear; data on gastric emptying times (Brett and Higgs, 1970) suggest that such fish are rarely post-absorptive. Murai et al. (1987) examined portal and hepatic uptake of amino acids in rainbow trout force-fed a meal containing either casein or an amino acid mixture. Hepatic uptake of amino acids amounted, for virtually all indispensable amino acids including branched chain amino acids, to almost half of those taken into the portal vein. At the same time ammonia was released from the liver to the hepatic vein showing that catabolism of amino acids occurs extensively soon after they have been assimilated.

In common with other vertebrates fish cannot store protein in the sense that glycogen or triacylglycerol is stored. Their tissues may contain variable amounts of cytosolic protein (composition data of Brett et al., 1969; see also Cowey and Sargent, 1972), and in this sense part of N retention is repletable or depletable, but any amino acids not used directly for protein synthesis are deaminated and the carbon skeletons used as a source of energy. Oxidative losses occur during feeding because amino acids are absorbed at a rate which is in excess of that at which they can be used for protein synthesis. Fish appear to have little ability to conserve indispensable amino acids even under conditions of dietary protein restriction. The activities of tissue enzymes which initiate the catabolism of amino acids show little adaptive decrease in response to these conditions (Cowey and Walton, 1989).

Apart from this use as an energy source, either directly or after gluconeogenesis or lipogenesis, amino acids may be lost to protein deposition in the body as a consequence of two other processes. First, they serve as precursors of a variety of other nitrogenous compounds and, secondly, they may be lost as a consequence of protein turnover. Table 2 lists some of the nitrogenous compounds formed from specific amino acids. Their quantitative significance in fish is difficult to assess, many of them are minor pathways, others (glycine usage for haem, purine bases and creatine) appear to involve appreciable amounts of the amino acid concerned.

The protein pool of the body is not static, synthesis and degradation occur even when N-balance is zero or negative. While turnover, of itself, does not necessarily lead to a loss of amino acids from body protein, nevertheless the amino acids liberated during turnover may be catabolised by the deaminating enzymes that are generally present, or even lost to minor pathways of N metabolism. Reduction in the concentration of one amino acid from the precursor pool would limit the extent to which other amino acids could be re-incorporated into protein.

Table 2. Nitrogenous compounds formed from amino acids

Amino acid	Products
Glycine	Creatine, purines, porphyrins
Methionine	Choline, creatine, polyamines
Glutamine	Purines, pyrimidines
Aspartic acid	Pyrimidines
Phenylalanine/ Tyrosine	Catecholamines, thyroid hormones
Tryptophan	Serotonin
Lysine	Carnitine
Histidine	Histamine

PROTEIN DEPOSITION

It is implicit in what has been said so far that protein deposition is the sum of two processes and gain or loss of body protein may result from a change in protein synthesis, protein degradation or in both processes. Reliable *in vivo* methods for the measurement of protein synthesis, such as the flooding dose phenylalanine method of Garlick et al. (1980) have been applied to several species of fish in recent years. Measurement of protein breakdown is less satisfactory and estimates have been obtained from the difference between protein accretion (usually measured over a period of weeks) and protein synthesis measured over a period of hours or minutes.

These measurements have shown the pattern of protein

deposition in fish tissues to be similar to that in mammals. Increased rates of growth and of protein deposition are a result of increased rates of both protein synthesis and breakdown, i.e. protein turnover rate increases - acceleration of synthesis occurring at a greater rate than that of degradation (Houlihan et al., 1986, 1988). Fractional rates of protein synthesis were shown to be much higher in tissues such as liver, gills, intestine and spleen than in skeletal muscle, especially white muscle. In fact, muscle protein synthesis in fish is seen as differing strikingly from that in mammals (Smith, 1981). His measurements show that trout growing at a rate of 0.25%/day have a fractional rate of muscle protein synthesis of 0.38%/day, this compares with a muscle protein synthesis rate of 4.9%/day in rats growing at a similar rate (0.29%/day). Houlihan et al. (1986) noted in addition, that in a growing fish, much of the muscle protein synthesised results in growth; approximately 76% of the protein synthesised in white muscle of trout remained as growth while the corresponding figure for gills was 3-5%. As white muscle accounts for about 60% of the body weight of fish, then the total amount of protein synthesised in this tissue is greater than in any other tissue. It may be inferred that whole body protein turnover is much lower than in mammals and losses of amino acids from the body as a consequence of turnover will be lower in fish than in mammals. A further consequence may be that fish are more immediately dependent on the dietary supply of amino acids for tissue protein synthesis than is the case in mammals.

Millward and Rivers (1988) introduced the concept of "anabolic drive" a principle tenet of which is that amino acids serve not only as substrates for the synthesis of protein and other molecules and as a source of energy, but also exert a regulatory influence on some of the individual pathways themselves. These regulatory influences would be primarily anabolic but might also stimulate protein breakdown and oxidative loss, some of the effects would be transient. The amino acids involved would be part of the normal requirement.

With this in mind, Millward (1989) reviewed the influence of dietary protein, insulin, tri-iodothyronine and insulin-like growth factor 1 on rates of protein turnover in muscle. He concluded that insulin may have a crucial role in sensing indispensable amino acid intakes and in initiating the anabolic drive.

Direct effects of amino acids on muscle protein deposition have been difficult to demonstrate. There is evidence that dietary concentrations of tryptophan, over and above those necessary for optimal growth of swine, are needed for maximal activity of muscle ribosomes - but this increased activity was not reflected in increased protein synthesis following simple recombination *in vitro* (Lin et al., 1988). Leucine has been shown to stimulate protein synthesis and reduce protein breakdown in muscle in many mammalian in vitro systems (Block, 1989). However, McNurlan et al. (1982) were unable to demonstrate any direct effect of leucine on the fractional rate of protein synthesis in tissues of intact rats in different nutritional states. Recent experiments on rainbow trout in this laboratory have similarly failed to demonstrate any proteogenic effect of dietary leucine. Choo (1990) fed diets containing increasing concentrations of dietary leucine beyond the requirements for maximal growth but no increase in carcass protein deposition occurred.

CONTROL OF AMINO ACID CATABOLISM

Little is known about the extent to which amino acid

catabolism is controlled in fish. Attempts to examine this imply a prior knowledge of the pathways involved. These are generally assumed to be similar to, if not identical with, those occurring in omnivorous mammals. That differences, at least of degree, occur is evident from observations on the branched chain amino acids (BCAA). In mammals the branched chain amino acid aminotransferase (BCAAT) is found almost entirely in skeletal muscle so that, alone of absorbed amino acids, BCAA (other than those used for protein synthesis in liver) by-pass this organ and the first steps in their catabolism occur in the muscle. Branched chain keto acids are later transported back to the liver. It was noted earlier (Murai et al., 1987) that BCAA are removed from the portal vein of trout by the liver. In line with this several tissues (including liver) contain appreciable BCAAT activity (Hughes et al., 1983) but activities in white muscle are very low (Teigland and Klungsoyr, 1983). High activities occur in kidney and in red muscle, but there is little red muscle in fish.

Another observation of interest is that BCAA antagonism, seen in mammals and birds after they have consumed an excessive amount of leucine, does not appear to occur in trout. The antagonism probably stems from the fact that the first two steps of BCAA catabolism are carried out by the same enzymes. The activities of these are enhanced in response to the high leucine intake and this leads to rapid depletion of the total body free pools of isoleucine and valine as well as of their keto acids, free pools of leucine remain high. By contrast, the concentrations of isoleucine and valine in the tissues of rainbow trout were not significantly affected when the fish were given diets containing up to 13% leucine (Table 3, Choo, 1990).

Some control of amino acid catabolism is exerted at the substrate level (control by K_m). When trout are fed diets (40% crude protein) containing graded levels of one or other indispensable amino acid (lysine, tryptophan, arginine) then little oxidation occurs at low levels of intake. As concentration in the diet increases,

Table 3. Concentrations of branched chain amino acids in plasma and liver of rainbow trout given diets with different concentrations of leucine

	Dietary leucine (%)			
	3.3	6.2	9.2	Pooled SEM
Plasma (nmoles/ml)				
Valine	1017	952	902	127
Leucine	480	600	970	370
Isoleucine	89	81	68	10
Liver (μmoles/g)				
Valine	6.62	8.38	8.29	0.61
Leucine	3.29	4.14	4.23	0.26
Isoleucine	0.87	1.25	1.12	0.08

Data of Choo (1990)

presumably leading to enhanced tissue concentrations, oxidation of the amino acid rises markedly (Walton et al., 1986). In much the same way dietary restriction of one amino acid will lead to enhanced oxidation of other indispensable amino acids present at normal levels in the diet (Were, 1989). In this instance protein synthesis is restricted by the limiting amino acid, and the "excess" of other indispensable amino acids, which cannot be used for protein synthesis, are oxidised.

Methionine catabolism in trout appears to follow the transmethylation - transulphuration pathway established in mammals (Walton et al., 1982). Again, no adaptive control is apparent at the enzyme level in response to restriction of methionine intake. Hepatic activities of methionine adenosyl transferase and cystathionine synthase, key regulatory loci, were unchanged in response to this treatment. Control was apparent at the substrate level, however. Thus serine concentration was elevated in both liver and plasma of methionine deficient trout but fell as dietary methionine was increased (Table 4). These changes are most reasonably interpreted as an indication of a decreased flux through cystathionine synthase when dietary methionine intake is lowered. Taurine concentrations, on the other hand, were elevated in these tissues as dietary methionine concentration rose (Table 4). These increases must likewise reflect enhanced levels of transulphuration at high methionine intake.

Table 4. Concentrations of serine and taurine in plasma and liver of trout fed diets containing different amounts of methionine and cystine

Diet (%)		Plasma (μmole/l)		Liver (μmole/l)	
Methionine	Cystine	Serine	Taurine	Serine	Taurine
0.40	0.16	528[a]	55.9[a]	3.18[a]	0.80[a]
0.80	0.16	462[ab]	301.4[b]	1.88[b]	10.35[b]
1.90	0.16	152[c]	458.4[b]	2.08[b]	28.02[c]
0.60	0.36	352[b]	309.3[b]	1.98[b]	9.89[b]

Values in the same column with different superscripts are significantly different ($P<0.05$). Unpublished data of Cowey and Cho.

ACKNOWLEDGEMENT

I thank Mrs. Andra Williams for expert help in the preparation of the manuscript.

REFERENCES

Block KP (1989) Interactions among leucine, isoleucine and valine with special reference to the branched chain amino acid antagonism. In: Absorption and Utilization of Amino Acids (M Friedman ed) Vol. 1 pp 229-244. CRC Press Inc. Boca Raton Florida

Boorman KN (1980) Dietary constraints on nitrogen retention. In: Protein Deposition in Animals (PJ Buttery, DB Lindsay eds) pp 147-166. Butterworths London

Bowen SH (1987) Dietary protein requirements of fishes - a reassessment. Can J Fish Aquat Sci 44:1995-2001

Brett JR, Higgs DA (1970) Effect of temperature on the rate of gastric digestion in fingerling sockeye. J Fish Res Bd Can 27:1767-1779

Brett JR, Shelbourne JE, Shoop CT (1969) Growth rate and body composition of fingerling sockeye salmon, *Oncorhynchus nerka*, in relation to temperature and ration size. J Fish Res Bd Can 26:2363-2394

Brett JR, Zala CA (1975) Daily pattern of nitrogen excretion and oxygen consumption of sockeye salmon (*Oncorhynchus nerka*) under controlled conditions. J Fish Res Bd Can 32:2479-2486

Cho CY, Kaushik SJ (1990) Nutritional energetics in fish: energy and protein utilization in rainbow trout (*Salmo gairdneri*). World Rev Nutr Diet 61: 132-172

Choo PS (1990) Leucine nutrition of rainbow trout. MSc thesis, University of Guelph, Ontario, 96 pp

Cowey CB, Sargent JR (1972) Fish nutrition. Adv Mar Biol 10:383-492

Cowey CB, Walton MJ (1989) Intermediary metabolism of carbohydrates and amino acids. In: Fish Nutrition (JE Halver ed) pp 259-329. Academic Press New York

Dabrowski K, Dabrowska H (1981) Digestion of protein by rainbow trout (*Salmo gairdneri* Rich.) and absorption of amino acids within the alimentary tract. Comp Biochem Physiol 69A:99-111

Garlick PJ, McNurlan MA, Preedy VR (1980) A rapid and convenient technique for measuring the rate of protein synthesis in tissues by injection of [^{3}H]phenylalanine. Biochem J 192:719-723

Houlihan DF, MacMillan DN, Laurent P (1986) Growth rates, protein synthesis and protein degradation rates in rainbow trout: effects of body size. Physiol Zool 59:482-493

Houlihan DF, Hall SJ, Gray C, Noble BS (1988) Growth rates and protein turnover in Atlantic cod, *Gadus morhua*. Can J Fish Aquat Sci 45:951-964

Hughes SG, Rumsey GL, Nesheim MC (1983) Branched chain amino acid aminotransferase activity in the tissues of lake trout. Comp Biochem Physiol 76B:429-431

Johnsen F, Wandsvik A (1991) The impact of high energy diets on pollution control in the fish farming industry. In: Nutritional Strategies in the Management of Aquaculture Waste (CB Cowey, CY Cho eds). University of Guelph, Ontario. In Press
Kim K-I, Kayes TB, Amundsen CH (1991) Purified diet development and re-evaluation of the dietary protein requirement of fingerling rainbow trout (*Oncorhynchus mykiss*). Aquaculture 96:57-67
Lee DJ, Putnam GB (1973) The response of rainbow trout to varying protein/energy ratios in a test diet. J Nutr 103:916-922
Lin FD, Smith TK, Bayley HS (1988) A role for tryptophan in regulation of protein synthesis in muscle. J Nutr 118:445-449
McNurlan MA, Fern EB, Garlick PJ (1982) Failure of leucine to stimulate protein synthesis in vivo. Biochem J 204:831-838
Millward DJ (1989) The nutritional regulation of muscle growth and protein turnover. Aquaculture 79:1-28
Millward DJ, Rivers J (1988) The nutritional role of indispensable amino acids and the metabolic basis for their requirements. Eur J Clin Nutr 42:367-393
Murai T, Ogata H, Hirasawa Y, Akiyama T, Nose T (1987) Portal absorption and hepatic uptake of amino acids in rainbow trout force fed complete diets containing casein or crystalline amino acids. Nippon Suisan Gakkaishi 53:1847-1859
Smith DA (1987) Studies on the protein requirement of rainbow trout (*Salmo gairdneri*). PhD thesis, University of Wisconsin - Madison, 242 pp
Smith MAK (1981) Estimation of growth potential by measurement of protein synthetic rates in feeding and fasting rainbow trout, *Salmo gairdneri* Richardson. J Fish Biol 19:213-220
Teigland M, Klungsoyr L (1983) Accumulation of α-ketoisocaproate from leucine in homogenates of tissues from rainbow trout (*Salmo gairdneri*) and rat. An improved method for determination of branched chain keto acids. Comp Biochem Physiol 75B:703-705
Walton MJ, Cowey CB, Adron JW (1982) Methionine metabolism in rainbow trout fed diets of differing methionine and cystine content. J Nutr 112:1525-1535
Walton MJ, Cowey CB, Coloso RM, Adron JW (1986) Dietary requirements of rainbow trout for tryptophan, lysine and arginine determined by growth and biochemical measurements. Fish Biochem Physiol 2:161-169
Were KE (1989) Determination of the tryptophan requirement of the juvenile rainbow trout using the indicator oxidation technique. MSc thesis, University of Guelph, Ontario, 199 pp

9

Nutrient Transport in Fish: Studies with Membrane Vesicles

C. Storelli and T. Verri

The study of intestinal nutrient absorption in fishes is of great interest both in order to increase our theoric knowledge of fish physiology and in order to suggest applications to fish farming. After Crane's Na^+-dependent hypothesis (Crane, 1962), the observation made by Smith (1966) that intestinal nutrient transport in fish was associated with Na^+-ion movement remains a mile-stone in the study of fish intestinal physiology.

As happened for studies on mammalian intestine, introduction of the isolated plasma membrane technique (Eveloff et al., 1980) dramatically increased our knowledge of fish intestinal nutrient absorption.

In this paper a summary of the results on sugar, amino acid and water soluble vitamin absorption, obtained using the isolated intestinal brush border and basolateral membrane of the european eel Anguilla anguilla is reported. In addition, some comparative and applied aspects of intestinal nutrient absorption mechanisms are discussed.

MATERIALS AND METHODS

European yellow eels (Anguilla anguilla) (150-200 g) were purchased from Ittica Ugento (Lecce, Italy) and were kept in seawater aquaria. Brush border membrane vesicles were routinely prepared using a Mg^{2+}

precipitation technique combined with differential centrifugation (Storelli et al., 1986). Basolateral membranes were prepared by centrifugation in a discontinous sucrose gradient, after intestinal cell isolation and homogenization, as described by Reshkin et al. (1988). Measurement were conducted on these osmotically active vesicles using both radioactive (Storelli et al., 1986) and fluorescent tracer techniques (Cassano et al., 1988). In the first case, by combining the use of labeled substrates and a rapid filtration technique, nutrient uptake (pmol/min x mg prot) was directly followed and plotted versus time (Storelli et al., 1986). When the fluorescent potential sensitive tracer DiS-C_2(5) was used, a transmembrane potential difference, inside negative, was imposed by the presence of a K^+ gradient plus valinomycin, and Na^+-linked substrate uptake was monitored following dissipation of the electrical potential (Cassano et al.,1988).

SUGAR TRANSPORT IN EEL BRUSH BORDER MEMBRANE VESICLES

Table I summarizes information on sugar transport in eel intestinal BBMV concerning Na+ dependence, electrogenicity, kinetic parameters, substrate specificity and stoichiometric ratios of Na^+-D-glucose transport and compares them with those of other vertebrate species.

Data shown in Table I suggest that all vertebrate species examined share basic characteristics of the Na^+-D-glucose cotransporter such as Na^+ dependence, electrogenicity, substrate specificity and substrate affinity scale.

Of particular interest is the observation that, using BBMV, two Na^+-dependent carriers of different kinetic characteristics have been described in mammalian intestine: a low affinity - high capacity transport system with a K_m ranging between 0.2 (Kessler et al., 1983) and 4.0 mM (Dorando et al., 1984) and a

Table I: Na^+-D-glucose cotransport in eel and other vertebrate brush border membranes

Characteristics	Eel	Other vertebrates	Ref.
Na^+-dependence	+	+	
Electrogenicity	+	+	
Substrate specificity	D-pyr. ring, C_2-OH, C_6; D-glucose> ß-Me-glucose> α-Me-glucose> D-galactose	Same	(1)
K_m	0.10	0.03 - 4.00	(2,5)
J_{max}	6.87	0.65 - 11.00	(2,3)
Na^+/substrate stoichiometric ratio	2:1	2:1 1:1	(2,4) (2)

K_m: mM; J_{max}: nmol/min x mg prot;
(1) Shimada et al., 1986; (2) Kaunitz et al., 1984; (3) Stevens et al., 1984; (4) Kimmich et al., 1984; (5) Dorando et al., 1984

high affinity - low capacity transport system with a K_m ranging between 0.03 and 0.08 mM (Kaunitz et al., 1984). K_m measured using other different experimental approaches have been reported to range between 0.6 mM in bull frog (Karasov et al., 1985) and 5.0 mM in guinea pig (Thomson et al., 1982).

Fixed stoichiometric Na^+-substrate ratios of both 1:1 and 2:1 seem to be present in mammalian intestine (Kaunitz et al., 1982; Hopfer et al., 1980; Kessler et al., 1983). In lower vertebrates a fixed stoichiometric ratio of 2:1 has been found in chick (Kimmich et a., 1984) and eel (Maffia et al., 1990).

On the other hand Reshkin et al. (1989), in the intestine of the teleost tilapia (<u>Oreochromis mossambicus</u>), observed a change in intestinal brush border Na^+-glucose cotransport stoichiometry from 1 Na^+: 1 glucose, under normal physiological conditions, to 2 Na^+:1 glucose under hormonal 17-α-methyltestosterone stimulation. This change seems to be due to a novel insertion of 2 Na^+:1 D-glucose carriers in the brush border membranes (Hazzard et al., in press). Two distinct Na^+-glucose cotransport systems with, respectively, 2:1 and 1:1 stoichiometric ratio, also seem to be present in fishes such as the Pacific

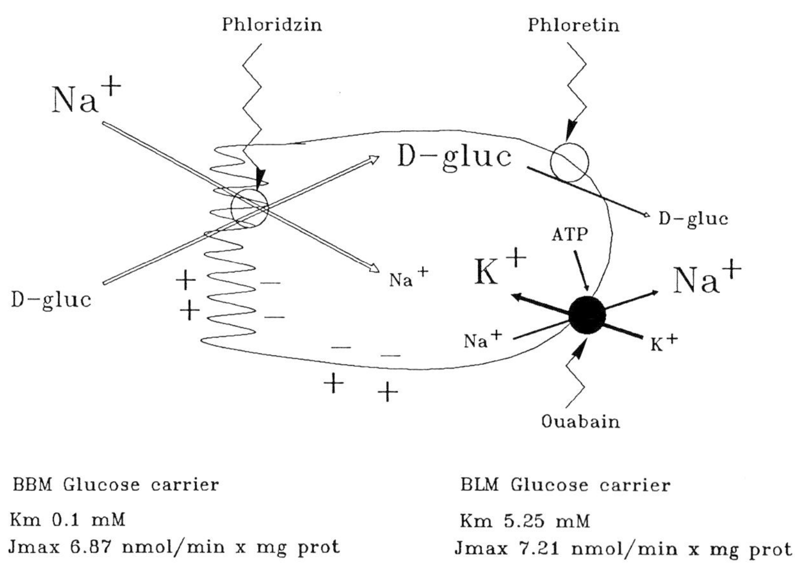

<u>Fig. 1</u>: Transephitelial D-glucose transport

copper rockfish (Sebastes caurinus) which have differentiated proximal (pyloric caeca) and distal (intestinal) portions of the gastrointestinal tract (Ahearn et al., submitted for publication).

Fig. 1 suggests a complete scheme of the transcellular D-glucose transport mechanism in eel intestine. As suggested in mammals, it is based on a) a secondary active transport located on BBM, which, using the Na+ electrochemical gradient, accumulates the sugar inside the cell; b) a Na^+-independent, phloretin inhibitable, mechanism, which translocates the sugar from the cell towards the extracellular space down its chemical gradient (Reshkin et al., 1988).

AMINO ACID TRANSPORT IN EEL BRUSH BORDER MEMBRANE VESICLES

The study of intestinal amino acid transport both in mammals and in fishes has been and still is a difficult matter because it is complicated by the existence of more than one carrier system with overlapping specificities. Those carriers can be, as a first step, operationally defined in terms of Na^+ dependence (or independence) and their criteria of saturation (kinetic parameters).

Na^+-dependence and Electrogenicity

Using both fluorescence and radioactive tracers, the dependence of amino acid transport across eel BBMV on Na^+ ion and electrical potential gradients was studied.

In Table II results obtained using amino acids belonging to the different classes are reported. As it can be noted the transport of all tested amino acids is strictly dependent on the presence of Na^+ ion in the extravesicular space. Furthermore, L-Ala, L-Pro, L-Lys and L-Glu were also studied using radiolabeled compounds. Among them, in addition to a Na^+-dependent system, L-alanine and L-lysine, also showed a Na^+-independent mechanism. Finally, the results obtained using a membrane potential sensitive fluorescent dye

Table II: Na^+-dependent and Na^+-independent amino acid uptake in eel BBMV

Amino acid	Fluorescent tracers	Radioactive tracers	
	Na^+-dep.	Na^+-dep.	Na^+-indep.
Gly	+		
N-Me-Gly	+		
L-Ala	+	+	+
N-Me-Ala	+		
ß-Ala	+		
L-Val	+		
L-Leu	+		
L-Met	+		
L-Phe	+		
L-Pro	+	+	-
α-AIB	+		
L-Lys	+	+	+
L-Glu	+	+	-
L-Asp	+		
L-Thr	+		

(DiS-C_2(5)) strongly suggests the existence of an electrogenic Na^+-amino acid translocation for all tested amino acids, since the addition of any of them in the extravesicular space, directly drives the movement of positive (Na^+-ion) electric charges inside the vesicles.

Kinetic parameters

Table III reports kinetic parameters of Na^+-dependent transport systems for some representative neutral, imino, basic and acidic amino acids. In addition the kinetic parameters of some methylated amino acids, useful for studying carrier specificity, are reported.

Table III: Kinetic parameters of Na^+-amino acid cotransport systems in eel BBMV

Substrate	Radioactive tracers		Fluorescent tracers	
	K_m	J_{max}	K_m	J_{max}
L-Ala	1.29	3.61	0.59	32.1
Gly	7.33	17.70	0.55	14.8
α-AIB	-	-	0.98	13.6
N-Me-Gly	-	-	1.22	11.7
N-Me-L-Ala	-	-	1.23	12.6
α-MeAIB	-	-	0.05	10.4
L-Pro	0.23	7.96	0.12	22.6
L-Phe	1.67	6.40	-	-
L-Lys	0.16	3.57	-	-
L-Glu	0.72	2.70	0.45	2.7

K_m: mM; J_{max}; nmol (radioactive tracers) or ΔF (fluorescent tracers)/min x mg prot; S.E. < 5 %

It can be observed that the J_{max} values, with the exception of glycine, varied in an homogenous range between 2.7 (L-glutamate) and 7.9 (L-proline) nmol/mg prot x min. Furthermore, with the exception of glycine, all K_m values measured under short circuit conditions (radioactive tracers) were of the same order of magnitude (0.16 mM; L-lysine - 1.67 mM; L-phenylalanine).

When, with the use of the fluorescence technique, measurements were made in the presence of a transmembrane electrical potential (internal negative) a maximal reduction of the K_m values was obtained for L-alanine, L-proline, glycine, L-glutamic acid.

Experiments using labeled glutamate or proline, carried out in the presence and in the absence of the electrochemical potential difference across the vesicles membrane, have confirmed that the presence of a transmembrane potential difference doubles the affinity constants of the carrier for the amino acids (Romano et al., 1989), thus acting as an additional driving force.

Another example of an additional driving force acting on fish intestinal brush border membrane amino acid translocation mechanisms, seems to be represented by the Cl^- electrochemical gradient (Bogé et al., 1985).

Romano et al. (1989) have shown that in the eel, both K^+ (intravesicular) and Cl^- (extravesicular) enhance the uptake of glutamate inside the vesicles. It has been also demonstrated that K^+ ion clearly decreases the K_m, while Cl^- significantly increases the J_{max} for glutamate. This effect has been observed in other fish intestines (Bogé et al., 1985) and other different tissues (Corcelli et al., 1983; Scalera et al., 1987) and, represents a very interesting aspect of the amino acid transport problem still opened to further investigations (Turner, 1986; Balon et al., 1991).

The roles of the Na^+ gradient, and other additional driving forces (membrane potential difference, Cl^-, K^+, H^+ gradients), are of fundamental importance in the intestinal transepithelial absorption mechanisms, because they affect the kinetic parameter values thus regulating the intracellular accumulation capacity.

To fully understand intestinal amino acid absorption mechanisms other aspects, such as the number of carriers, the specificities and overlapping transport capacities and the nature of mutual inhibition, also need to be elucitated.

Among the neutral amino acids transport systems the characterization of the imino acid L-proline transporter, in eel BBMV, has prooved to be of particular interest to face this intriguing matter.

Using the scheme suggested by Scriver (1964) experiments were carried out among L-proline, L-alanine and L-α-(methylamino)-isobutyric acid (MeAIB) in order to establish which kind of inhibition was present among them. With this kind of information it was possible to establish whether a carrier was shared or not by different substrates.

Table IV summarizes current information on the eel BBM L-proline carrier and compares this information to the data from the mammalian rabbit intestinal L-proline carrier reported by Peerce et al. (1984).

Although kinetic parameters of the eel and rabbit intestinal carriers for L-proline, K_m and J_{max}, seem to be comparable, the eel L-proline carrier shows significantly different substrate characteristics. Eel BBMV transports MeAIB and other N-methylated amino acids (N-MeAA), while (with the exception of MeAIB) N-MeAA are not transported by the rabbit BBMV L-proline carrier.

Furthermore, it has been found in the eel that both the stoichiometric ratio (Vilella et al., 1988) and the translocation mechanism (Maffia et al., 1990) are different than those found in mammals. The proline carrier in eel first binds L-proline and then Na^+, while in rabbit the proline carrier sequentially binds two Na^+ ions and then L-proline according to an iso-ordered ter-ter mechanism (Stevens and Wright, 1987). Data obtained studying amino acid transport mechanisms of different amino acid groups in eel intestine suggested that, with regard to Na^+-dependent transport systems, the molecular organization found in the eel intestinal BBM seems to be very near to those of the

Table IV: Na^+-L-proline transport system in eel and rabbit BBMV

Characteristics	Eel	Rabbit
Na^+-dependence	+	+
Electrogenicity	+	+
K_m	0.23 ± 0.04	0.55 ± 0.05
J_{max}	7.96 ± 0.87	9.96 ± 0.01
Transported substrates	- Pro, MeAIB, other N-MeAA;	- Pro, MeAIB
Substrate inhibition	- Inhibited by Ala (completely) and Phe (uncompletely)	- Inhibited by Ala (uncompletely) and Phe (completely)
Stoichiometric ratio	1:1	2:1

K_m: mM; J_{max}: nmol/min x mg prot

mammalian gut. Four different Na^+-dependent carrier processes were identified: a) a transport process for anionic L-glutamic and L-aspartic acid; b) a specific mechanism for cationic L-lysine and L-arginine amino acids; c) an imino acid carrier system for L-proline and MeAIB, which probably also accepts L-phenylalanine;

d) a second transport system for neutral amino acids, with a broader specificity for short-chain substrates (L-alanine, L-serine, glycine).

Furthermore, the molecular organization found in the eel BBM seems to have the following specificities: a) a transport system specific for L-proline but with many characteristics (stoichiometric ratio, translocation mechanism, substrate specificities) quite different from those found in mammals (Vilella et al., 1988; Maffia et al., 1990); b) the probable absence of a distinct transport system for L-phenylalanine (PHE system); c) the presence of both Na^+-dependent and Na^+-independent L-lysine carriers with similar kinetic parameters (Vilella et al., 1990).

UPTAKE OF AMINO ACIDS ON BASOLATERAL MEMBRANE

Results obtained studying amino acid transport in eel basolateral membrane vesicles have confirmed the observations made with mammal enterocytes that most of the amino acids accumulated inside the cell leave the intracellular compartment via Na^+-independent facilitated mechanisms (Reshkin et al., 1988). Of interest is the finding reported by the same authors that both L-glutamate and L-proline possess a specific Na^+-dependent transport system. Similar Na^+-dependent mechanisms have been reported for L-glutamate (Scalera et al., 1984) and L-alanine (Mircheff et al., 1980) in rat basolateral membranes.

UPTAKE OF WATER SOLUBLE VITAMINS

Data obtained on the uptake of water soluble vitamins in eel brush border and basolateral membrane vesicles suggest that the model described in Fig. 1 seems to be suitable for a great variety of nutrients present in the intestinal lumen. It has been reported that water soluble vitamins, such as inositol (Vilella et al., 1989) and ascorbic acid (Maffia et al., 1990), are transported across eel BBM vesicles in a Na^+-dependent, electrogenic manner, while the transport throughout the BLM is obtained by means of Na^+-independent transport systems as found for inositol (Reshkin et al., 1989). With regard to carrier specificity, it has been reported that inositol was fully non-competitively inhibited by D-glucose and by phloridzin, while no inositol inhibitory effect was found on D-glucose transport (Vilella et al., 1989). Interestingly, this effect of D-glucose on inositol tranport is also present in hamster (Caspary et al., 1970) and in chicken (Lerner et al., 1979) small intestine. On the other hand ascorbic acid was inhibited competitively by isoascorbic acid, suggesting the use of the same carrier (Maffia et al., 1990) by the two molecules.

DISCUSSION AND CONCLUSIONS

The study of fish intestinal brush border transport systems by which metabolites are absorbed from the gut lumen, has become a very important research field since, in addition to representing a powerful means for understanding intestinal physiology at the molecular level, it can also suggest useful applications in the preparation of artificial diets for fish farming.

Using membrane vesicles as a tool for studying nutrient transport processes, it is becoming evident that the molecular organization of intestinal absorption in teleost fishes has reached a level quite near to that of mammals.

Features of the eel intestinal brush border D-glucose (and other monosaccaride) transport system (i.e. Na^+-dependence, phloridzin sensitivity; substrate specificity) seems to be similar to those described in other vertebrates, thus supporting the idea that a phylogenetically identical carrier mechanism for sugar transport is present in the intestinal BBM of most vertebrates.

Comparison of characteristics such as Na^+-D-glucose stoichiometric ratio or binding activity (K_m) strongly suggest the possibility that in fishes, as described for mammals, different forms of the sugar transport have evoluted under different environmental or physiological conditions. In fact, while a single 2 Na^+/1 D-glucose transporter was found in the eel, two distinct D-glucose carriers (2:1 and 1:1) seems to be present in fishes with anatomically differentiated gastrointestinal tracts (Ahearn et al., submitted for publication) or in fishes under different hormonal status (Hazzard et al., 1991). In this regard the eel D-glucose cotransporter seems definitely to be more comparable to that of the upper intestinal region of the carnivorous rockfish Sebastes caurinus (K_m 0.15 mM; Na^+-D-glucose stoichiometric ratio 2:1) than to the one described in the more proximal pyloric caeca of the same fish (K_m 0.24 mM; Na^+-D-glucose stoichiometric ratio 1:1) (Ahearn et al., submitted for publication) or to that reported in the intestine of the herbivorous teleost tilapia (Oreochromis mossambicus) (K_m 0.6 mM; Na^+-D-glucose stoichimetric ratio 1:1).

Amino acids in eel intestinal BBMV are transported by at least four different Na^+-amino acid coupled mechanisms: two for neutral, one for cationic and one for anionic amino acids. This molecular organization is similar to that described for rabbit BBMV (Stevens et al., 1984) with some notable differences mainly regarding the lack of a specific system transporting L-phenylalanine (PHE), which in eel intestine could be transported either by L-proline (IMINO) or L-alanine (L-alanine, L-serine, glycine) carrier systems.

Another special feature shown by the strictly carnivorous eel is the simultaneous presence of both Na^+-dependent and Na^+-independent L-lysine-L-arginine intestinal transport systems. This finding could be explained by the high level of protein in the diet of this fish. A Na^+-dependent cotransport system for L-lysine has been described in rat only when animals were fed on a high-protein diet (Wolffram and Scarrer, 1984).

Na^+-dependent transport mechanisms for water soluble vitamins, such as ascorbic acid, have been reported in rainbow trout (Rose et al., 1990). Our preliminary results obtained in eel BBMV (Maffia et al., 1990) confirm these results suggesting that isoascorbic acid is also transported by the same carrier. This seems to be a common characteristic among the vertebrates as the same evidence has also been obtained in guinea pig small intestine (Siliprandi et al., 1979).

With respect to the basolateral membrane, common transporting characteristics found in the eel are a Na^+-independent, phloretin-inhibitable carrier mediated system for D-glucose and Na^+-independency of carrier systems for most amino acids. On the other hand peculiar features of the eel are a Na^+-dependent transporter for both aspartic acid and L-proline and a Na^+-independent, phloretin-inhibitable carrier system for the water soluble vitamin inositol.

The information presented here suggest that in fishes, as in mammals, the first step in nutrient intestinal transepithelial absorption is mediated by the presence of specific Na^+-substrate cotransport systems. These cotransporters, by using the Na^+ electrochemical gradient present across the brush border membrane can specifically bind and accumulate sugar, amino acid, water soluble vitamins and probably other nutrients inside the enterocytes. The amount of nutrients which is actually translocated through the apical cell membrane and consequently accumulated inside the cell depends on many different factors: i) the kinetic parameter values of the Na^+-cotransporters; ii) the stability of the driving forces present across the membrane; and iii) the interactions among different nutrients (competitive, non competitive inhibition) at the substrate binding sites.

Transport inhibitions have been observed either among nutrient molecules of different chemical species (sugars - amino acids; sugars - water soluble vitamins) or among molecules (amino acids) of the same group (neutral - imino; neutral - basic). Murer et al. (1975) demonstrated that the uptake inhibition between sugar and amino acids, observed in rat BBMV, was due to competition for the Na^+-electrochemical gradient, while in eel BBMV competitive (glutamate - aspartate; ascorbate - isoascorbate) or non-competitive inhibition (proline - alanine; D-glucose - inositol) for substrate binding sites was found to be the explanation.

FINAL REMARKS

Since the amount of sea food coming from fishing has probably reached the maximal level, both because of overfishing and environmental pollution problems, the practice of fish aquaculture is assuming an increasing importance in world protein production. In this context, altough many basic problems remain to be solved, the use of artificially compuonded food instead of living animals, seems to be a very promising means for metabolic energy supply in fish nutrition. Furthermore, the observation that some essential amino acids and vitamins need to be present in the diet for a regular fish growth and development, suggest that results obtained studying nutrient absorption in fish intestine, could be of great practical interest.

Proteins, sugars, fats, vitamins are fundamental components in fish nutrition and the knowledge of their intestinal absorbing characteristics at the molecular level could be used in order to optimize their utilization in the diet. In this regard, information on kinetic parameters, number, and specificities of the carriers for different nutrient groups observed in eel intestine could explain and solve problems related to the mutual inhibition observed in nutrient absorption (Murer et al., 1975; Vilella et al., 1989; Vilella et al., 1990; Primbs et al., 1971). Moreover an important role in amino acid absorption could be played by the presence of dipeptide or oligopeptide in the diet, since, as reported in tilapia (Reshkin et al., 1991) and eel (Verri et al., 1990) intestines, they can be specifically transported across the apical cell membrane. Finally, very interesting, even if very complex, are the information regarding adaptive nutrient transport regulation. In fact comprehension of different pattern of adaptation due to dietary composition, starvation or hyperphagia, time of day or different seasons and aging (Karasov and Diamond, 1983) could play a fundamental role in optimizing fish growth and development.

REFERENCES

Ahearn GA, Behnke RD, Zonno V, Storelli C (1991) Kinetic heterogeneity of Na^+/D-glucose cotransport in absorptive organs of the teleost gastrointestinal tract. (submitted for publication)

Balon LM, Ahearn GA (1991) Both Na^+ and Cl^- gradients energize NaCl/L-glutamate cotransport in lobster hepatopancreatic brush border membrane vesicles. Biochim Biophys Acta 1067:123-130

Bogé G, Roche H, Peres G (1985) Role of chloride ions in glycine transport in a sea fish (Dicentrarchus labrax). Biochim Biophys Acta 820:122-130

Caspary WF, Crane RK (1970) Active transport of myo-inositol and its relation to the sugar transport system in hamster small intestine. Biochim Biophys Acta 203:308-316

Cassano G, Maffia M, Vilella S and Storelli C (1988) Effects of membrane potential on Na^+ cotransports in eel intestinal brush-border membrane vesicles: studies with a fluorescent dye. J Membrane Biol 101:225-236

Corcelli A, Storelli C (1983) The role of potassium and chloride ions on the Na/acidic amino acid cotransport system in rat intestinal brush border membrane vesicles. Biochim Biophys Acta 732:24-31

Crane RK (1962) Hypothesis for mechanism of intestinal active transport of sugars. Fed Proc 21:891-895

Dorando FC, Crane RK (1984) Studies of the kinetics of Na^+ gradient-coupled transport as found in brush border membrane vesicles from rabbit jejunum. Biochim Biophys Acta 772:273-287

Eveloff J, Field M, Kinne R, Murer H (1980) Sodium-cotransport systems in intestine and kidney of the winter flounder. J Comp Physiol 135:175-182

Hazzard EC and Ahearn GA (1991) Rapid stimulation of D-glucose transport in teleosts by 17-α-methyltestosterone: possible involvment of microtubular action. Am J Physiol (in press)

Hopfer U (1987) Membrane transport mechanisms for hexoses and amino acids in the small intestine. In: Physiology of the gastrointestinal tract, second edition. Edited by Leonard R. Johnson, Raven Press, New York, 1499-1526

Hopfer U, Groseclose R (1980) The mechanism of Na-dependent D-glucose transport. J Biol Chem 255:4453-4462

Karasov WH (1988) Nutrient transport across Vertebrate intestine. In: Advances in Comparative and Enviromental Physiology 2:131-172

Karasov WH, Buddington RK, Diamond JM (1985) Adaptation of intestinal sugar and amino acid transport in vertebrate evolution. In: Transport processes

processes, iono- and osmoregulation, Springer, Berlin Heidelberg New York, 227-239
Karasov WH, Diamond JM (1983) Adaptative regulation of sugar and amino acid transport by vertebrate intestine. Am J Physiol 245:G443-G462
Kaunitz JD, Gunther R, Wright EM (1982) Involvment of multiple sodium ions in intestinal D-glucose transport. Proc Natl Acad Sci USA 79:2315-2318
Kaunitz JD, Wright EM (1984) Kinetics of sodium D-glucose cotransport in bovine intestinal brush border membranes. J Membr Biol 79:41-51
Kessler M, Semenza G (1983) The small intestinal Na^+ D-glucose cotransporter: An asimmetric gated channel (or pore) responsive to $\Delta\psi$. J Membr Biol 76:27-56
Kimmich GA, Randles J (1984) Sodium-sugar coupling stoichiometry in chick intestinal ephitelia cells. Am J Physiol 247:C74-C82
Lerner J, Smagula RM (1979) Myo-inositol transport in the small intestine of the domestic fowl. Comp Biochem Physiol 62A:939-945
Maffia M, Cassano G, Marcucci D, Vilella S, Storelli C (1990) The Na^+-dependent proline carrier, of eel intestinal brush-border membrane, sequentially binds proline and then Na^+. Biochim Biophys Acta 1027:8-16
Maffia M, Vilella S, Zonno V, Storelli C (1990) Il trasporto di aciddo ascorbico sull'orletto a spazzola dell'enterocita di Anguilla anguilla. XLII Meeting della Società Italiana di Fisiologia
Mircheff AK, van Os CH, Wright EM (1980) Pathways for alanine transport in intestinal basal lateral membrane vesicles. J Membr Biol 52:83-92
Murer H, Sigrist-Nelson K, Hopfer U (1975) On the mechanism of sugar and amino acid interaction in intestinal transport. J Biol Chem 250:7392-7396
Peerce BE, Wright EM (1984) Sodium-induced conformational changes in the glucose transporter of intestinal brush borders. J Biol Chem 259:14105-14112
Primbs ERJ, Sinhuber RO, Warren CE (1971) Hypervitaminosis in rainbow trout: counteraction by vitamin C. Int J Vit Nutr Res 41:331-338
Reshkin SJ, Ahearn GA (1991) Intestinal glycil-L-phrnylalanine and L-phenylalanine transport in a eurialine teleost. Am J Physiol 260:R563-R569
Reshkin SJ, Grover ML, Howerton WD, Grau EG, Ahearn GA (1989) Dietary hormonal modification of growth, intestinal ATPase and glucose transport in tilapia. Am J Physiol 256:E610-E618
Reshkin SJ, Vilella S, Ahearn GA, Storelli C (1989) Basolateral inositol transport by intestines of carnivorous and herbivorous teleosts. Am J Physiol 256:G509-G516

Reshkin SJ, Vilella S, Cassano G, Ahearn GA, Storelli C (1988) Basolateral amino acid and glucose transport by the intestine of the teleost, Anguilla anguilla. 91A:779-788

Rose RC, Choi JL (1990) Intestinal absorption and metabolism of ascorbic acid in rainbow trout. Am J Physiol 258:R1238-R1241

Scalera V, Corcelli A, Frassanito A, Storelli C (1987) Chloride dependence of the sodium dependent glycine transport in pig kidney cortex brush border membrane vesicles. Biochim Biophys Acta 903:1-10

Scalera V, Corcelli A, Vilella S, Storelli C (1984) Acidic amino acid transport through brush border and basolateral plasma membranes of rat intestinal absorbing cells. Ital J Biochem 33, 205A-207A

Scriver CR, Wilson OH (1964) Possible locations for a common gene product in membrane transport of imino-acids and glycine. Nature 202:92-93

Shimada T, Hoshi T (1986) A comparative study of specificity of the intestinal Na^+/sugar cotransport among vertebrates. Comp Biochem Physiol 84A:365-370

Siliprandi L, Vanni P, Kessler M, Semenza G (1979) Na^+-dependent electroneutral L-ascorbate transport across brush border membrane vesicles from guinea pig small intestine. Biochim Biophys Acta 552:129-142

Smith MW (1966) Time course and nature of temperature-induced changes in sodium-glucose interactions of the gold fish intestine. J Physiol (London) 183:649-657

Smith MW (1970) Selective regulation of amino acid transport by the intestine of goldfish (Carassius auratus). Comp Biochem Physiol 35:387-401

Stevens BR, Kaunitz JD, Wright EM (1984) Intestinal transport of amino acids and sugars: advances using membrane vesicles. Ann Rev Physiol 46:417-433

Stevens BR, Ross HJ, Wright EM (1982) Multiple transport pathways for neutral amino acids in rabbit jejunal brush border vesicles. J Membr Biol 66:213-225

Stevens BR, Wright EM (1987) Kinetics of the intestinal brush border (IMINO) carrier. J Biol Chem 262:6546-6551

Storelli C, Vilella S, Cassano G (1986) Na^+-dependent D-glucose and L-alanine transport in eel intestinal brush border membrane vesicles. Am J Physiol 251:R463-R469

Thomson ABR, Hotke CA, Weinstein WM (1982) Comparison of the kinetic constants of hexose uptake in four animal species and man. Comp Biochem Physiol 72A:225-236

Turner RJ (1986) ß-amino acid transport across the renal brush-border membrane is coupled to both Na^+ and Cl^-. J Biol Chem 261:16060-16066

Verri T, Cassano G, Storelli C (1990) Il cotrasporto glicil-glicina/Na^+. Studi con vescicole dell'orletto a spazzola di intestino di anguilla. XLII Meeting della Società Italiana di Fisiologia

Vilella S, Ahearn GA, Cassano G, Maffia M, Storelli C (1990) Lysine transport by brush border membrane vesicles of eel intestine: interactions with neutral amino acids. Am J Physiol 259:R1181-R1188

Vilella S, Ahearn GA, Cassano G, Storelli C (1988) Na^+-dependent L-proline transport by eel intestinal brush-border membrane vesicles. Am J Physiol 255:R648-R653

Vilella S, Reshkin SJ, Storelli C, Ahearn GA (1989) Brush border inositol transport by intestines of carnivorous and herbivorous teleosts. Am J Physiol 256:G501-G508

Wolffram S, Scharrer E (1984) Effect of feeding a high protein diet on amino acid uptake into rat intestinal brush border membrane vesicles. Pfluegers Arch 400:34-39

10

Development of Fish Larvae and Rearing Conditions in Hatcheries

F.J. Gatesoupe

INTRODUCTION

The early life of fish is a train of dramatic events under pressure of a growth efficiency close to its possible maximum (Kiorboe, 1989). For instance, Hansen et al. (1989) found that protein synthesis in cod larvae, *Gadus morhua*, was twenty times higher than in cod muscle. If growth is generally described by simple mathematical models, the theory of saltatory ontogeny of Balon (1984) is more representative of development, where periods of slow preparation alternate with stages of rapid changes. The saltatory ontogeny is particularly obvious in flatfish, the metamorphosis of which is spectacular.

For instance, the development of turbot, *Scopthalmus maximus*, was studied by Al-Maghazachi and Gibson (1984, Fig. 1). The main morphological changes during the climax were notochord bending, right eye migration and swim bladder disappearance. The other changes were smoother and could be also observed in marine roundfish. Exogenous feeding started few days before complete yolk absorption, while food must be digested in a primitive gut. Gastric glands, bile, pancreas and liver were fully operationing only by the end of the first month. However, endocellular processes of digestion were observed at start feeding. Iwai (1969) described the pinocytosis of proteins and fat droplets in early feeding larvae of several species, and, recently, Hansen and Olafsen (1990) showed the endocytosis of bacteria in cod at the yolk sac stage. Consequently, non specific immunity seemed to be initiated very early.

Stage	1a	1c	1d	2	3	4a	4c	5d
Day	0	2	3	6	10	20	27	120
mm	3	3	4	5	6	9	14	38

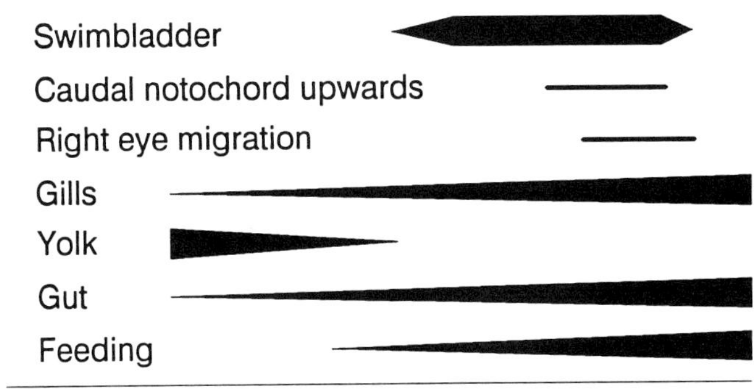

Fig.1- Developmental events during the metamorphosis of turbot, reared at 14-17°C, according to the description of Al-Maghazachi and Gibson (1984).

Larval rearing is generally the most difficult phase of fish farming, and so far, most of the technique aimed at improving the growth and survival rates. Development was taken into account only when abnormalities were observed, but the most serious developmental failures led to quick death and they were shaded off into the survival rates. According to Kiorboe (1989), "big is safe and growth improves survival", but does the highest growth rate correspond to the best development? The quality of hatchery-reared juveniles is more important than their mean weight at a given age, and the rearing conditions need to be considered in the light of their effects not only on growth and survival, but also on development. To this end, the present contribution is an attempt to review available litterature dealing with the relationship between growth, development, and rearing conditions. Then the biochemical and pathological aspects will be further examined.

GROWTH AND DEVELOPMENT

Winter flounder, *Pseudopleuronectes americanus*, started to feed during yolk absorption at low temperature (2-5°C), and after the complete utilization of yolk at higher temperature (7-10°C) (Buckley, 1982). This is a first indication that the developmental events are not necessarily synchronized between different temperatures. Moreover, Seikai et al. (1986) observed that the mean length at the last larval stage of Japanese flounder, *Paralichthys olivaceus*, decreased when temperature increased (13.5 mm at 13°C versus 11 mm at 19°C). Practically, it may seem interesting to shorten the larval stages as far as possible, but the size of fish at metamorphosis is an important criterion of quality, thus maximum and optimum growth rates should not be confused.

If the environmental factors like temperature may act directly on metabolism, their effect may also be mediated by the endocrine system. In particular, the role of thyroxine on fish metamorphosis was clearly demonstrated by Miwa et al. (1988), who observed a peak of thyroxine during the climax. Parker and Specker (1990) observed an increase of thyroxine in larval striped bass, *Morone saxatilis*, by increasing salinity, and they suggested that the beneficial effect of salinity on growth and survival may be mediated by thyroxine. Some authors tried direct hormonal treatments either by maternal injection (Brown et al., 1988) or by immersion of the larvae (e.g. Lam, 1980). In these experiment performed on several species of round fish, the acceleration of both metamorphosis and growth rate was observed. On the other hand, an experiment of thyroxine immersion of Japanese flounder by Inui and Miwa (1985) gave also an accelerated metamorphosis, but the growth rate was not increased, and these authors obtained dwarf juveniles with this treatment, while they obtained giant larvae treated with thiourea, an antithyroid drug.

Table 1 - Effect of delayed feeding on the specific growth rate of sea bass larvae, *Dicentrarchus labrax*.

Age at start feeding (days)	5	6	7	8	9
Specific growth rate *	15	17	16	13	12

(*) computed from Gatesoupe (1986), according to the formula $100*(\ln(W_t)-\ln(W_0))/t$, where W_0 and W_t corresponded to the formalin-fixed weight (mg) at days 10 and 20, respectively, and t was 10 days.

An extreme example of disconnection between growth and development was provided by starvation experiments, where there was no growth at all, but abnormal development (Ehrlich et al., 1976). Intermediate treatments of delayed feeding led to inferior growth rate, though it might be compensated for afterwards with an increased specific growth rate (Table 1). Dabrowski et al. (1986) raised the hypothesis that it was due to recovery growth, but delayed feeding may cause a selective mortality of fish with low growth efficiency, as well. The issue of recovery growth in fish larvae will remain under discussion as long as we shall not be able to measure individual growth rates.

BIOCHEMICAL ASPECTS

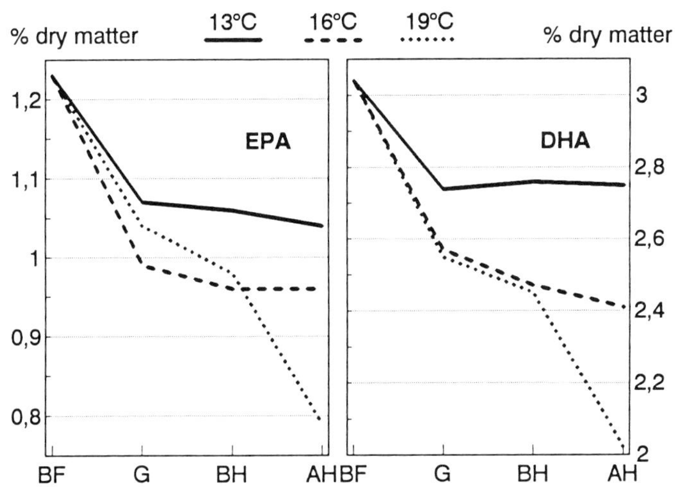

Fig.2- Influence of incubation temperature on the EPA and DHA contents of turbot eggs from fertilization to hatching, according to the data of Rubio Rincon (1986). (BF: before fertilization - G: gastrula - BH: before hatching - AH: after hatching)

A lot of data showing the effect of live food organisms on the fatty acid content of fish larvae are available. However, the interest of this biochemical criterion should not be restricted to dietary aspects. Rubio Rincon (1986) showed that the fatty acid content of turbot embryo was affected by incubation temperature. The preservation of the essential fatty acids, eicosa-pentaenoic (EPA) and docosahexaenoic (DHA) was more efficient at 13°C than at higher temperature (Fig 2). Nevertheless, it is important to consider separately polar and neutral lipids, as the difference between the amounts of DHA in the polar lipids of newly hatched larvae at 13 and 19°c (1.7 and 1.3% of dry matter, respectively) was smaller than the difference observed with the DHA contents of the neutral fractions (1 and 0.5 % of dry matter, respectively).

The case of the free amino acids is more complex. First, they are involved in the adaptation of the larvae to salinity and Marangos et al. (1985) showed that the free amino acid content of newly hatched sea bass larvae increased with salinity. Taurine and the non-essential amino acids were more affected than the essential amino acids. Free amino acids are also useful for assessing the nutritional status. Fyhn and Serigstad (1987) observed a sharp decrease of the free amino acid content in cod and halibut, *Hippoglossus hippoglossus*, before first feeding, and these authors showed that it was due to the use of the free amino acids as energy source. However in sea bass, the Essential Amino Acids (EAA) were much more affected than the non-essential ones (NEAA) and furthermore, the EAA/NEAA ratio was affected by starvation in the same way as the EAA content (Fig. 3). In energy catabolism, the non-essential nutrients are used as priority and it is likely that amino acids are quickly used for protein synthesis, which is very intense in the early stages (Fauconneau, 1985, Hansen et al., 1989). Further studies are necessary with a view to specify the dynamic relation between yolk, free amino acids, protein synthesis and energy production.

The RNA/DNA ratio may also be used as an index of protein synthesis. This ratio was quickly decreased by starvation (e.g. Wright and Martin, 1985), but it was poorly affected by temperature (Buckley, 1982), whereas protein synthesis is strongly dependent on temperature (Fauconneau, 1985). Therefore, the biochemical criteria are helpful, but they should be interpreted with caution in regard to the effect of rearing conditions, especially temperature.

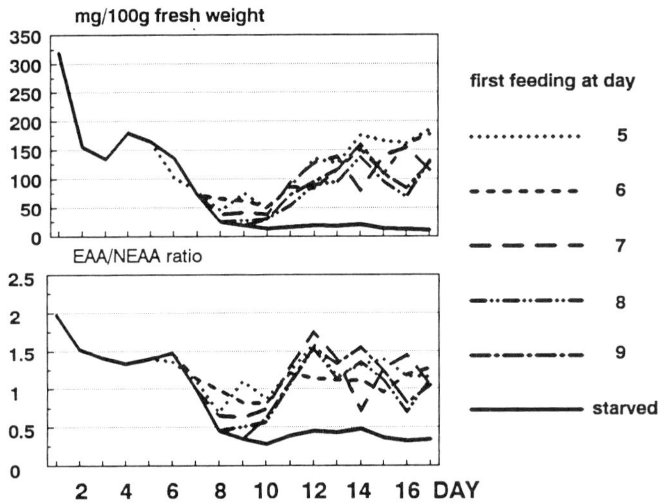

Fig.3- Effect of starvation and delayed feeding on the total free EAA content (mg/100g) and the EAA/NEAA ratio in sea bass larvae (Gatesoupe, 1986).

ABNORMAL DEVELOPMENT

Many enviromental factors may cause abnormal development. For instance, the exarcerbation of metabolism by temperature increase may lead to abnormal development and mortality. Considering again the effect of high temperature on the incubation of turbot eggs (Rubio Rincon, 1986), not only the essential fatty acid content was decreased, but the hatching rate was only 23 % at 19°C, versus 78 % at 13°C, and the rate of abnormal larvae was increased (21 and 5 % at 19 and 13°C, respectively). In sea bass larvae, Johnson and Katavic (1984) observed that high temperature or high salinity increased the rate of swim bladder hypertrophy. The best quality of fry without mortality and abnormality was obtained with the lowest salinity (25 p.p.t.) and temperature (12°C), and it corresponded to the lowest mean length. Among the intermediate treatments, if the proportion of abnormal larvae with Hypertrophied Swim Bladder (HSB) was considered while disregarding the suvival rates, one could think that the combination of 12°C and 36 p.p.t. gave a better result than the opposite treatment with 18°C and 25 p.p.t. (33 and 50 % of HSB, respectively). If the survival rate is considered together with the normality rate it appears that the contrary is true (10 and 25% of HSB, 70 and 50% of dead, respectively). That shows how important it is to keep in mind that mortality and abnormality are not independent. This may seem obvious in this case, but the same rule to consider survival rate together with normality rate should be applied even when the abnormality does not seem lethal, like albinism for instance. On the other hand, the lack of swim bladder inflation led to lordosis in Perciforms, as originally described on red sea bream, *Pagrus major*, by Kitajima et al. (1981). The environmental factors hindering the larvae from gulping air at the surface caused the lack of swim bladder inflation. Many other parameters of water quality might induce abnormal developments and they need further investigations.

An inadequate diet may also cause abnormality. Ehrlich et al. (1976) showed that starvation caused the atrophy of gut and other diformities like flat head with big eyes in herring, *Clupea harengus*, and plaice, *Pleuronectes platessa*. Watanabe et al. (1989) related the occurence of hydrops in red sea bream to the DHA deficiency of live food. Kanazawa (1991) showed that the albinism of flatfish might be caused by a complex deficiency of DHA, vitamin A and lecithin, which are precursors of rhodopsin in retina, the lack of which might hinder the secretion of the melanocyte stimulating hormone. However, this theory does not explain why Brazilian *Artemia* were more detrimental for the pigmentation of Japanese flounder than other strains of *Artemia* (Seikai et al., 1987a). Furthermore, Seikai et al. (1987b) showed that the amount of abnormal pigmentation was increased by the duration of feeding with Brazilian *Artemia* nauplii from day 10, while the growth rate was increased in the same way. Again, there was a discrepancy between the maximum growth rate and the quality of fish. The deficiency of lecithin induced also scoliosis in Ayu, *Plecoglossus altivelis* (Kanazawa et al., 1983). This phospholipid requirement would be specific of the larval stages and led to the concept of the ontogeny of nutritional requirements. Last, Dabrowski et al. (1988) observed symptoms of vitamin C deficiency in larval carp, *Cyprinus carpio*, like gill deformation and caudal fin erosion. The description of abnormalities due to nutritional deficiencies in fish larvae is rapidly increasing, due to the progress of compound diets for larval stages which make easier the experimental deficiencies. The risk to be deficient is higher for compound diets than for live food organisms, but on the other hand, technical solutions can be found while we don't know how to correct the dietary value of *Artemia* against faltfish albinism, for instance.

The last causes of abnormal development are microbes, the proliferation of which may be induced by food and environmental factors. The description of microbial infections in fish larvae is recent. Only mortalities were reported in black sea bream, *Acanthopagrus schlegeli*, (Kusuda et al., 1986) and turbot (Gatesoupe, 1991a), but abnormal developments were described in other papers concerning flatfish. Iida et al. (1989) studied a virus which caused epidermial hyperplasia in the Japanese flounder, while Masumura et al. (1989)

described in the same species intestinal necrosis caused by *Vibrio* sp. Bergh et al. (1991) showed that the embryogenesis of halibut was blocked by *Flexibacter* sp., while *Vibrio anguillarum* induced later larval deformations. Optional treatments against bacteria with antibiotics and probiotics are in progress (Gatesoupe, 1991b). These products may also affect directly the metabolism of fish, and probiotics could be particularly helpful from this point of view, since in higher Vertebrates, they may act against toxins and stimulate non-specific immunity.

CONCLUSIONS

Further knowledge about the effect of rearing conditions on development should be expected from hormonal studies. We have already mentioned the role of thyroxine in the metamorphosis of Japanese flounder. These works were recently completed by the observation of a peak of cortisol during the climax, which may indicate a synergistic effect of cortisol on thyroxine (de Jesus et al., 1991). But the role of hormones in larval development seems more general, as suspected by Cambré et al. (1990) who detected several pituitary hormones in newly hatched sea bass. The study of hormonal regulation would help the understanding of the effect of rearing conditions. For instance, we have not quoted here the results concerning lighting conditions (light intensity and photoperiod), because of their obvious effect on feeding activity. However, it seems that light may increase respiration and activity even in starving fish (Dabrowski, 1976), indicating a direct action of light on metabolism. The theory of Kanazawa (1991) about the role of lipid deficiencies on flatfish albinism via the secretion of melanocyte stimulating hormone illustrated the relationship between diet and endocrine system, but many other relations should be found, for instance the secretion of digestive enzymes under hormonal control, or the induction of feeding behaviour after the perception of food by vision or chemoreception.

We have seen that the optimum growth rate corresponding to the best quality of fry was sometimes far from the maximum growth rate observed. Further improvements may be expected by reducing the period of feeding with *Artemia*, which are good for growth but not necessarily for quality. However the compound diets need improvements too, especially concerning their digestibility. Investigations should be continued on possible specific requirements of the larval stages, as it was demonstrated with phopholipids. Hormonal treatments might also increase the growth rate, but they are conditional on the quality of the diet. Pathology should be reduced by a better diet and better environmental conditions. To this end, abnormal levels of hormones in the larvae might indicate stress induced by wrong rearing conditions. Last but not least, probiotics are promising for reducing microbial pathology.

REFERENCES

Al-Maghazachi SJ, Gibson R (1984) The developmental stages of larval turbot, *Scophthalmus maximus* (L.). J. Exp. Mar. Biol. Ecol. 82:35-51

Balon EK (1984) Reflections on some decisive events in the early life of fishes. Trans. Amer. Fish. Soc. 113:178-185

Bergh O, Hansen GH, Jelmert A, Skiftesvik AB, Taxt RE (1991) Bacterial diseases of eggs and yolk sac larvae of halibut (*Hippoglossus hippoglossus* L.). Eur. Aquacult. Soc., Spec. Publ. 15:389-391

Brown CL, Doroshev SI, Nunez JM, Hadley C, Vaneenennaam J, Nishioka RS, Bern HA (1988) Maternal triiodothyronine injections cause increases in swimbladder inflation and survival rates in larval striped bass, *Morone saxatilis*. J. Exp. Zool. 248:168-176

Buckley LJ (1982) Effects of temperature on growth and biochemical composition of larval winter flounder *Pseudopleuronectes americanus*. Mar. Ecol. Prog. Ser. 8:181-186

Cambré M, Mareels G, Corneillie S, Moons L, Ollevier F, Vandesande F (1990) Chronological appearance of the different hypophysial hormones in the pituitary of sea bass larvae *Dicentrarchus labrax* during their early development: an immunocyto-chemical demonstration. Gen. Comp. Endocrinol. 77:408-415

Dabrowski K (1976) An attempt to determine the survival time for starving fish larvae. Aquaculture 8:189-193

Dabrowski K, Takashima F, Strüssmann C (1986) Does recovery growth occurs in larval fish? Bull. Jap. Soc. Sci. Fish. 52:1869

Dabrowski K, Hinterleitner S, Sturmbauer C, El-Fiky N, Wieser W (1988) Do carp larvae require vitamin C? Aquaculture 72:295-306

Ehrlich KF, Blaxter JHS, Pemberton R (1976) Morphological and histological changes during the growth and starvation of herring and plaice larvae. Mar. Biol. 35:105-118

Fauconneau B (1985) Protein synthesis and protein deposition in fish. In: Cowey CB, Mackie AM and Bell JG (eds) Nutrition and feeding in fish. Academic press London 17-45

Fyhn HJ, Serigstad B (1987) Free amino acids as energy substrate in developing eggs and larvae of the cod *Gadus morhua*. Mar. Biol. 96:335-341

Gatesoupe FJ (1986) The effect of starvation and feeding on the free amino acid composition of sea bass larvae, *Dicentrarchus labrax*. Océanis 12:207-222

Gatesoupe FJ (1991a) Experimental infection of turbot, *Scophthalmus maximus* (L.), larvae with a strain of *Aeromonas hydrophila*. J. Fish Dis. 14:

Gatesoupe FJ (1991b) The use of probiotics in fish hatcheries: results and prospect. Int. Cons. Explor. Sea, Mariculture Committee F:37

Hansen GH, Olafsen JA (1990) Endocytosis of bacteria in yolksac larvae of cod (*Gadus morhua* L.). In: Lésel R (ed) Microbiology in poecilotherms. Elsevier Amsterdam New York Oxford 187-191

Hansen PE, Lied E, Borresen T (1989) Estimation of protein synthesis in fish larvae using an in vitro polyribosome assay. Aquaculture 79:85-89

Iida Y, Masumura K, Nakai T, Sorimachi M, Matsuda H (1989) A viral disease in larvae and juveniles of the Japanese flounder *Paralichthys olivaceus*. J. Aquat. Anim. Health 1:7-12

Inui Y, Miwa S (1985) Thyroid hormone induces metamorphosis of flounder larvae. Gen. Comp. Endocrinol. 60:450-454

Iwai T (1969) Fine structure of gut epithelial cells of larval and juvenile carp during absorption of fat and protein. Arch. Histol. Jap. 30:183-199

de Jesus EG, Hirano T, Inui Y (1991) Changes in cortisol and thyroid hormone concentrations during early development and metamorphosis in the Japanese flounder, *Paralichthys olivaceus*. Gen. Comp. Endocrinol. 82:369-376

Johnson DW, Katavic I (1984) Mortality, growth and swim bladder stress syndrome of sea bass (*Dicentrarchus labrax*) larvae under varied environmental conditions. 38:67-78

Kanazawa A (1991) Nutritional mechanisms causing abnormal pigmentation in cultured marbled sole larvae, *Limanda yokohamae* (Heterosomata). Eur. Aquacult. Soc., Spec. Publ. 15:20-22

Kanazawa A, Teshima S, Kobayashi T, Takae M, Iwashita T and Uehara R (1983) Necessity of dietary phospholipids for growth of the larval Ayu. Mem. Fac. Fish. Kagoshima Univ. 32:115-120

Kiorboe T (1989) Growth in fish larvae: are they particularly efficient? Rapp. P.-v. Réun. Cons. int. Explor. Mer 191:383-389

Kitajima C, Tsukashima Y, Fujita S, Watanabe T, Yone Y (1981) Relationship between uninflated swim bladders and lordosis deformity in hatchery-reared red sea bream *Pagrus major*. Bull. Jap. Soc. Sci. Fish. 47:1289-1294

Kusuda R, Yokoyama J, Kawai K (1986) Bacteriological study on cause of mass mortalities in cultured black sea bream fry. Bull. Jap. Sci. Fish. 52:1745-1751

Lam TJ (1980) Thyroxine enhances larval development and survival in *Sarotherodon* (*Tilapia*) *mossambicus* Ruppell. Aquaculture 21:287-291

Marangos C, Yagi H, Ceccaldi HJ (1986) Rôle de la température et de la salinité sur le taux de survie et la morphogénèse au cours du développement embryonnaire chez les oeufs du loup de mer *Dicentrarchus labrax* (Linnaeus, 1758)(Pisces, Teleostei, Serranidae). Aquaculture 54:287-300

Masumura K, Yasunobu H, Okada N, Muroga K (1989) Isolation of a *Vibrio* sp., the causative bacterium of intestinal necrosis of Japanese flounder larvae. Fish. Pathol. 24:135-141

Miwa S, Tagawa M, Inui Y, Hirano T (1988) Thyroxine surge in metamorphosing flounder larvae. Gen. Comp. Endocrinol. 70:158-163

Parker SJ, Specker JL (1990) Salinity and temperature effects on whole-animal thyroid hormone levels in larval and juveniles stiped bass, *Morone saxatilis*. Fish Physiol. Biochem. 8:507-514

Rubio Rincon EA (1986) Evolution de la composition en acides gras de l'ovocyte à la larve du turbot (*Psetta maxima* L.) en fonction du régime alimentaire des reproducteurs et des larves, ainsi que de la température d'incubation. Thèse 3° cycle, Univ. Bretagne Occident. Brest (France)

Seikai T, Watanabe T, Shimozaki M (1987a) Influence of three geographically different strains of *Artemia* nauplii on occurence of albinism in hatchery-reared flounder *Paralichthys olivaceus* Nippon Suisan Gakkaishi 53:195-200

Seikai T, Shimozaki M, Watanabe T (1987b) Estimation of larval stage determining the appearance of albinism in hatchery-reared juvenile flounder *Paralichthys olivaceus*. Nippon Suisan Gakkaishi 53:1107-1114

Seikai T, Tanangonan JB, Tanaka M (1986) Temperature influence on larval growth and metamorphosis of the Japanese flounder *Paralichthys olivaceus* in the laboratory. Bull. Jap. Soc. Sci. Fish. 52:977-982

Watanabe T, Izquerdio MS, Takeuchi T, Satoh S, Kitajima C (1989) Comparison between eicosapentaenoic and docosahexaenoic acids

in terms of essential fatty acid efficacy in larval red seabream. Nippon Suisan Gakkaishi 55:1635-1640
Wright DA, Martin FD (1985) The effect of starvation on RNA:DNA ratio and growth of larval striped bass, *Morone saxatilis*. J. Fish Biol. 27:479-485

11

Effect of GH Treatment on Salmonid Growth: Study of the Variability of Response

P.-Y. Le Bail, J. Pérez-Sanchez, K. Yao, and G. Maisse

INTRODUCTION

Pickford showed for the first time in 1948 that mammalian growth hormone was active in teleost fish such as Fundulus (*Fundulus heteroclitus*). Since this date, and until the eighties, the GH used in supplementation experiments in fish were all of bovine origin (Donaldson *et al.* 1979). Following this, scientists started treating animals with purified fish GH to estimate the biological activity of this preparation. The industrial production of massive amounts of recombinant fish GH by several different companies has enabled a large increase in the number of experiments using these homologous hormones.

These experiments demonstrated the pleïotropic effects of this hormone, which plays an important role in osmoregulation, reproduction and in both protein and lipid metabolism. However, it is mainly the effect on length and weight gain which has been the object of numerous experiments. The results show that the degree of response is very variable between the different experiments with exogenous GH; however, few researchers have analysed this variability.

Based on physiological data available and by analysing (multifactorial analysis) data collected from salmonids, the factors which may modify the response of fish to growth hormone treatment are listed. These variability factors are of two types:

- those linked to the hormonal preparation used and its mode of administration.
- those linked to the physiology of treated fish.

We have tried to situate the different levels of endocrine regulation which might be involved in this variability, based on a hypothetical scheme of the mechanisms involved.

I) HORMONE ACTIVITY AND ADMINISTRATION

a) Measuring techniques

In order to compare the effectiveness of different GH preparations on growth and on other functions, it is essential that the amounts administered are precisely known. Several techniques are available in order to measure this, each with its advantages and inconveniences.

Immunological assays (ELISA, RIA) give a very good quantification of the GH present. These have been developed in several teleost species such as carp, *Cyprinus carpio* (Cook *et al.* 1983), salmonids (Bolton *et al.* 1986, Wagner *et al.* 1986, Furuya *et al.* 1987, Le Bail *et al.* 1991, Farbridge and Leatherland 1991, Takahashi *et al.* 1991a), the Japanese eel, *Anguilla japonica* (Kishida and Hirano, 1988), tilapias (Hirano *et al.* personal communication, Ricordel *et al.* unpublished), and sea-bream, *Sparus aurata* (Le Bail *et al.* unpublished). The large majority of these assays use polyclonal antibodies. However, the immunological activity of the GH measured in this way is not generally identifiable with the biological activity, unless it has been specifically studied (Le Bail *et al.* 1991, Niu 1990, Smal personal communication). The use of monoclonal antibodies directed against the active site of the hormone molecule may lead to further research possibilities.

Several assays using homologous receptors are available for tilapia, *Oreochromis mossambicus* (Fryer 1979, Mori *et al.* in press), Coho salmon, *Oncorhynchus kisutch* (Gray *et al.* 1990) Rainbow trout, *Oncorhychus mykiss* (Yao *et al.* 1991) Japanese eel (Hirano 1991) and sea-bream (Perez-sanchez *et al.* 1991a). Radioreceptor assay is directed against the binding site of the GH; it differs from radioimmunoassay which can be directed against epitopes dispersed over the whole molecule. The binding activity of the GH to its receptor is a necessary condition for obtaining biological activity. However, even if these two activities are very closein character, they cannot be confused, at least in theory, (see the antagonist case). Thus, this is not strictly speaking a biological test. However, the precision and the repeatibility of radioreceptorassay are superior to that of *in vivo* tests and by using a reference preparation it is possible to attribute and compare the binding activity of each preparation tested.

True biological tests or assays can be carried out either *in vivo* or *in vitro*, these two approaches being often complementary.

Certain authors have suggested an *in vitro* bioassay which is inspired from the rat tibia test. This involves measuring the quantity of radioactive sulphate incorporated into the ceratobranchial cartilages after stimulation by serum from animals treated *in vivo* with GH (Ash 1977, Komourdjian and Idler 1978, Duan and Inui 1990). However, NIH researchers have now abandonned this complicated technique. To standardized the biological activity of their GH preparations, they now measure weight gain in hypox female rats injected with GH. A similar test, based on the reoccurance of growth after GH treatment in animals previously hypophysectomized, was the first *in vivo* biological test used in fish (Pickford 1954, Komourdjian *et al.* 1978, Kayes 1977). As hypophysectomy is difficult to perform in teleosts and as other means of characterizing the GH have become available, most authors have limited themselves to measuring length and and weight gain in entire animals (Table 1).

While this *in vivo* approach is necessary to demonstrate the real GH biological activity of a preparation, it presents three disadvantages:

- effects are difficult to quantify and reproduce because they depend on the physiological variability of the target animal as (genetic line, stage, stress, aquaculture conditions etc.). Moreover, diffusion and degradation of the injected hormone are difficult to control. In the majority of cases, these tests only give an indication as to the biological activity of the preparation.
- injected quantities are large (in general from 0.1 to 1.0mg/kg live weight).
- response time is at least approximately 10 days.

Simplification is thus necessary for making general use of reliable biological tests. Cell culture techniques are being increasingly used by scientists working on fish, and thus it is probable that in the near future a true biological *in vitro* test will be developed. At the present time, radioreceptor assays are the most adapted tool for estimating biological activity of fish GH. However, there is no real standard test between laboratories. Thus, one should remain prudent when assessing the amounts of GH injected in experiments on fish (except in the case of NIH mammalian GH, as specific activities are well characterized).

b) Hormone quality

GH preparations, purified or recombinant, are rarely composed of only one biochemical entity. They may be composed of forms of different size (monomers, polymers), different charge or different secondary modifications (deaminated, phosphorylated, glycosylated etc.). These forms may exist in the natural state or be generated during manipulation. In mammals, these forms can have different biological activity (Charrier and Martal 1988). However, a recent study has demonstrated that, in trout, the different forms detected seem to have similar binding activity to the receptor (Niu 1990).

TARGET SPECIE / GH USED	O. mykiss	O. kisutch	O. nerka	O. keta	O. rhodorus	S. salar	S. trutta	S. fontinalis	Esox americanus	Cyprinus carpio	Carassius auratus	Tilapia mossanbica	Ophicephalus striatus	Gillichthys sp	Ictalurus melas	Fundulus heteroclitus	Anguilla anguilla	Anguilla rostrata	Anguilla japonica	AUTHORS
mammals							+ a						+ b			+ c				a) Swift 1954, b) Venugopalan 1967, c) Pickford *et al.* 1948
human	a + c d							+ c						+ b						a) Cotten 1987, b) Doneen 1976, c) Le Bail *et al.*, unpublished, d) Maclatchy *et al.* 1990, e) Skyrud *et al.* 1989
bovine	b + e k, l n, t	g + h m, q r	+ s						u	+ a	+ d p	+ c			+ j	+ o		+ f	+ i	a) Adelman 1977, 1982, b) Chartier-Barbaduc 1959, c) Clarke *et al.* 1977, d) Cook *et al.* 1983, e) Danzman *et al.* 1990, f) Degani *et al.* 1985, g) Gill *et al.* 1985, h) Higgs *et al.* 1975, 1976, 1977, 1978, i) Inui *et al.* 1985a, j) Kaynes 1977, 1978, k) Leatherland *et al.* 1981, l) Le Bail *et al.*, unpublished, m) Markert *et al.* 1977, n) Niu 1991, o) Pickford *et al.* 1954, 1957, 1959, 1972, 1973, p) Prack *et al.* 1980, q) Richman III *et al.* 1987, r) Sheridan 1986, s) Swift 1954, t) Weatherley *et al.* 1980, 1982, u) Weatherley *et al.* 1987
r bovine		+																		Down *et al.* 1988, 1989, Gill *et al.* 1985, Mc Lean *et al.* 1990
r bovine 21 K	+ b	+ a																		a) Down *et al.* 1989, b) Schulte *et al.* 1989
ovine	+ b d, f	+ c, n			+ l	+ a	+ k									+ h	+ e		+ i	a) Bœuf *et al.* 1990, b) Bolton *et al.* 1987, c) Björnsson *et al.* 1987, d) Collie *et al.* 1989, e) de Luze *et al.* 1984, 1987, f) Farbridje *et al.* 1988, g) Foster *et al.* 1991, h) Grau *et al.* 1979, i) Inui *et al.* 1985a et b, j) Kishida *et al.* 1987, k) Madsen 1990, l) Miwa *et al.* 1985, m) Wagner *et al.* 1985, n) Young 1988
porcine	+ a					+ b										+ c				a) Cheema *et al.* 1978, b) Komourdjian *et al.* 1976 a et b, c) Pickord *et al.* 1959
rat																	+			de luze *et al.* 1987
turkey, duck, ostrich																	+			de Luze *et al.* 1987
r chicken		+																		Gill *et al.* 1985
turtles																	+			de Luze *et al.* 1987
frog																	+			de Luze *et al.* 1987

TARGET SPECIE / GH USED	O. mykiss	O. kisutch	O. nerka	O. keta	O. rhodorus	S. salar	S. trutta	S. fontinalis	Esox americanus	Cyprinus carpio	Carassius auratus	Tilapia mossanbica	Ophicephalus striatus	Gillichthys sp	Ictalurus melas	Fundulus heteroclitus	Anguilla anguilla	Anguilla rostrata	Anguilla japonica	AUTHORS
shark																+				Lewis *et al.* 1972
teleosteen																+				Pickford 1954
tilapia			+ a									+ a					+ b			a) Clarke *et al.* 1977, b) de Luze *et al.* 1984, 1987
bonite	+																			Noso *et al.* 1988
seriole	+																			Kawazoe *et al.* 1988
hake																+				Lewis *et al.* 1972
pollack																+				Pickford *et al.* 1959
cod	+																			Rand-Weaver *et al.* 1982
carp											+									Cook *et al.* 1983, Van der Kraak *et al.* 1990
eel	+																			Kishida *et al.* 1987
r eel																			+	Duan *et al.* 1990
trout	+																			Agellon *et al.* 1988, Danzman *et al.* 1990, Pector *et al.* 1990, Yao *et al.* unpublished
chum	+																			Bolton *et al.* 1987, Collie *et al.* 1989, Kawauchi *et al.* 1986, Komourdjian *et al.* 1979, Wagner *et al.* 1985
r chum	+ a, c	+ b		+ b							+ d									a) Kawauchi *et al.* 1986, b) Moriyama *et al.* 1990, c) Sekine *et al.* 1985, Singh *et al.* 1988
chinook	+																			Le Bail *et al.* 1989, Le Gac *et al.* in press.

TABLE 1: ACTIVITY OF GH FROM DIFFERENT ZOOLOGICAL ORIGINS ON FISH BIOASSAY.

Other than this polymorphism, the same form can be denatured to different degrees (cleavage, inadapted tertiary structure etc.) which renders it inactive. In the case of purified GH, the quality of the pituitaries (conservation etc.) or the choice of the techniques used (hydrophobic conditions, mechanic constraints, oxydation, enzymatic attack etc.) may be the cause of this inactivation.

Gray *et al.* (1990) have demonstrated that the binding activity of salmonid GH can vary from 1 to 25 depending on the method of preparation of the hormone. Even when the purification technique is standardized, the binding activity can vary from one preparation to another (Yao *et al.* 1991).

In the case of recombinant hormone produced by *E. coli*, the molecule did not undergo maturation and its tertiary structure does not conform to that of natural GH. A further step is thus necessary: renaturation. This step, which is in fact the key step in genetic engineering production, determines, for the greater part, the quality of recombinant hormones. Le Bail and Smal (unpublished) have found that, before renaturation, recombinant trout GH did not bind to liver GH receptors; while after adequate renaturation, this GH presented a binding activity similar to that of purified natural GH. Depending on the conditions of renaturation, recombinant GH can thus have different biological activity.

It should be noted that in the case of recombinant hormones produced by *E. coli*, contamination by proteins of bacterial origin could result in a toxic preparation which slows down growth.

c) Zoological specificities

Immunological similarities of GH from different zoological origins are variable. Immunodiffusion techniques (Hayashida 1970) or RIA developed for fish such as chum salmon (Wagner and Mckeown 1986, Bolton *et al.* 1986), chinook salmon (Le Bail *et al.* 1991), carp (Cook *et al.* 1983), eel (Kishida and Hirano 1988), sea bream (Le Bail *et al.* unpublished) and tilapia (Ricordel *et al.* unpublished), do not cross react with mammalian GH. Generally, immunological cross reactivity is total within the same family, partial within the same order and very weak between orders. However, in some cases, fish hormones from very different groups may be partially recognised by the antibody directed against the GH of the reference species (Le Bail *et al.* unpublished). These results confirm those obtained from GH cDNA, which shows that sequence homology between fish groups is partial (Kawauchi *et al.* 1990). So, injections of heterologous hormone risks producing an immune reaction in the treated animal which would diminish the effectiveness of the treatment, even to the extent of inducing auto-immunisation against its own GH that would block growth.

Table 2 summarises the effects of GH treatment in fish, using hormones from different zoological origins. It shows that mammalian GH are active in all the species tested. Further, all fish GH are active in all the fish species tested. In view of these results, it seems that vertebrate GH have similar biochemical structures which give them similar biological activities.

In order to analyse the relative biological activities of GH preparations, we carried out a multifactorial analysis on data obtained from GH supplementation experiments in Rainbow trout, described in the literature or obtained in our laboratory (Table 2). It shows (Figure 1a) that the greatest weight gains were obtained with salmonid GH and that other fish GH appear more active than mammalian GH. It should, however, be noted that this analysis brings together the greatest variability of results (Table 2) and does not take the doses of hormones used into account.

Comparison of relative potency of GH from different zoological origins during the same experiment are few. They show that in salmonids, human GH, bovine GH (natural and recombinant), ovine and chicken GH have comparable activities (Gill *et al.* 1985, Kishida *et al.* 1987, Le Bail *et al.* unpublished). Paradoxically, tilapia or salmon GH appear to be equipotent (Clarke *et al.* 1977, Wagner and McKeown 1985) or less active (Danzman *et al.* 1990) than mammalian GH tested in trout. These results, which are in apparent contradiction with those obtained in radioreceptor assays (see the following paragraph), could be explained by the injected doses which were too large and saturating or by denatured fish hormone preparation or by toxic fish hormone preparations, as may be the case in some experiments.

The biological activity on fish of vertebrate GH is confirmed by radioreceptor assays. All the GH tested were capable of binding to the salmonid receptor (Fryer 1979, Fryer and Bern 1979, Tarpey and Nicol 1985, Le Bail *et al.* 1989a, Niu 1990, Yao *et al.* 1991, Gray *et al.* 1990, Perez-Sanchez *et al.* 1991b), tilapia (Fryer 1979, Mori *et al.* in press), eel (Hirano 1991) and sea bream (Perez-Sanchez *et al.* 1991a).

The use of radioreceptor assays using hepatic membranes demonstrates the differences in activity between the hormones. These differences vary with the reference system used (hormone, receptor). With the exeption of ovine GH, that binds eel receptors with high affinity (Table 3), mammalian GH have only a few percent of the binding activity of homologous fish GH. Converesly, heterologous fish GH has an activity at least equal to 30% of the homologous GH in all fish reference systems used. Tilapia GH (purified and recombinant) is a particular case, as it is practically inactive in eel or in salmonids. However, taken together, these binding study results confirm the results obtained in *in vivo* experiments, ie: as a rule, fish GH are generally more active than mammalian GH.

SPECIE	STRAIN	AGE	W	T°	PHO	SAL	NUT	GH	DOSE	ADM	TIME	CGR	GAIN	AUTORS
O. mykiss			25	12.5		0		Bovine	14	IP	14	0.860	1.035	Chartier-Baraduc 1957
O. mykiss			6	12	16	0	E	Bovine	20	IM	56	2.059	1.687	Weatherley et al 1980
O. mykiss			5	12	16	0	E	Bovine	10	IM	168	1.175	1.333	Weatherley et al 1982
O. mykiss			5	12	16	0	R	Bovine	10	IM	168	2.210	1.341	
O. mykiss			5	12	12	0	E	Ovine	2	IP	24	1.925	1.238	Wagner et al 1985
O. mykiss			5	12	12	0	E	Chum1	2	IP	24	1.925	1.202	
O. mykiss			5	12	12	0	E	Chum2	2	IP	24	1.925	1.202	
O. mykiss		0	10	9	12	0	E	Chum	0.1	IP	49	0.687	1.253	Sekine et al 1985
O. mykiss		0	10	9	12	0	E	rChum	0.1	IP	49	0.687	1.351	
O. mykiss			15	14	A	0	E	Chum	0.018	IP	24	0.330	1.113	Kawauchi et al 1986
O. mykiss			15	14	A	0	E	Chum	0.18	IP	24	0.330	1.434	
O. mykiss			15	14	A	0	E	Chum	0.1	IP	91	0.597	1.504	
O. mykiss			15	14	A	0	E	rChum	0.1	IP	91	0.597	1.542	
O. mykiss			13	15		0	R	Ovine	1.4	IP	25	1.723	1.150	Kishida et al 1987
O. mykiss			13	15		0	R	Eel	1.4	IP	25	1.723	1.165	
O. mykiss			13	15		0	R	Eel	1.4	IP	25	1.723	1.207	
O. mykiss	Klerk	0	20	13	P	0	E	Human	7.5	IP	56	1.316	1.390	Cotten 1987
O. mykiss	Klerk	0	20	13	P	0	E	Human	8.8	IMP	56	1.316	1.431	
O. mykiss			10	15.5	E	0	E	Bonite	0.1	IP	21	0.970	1.111	Noso et al 1988
O. mykiss			10	15.5	E	0	E	Bonite	1	IP	21	0.970	1.139	
O. mykiss	Normandale		110	10	12	0	E	Ovine	0.24	IP	28	1.629	1.064	Farbridje et al 1988
O. mykiss	Spring	0	1	15	12	0	E	rTrout	50	IMM	35	3.103	1.442	Agellon et al 1988
O. mykiss	Spring	0	1	15	12	0	E	rTrout	500	IMM	35	3.103	1.325	
O. mykiss	Spring	0	80	15	12	0	E	rTrout	0.2	IP	35	1.010	1.380	
O. mykiss	Spring	0	80	15	12	0	E	rTrout	1	IP	35	1.010	1.414	
O. mykiss			10	13	H	0	E	Cod	0.01	IP	20	1.013	1.134	Rand-Weaver et al 1989
O. mykiss			10	13	H	0	E	Cod	0.1	IP	20	1.013	1.169	
O. mykiss	Fraser	0	2	10	H	0	E	Bovine21	100	IMM	56	0.516	1.150	Schulte et al 1989
O. mykiss	Fraser	0	2	10	H	0	E	Bovine21	5	IP	56	0.783	1.710	
O. mykiss	Cornec	1	20	13	H	0	E	Chinook	0.1	IP	56	1.258	1.155	Le Bail et al 1989
O. mykiss	Reynold		40	17	12	0	E	Bovine	0.35	IP	33	1.677	1.160	Danzman et al 1990
O. mykiss	Reynold		40	17	12	0	E	rTrout	0.25	IP	33	1.677	1.050	
O. mykiss	Reynold	0	12	17	12	0	E	Bovine	1	IP	56	2.546	1.125	
O. mykiss	Reynold	0	12	17	12	0	E	rTrout	1	IP	56	2.518	0.970	
O. mykiss	Reynold	0	12	17	12	0	E	rTrout	0.1	IP	56	2.518	0.969	
O. mykiss	Reynold	0	12	17	12	0	R	Bovine	1	IP	56	2.349	0.982	
O. mykiss	Reynold	0	12	17	12	0	R	rTrout	1	IP	56	2.303	0.994	
O. mykiss	Reynold	0	12	17	12	0	R	rTrout	0.1	IP	56	2.303	0.939	
O. mykiss		0	50	11	P	0	R	Ovine	5	IMP	21	1.799	1.229	Foster et al 1991
O. mykiss	Cornec	1	140	15	P	0	E	Bovine	2	IP	28	1.327	1.144	Niu 1991
O. mykiss	Cornec	1	40	12	P	0	E	Bovine	2	IP	46	2.147	1.093	Le Bail et al unpublished
O. mykiss	Cornec	1	40	12	P	0	E	Human	2	IP	46	2.147	1.096	
O. mykiss	Mirwar		60	6.5	H	0	R	rTrout	0.1	IP	42	0.531	1.285	Smal et al unpublished
O. mykiss	Mirwar		60	6.5	H	0	R	rTrout	1	IP	42	0.531	1.401	
O. mykiss	Mirwar		60	7	H	0	R	rTrout	1	IP	42	0.605	1.420	
O. mykiss	Cornec	1	40	8	P	0	R	rTrout	0.5	IP	42	0.972	1.099	Yao et al unpublished
O. mykiss	Cornec	1	40	12	P	0	R	rTrout	0.5	IP	42	1.152	1.125	
O. mykiss	Cornec	1	40	16	P	0	R	rTrout	0.5	IP	42	1.159	1.127	
O. kisutch		1	10	10	12	0	E	Bovine	3.3	IMP	56	0.768	1.349	Higgs et al 1975
O. kisutch		1	10	10	12	0	E	Bovine	33	IMP	56	0.768	1.564	
O. kisutch		1	10	10	12	0	E	Bovine	20	IP	56	0.560	1.758	
O. kisutch		1	10	10	12	0	E	Bovine	100	IP	56	0.560	1.757	
O. kisutch		0	25	8	P	30	E	Bovine	10	IM	70	0.120	1.172	Higgs et al 1976
O. kisutch		1	100	8	P	30	E	Bovine	10	IM	70	0.284	1.193	
O. kisutch		0	8	10	12	0	E	Bovine	10	IP	84	0.651	4.065	
O. kisutch		0	8	10	12	0	E	Bovine	10	IM	84	0.525	4.915	
O. kisutch		0	8	10	12	0	E	Bovine	10	IM	84	0.423	4.056	
O. kisutch		0	8	10	12	0	E	Bovine	10	IM	84	0.513	3.487	
O. kisutch		0	8	10	12	0	E	Bovine	10	IP	84	0.588	4.005	
O. kisutch		0	8	10	12	0	E	Bovine	30	IP	84	0.588	4.176	
O. kisutch		1	12	10	12	0	E	Bovine	10	IM	59	0.650	2.032	Higgs et al 1977
O. kisutch		1	25	10	P	0	R	Bovine	10	IM	56	0.575	1.301	Markert et al 1977
O. kisutch		1	25	10	P	0	R	Bovine	10	IM	56	0.803	1.162	
O. kisutch		1	25	10	P	0	E	Bovine	10	IM	56	1.114	1.269	
O. kisutch		1	25	10	P	0	E	Bovine	10	IM	56	0.971	1.454	
O. kisutch		1	15	10	12	0	E	Bovine	0.1	IM	70	0.647	1.061	Higgs et al 1978
O. kisutch		1	15	10	12	0	E	Bovine	0.32	IM	70	0.647	1.202	
O. kisutch		1	15	10	12	0	E	Bovine	1	IM	70	0.647	1.271	
O. kisutch		1	15	10	12	0	E	Bovine	3.2	IM	70	0.647	1.364	
O. kisutch		1	15	10	12	0	E	Bovine	10	IM	70	0.647	1.658	
O. kisutch	Capilano	0	5	10	E	0	E	Bovine	1	IP	42	1.578	1.214	Gill et al 1985
O. kisutch	Capilano	0	5	10	E	0	E	Bovine	5	IP	42	1.578	1.329	
O. kisutch	Capilano	0	5	10	E	0	E	rBovine	1	IP	42	1.578	1.294	
O. kisutch	Capilano	0	5	10	E	0	E	rBovine	5	IP	42	1.578	1.400	
O. kisutch	Capilano	0	5	10	E	0	E	rChicken	1	IP	42	1.578	1.191	
O. kisutch	Capilano	0	5	10	E	0	E	rChicken	5	IP	42	1.578	1.386	
O. kisutch	Capilano	2	130	7.4	H	34	E	rBovine	0.5	IP	56	0.260	1.203	Down et al 1988
O. kisutch	Capilano	2	130	7.4	H	34	E	rBovine	5	IP	56	0.260	1.272	
O. kisutch	Capilano	2	130	7.4	H	34	E	rBovine	5	IP	56	0.260	1.182	
O. kisutch	Capilano	2	130	7.4	H	34	E	rBovine	0.5	IMP	56	0.260	1.189	
O. kisutch	Capilano	2	130	7.4	H	34	E	rBovine	5	IMP	56	0.260	1.112	
O. kisutch	Capilano	2	130	7.4	H	34	E	rBovine	0.5	PO	56	0.260	1.244	
O. kisutch	Capilano	0	3	11	E	0	E	rBovine	0.1	IP	56	2.109	1.034	Down et al 1989
O. kisutch	Capilano	0	3	11	E	0	E	rBovine	1	IP	56	2.109	1.112	
O. kisutch	Capilano	0	3	11	E	0	E	rBovin21	0.1	IP	56	2.109	1.137	
O. kisutch	Capilano	0	3	11	E	0	E	rBovin21	1	IP	56	2.109	1.316	
O. kisutch			4	12	E	0	E	rChum	3000	IMM	56	1.520	1.122	Moriyama et al 1990
O. kisutch			4	12	E	0	E	rChum	30000	IMM	56	1.520	1.207	
O. kisutch			30	10	E	0	E	rBovine	2.5	IP	49	1.859	1.551	McLean et al 1990
O. kisutch			30	10	E	0	E	rBovine	12.5	O	49	2.007	1.200	
O. kisutch			30	10	E	0	E	rBovine	12.5	O	49	2.007	1.293	
O. keta			2	7	H	0	E	rChum	5250	IMM	24	2.416	1.360	Moriyama et al 1990
O. keta			2	7	H	0	E	rChum	52500	IMM	24	2.416	1.500	
O. rhodorus	Kamazu	0	15	16	16	0	R	Ovine	2	IM	70	0.641	1.532	Miwa et al 1985
S. salar	St John	2	18	11.5	H	0	E	Porcine	3.5	IP	28	0.924	1.071	Komourdjian et al 1976
S. fontinalis		0	6	12	12	0	E	Human	10	IM	49	2.083	1.851	Skyrud et al 1989

TABLE 2: EFFECTS OF GH TREATMENT ON SALMONID WEIGHT INCREASE. **W**(animal weight in grams), **T°** (breeding temperature), **PHO**(winter (**W**), spring(**Sp**), summer(**Su**), and autumn(**A**) photoperiod), **SAL**(salinity), **NUT**(food in excess(**E**) limited(**R**)), **GH**(GH origin), **DOSE**(dose of GH in μg/gram of wet weight), **ADM**(administration way, **IP**(intraperitoneal), **IM**(intramuscular), **IMM**(immersion), **PO**(osmotic pump), **O**(oral)), **CGR**(control growth rate), **GAIN**(treated growth rate/control growth rate).

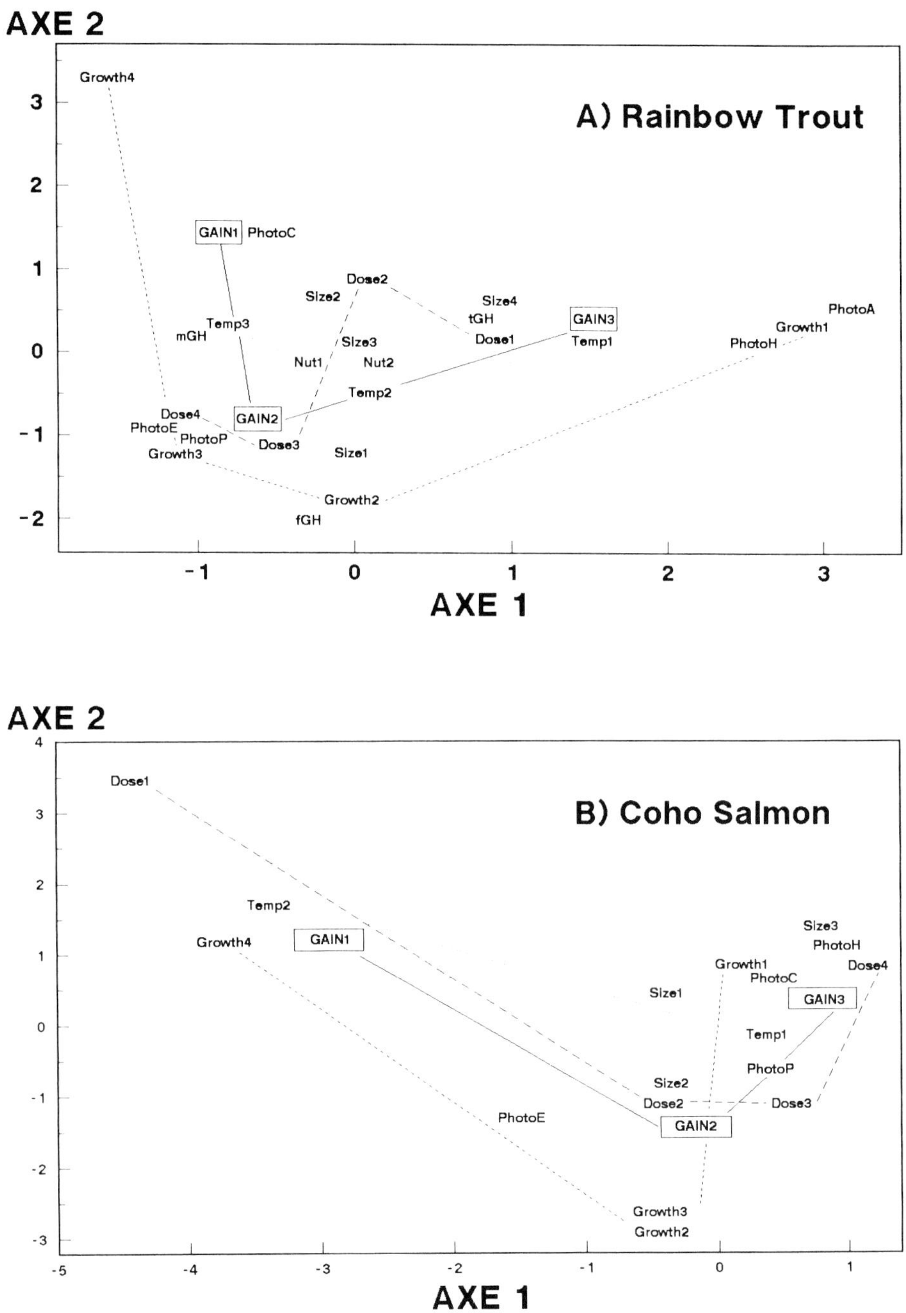

FIGURE 1: **MULTIFACTORIAL ANALYSIS OF PARAMETERS WHICH MODULATE THE SALMONID RECEPTIVITY (GAIN) TO GH TREATMENT (Data from TABLE 2).**
Numbers (1,2...) correspond to an increasing of the parameter value. **GAIN**(treated growth rate/control growth rate), **Growth**(control growth rate), **Size**(animal size at the beginning of the experiment), **Temp**(breeding temperature), **Photo**(winter(**H**), spring(**P**), summer(**E**) and autumn(**A**) photoperiod), **m**(mammalian), **f**(fish) and **t**(trout) **GH**, **Dose**(dose of the injected GH), **Nut**(food quantity).

TABLE 3: BINDING ACTIVITY TO LIVER GH RECEPTOR OF GH FROM DIFFERENT ZOOLOGICAL ORIGINS

human (Hum), bovine (Bov), ovine (Ov), sturgeon (St), tilapia (Til), recombinant tilapia (rTil), Sea bream (Sb), Eel (Eel), recombinant Eel (rEel), Chum salmon (Chum), recombinant Chum salmon (rChum), Chinook salmon (Chinook), recombinant Trout (rTrout)

RECEPTOR	LABELLED HORMONE	BINDING POTENCIALITY AS PERCENT OF HOMOLOGOUS GH													AUTHORS
		Hum	Bov	Ov	St	Til	rTil	Sb	Eel	rEel	Chum	rChum	Chinook	rTrout	
Rabbit	Bovine	103	100		<0,4										Tarpey and Nicoll 1985
Tilapia	Tilapia		7	1-3		100				60		8			Fryer 1979
Sea Bream	Sea Bream	1	2	2			4	100						40	Perez et Le Bail unpublished
Eel	Eel			230		0,4			100	100		100			Hirano 1991
Coho	Chum	3	3			3					100	30	500		Gray *et al.* 1990
Trout	Chinook	3	3	3				50					100	100	Yao *et al.* 1990 Le Bail *et al.* unpublished

d) Mode of administration

Numerous experiments on exogenous GH application, particularly in salmonids (Table 2), have used different mode of administration. Repeated intraperitoneal or intramuscular injection and implantation methods are the most frequently used, as these need relatively small amounts of the hormone and the effectiveness is well documented. However, these methods are often incompatible with modern aquaculture techniques.

Other methods, such as oral and immersion administration have, thus, been explored. Several studies using these methods of administration have shown promising results (see review of McLean and Donaldson, 1990). However, absorption levels are low. Paradoxically, few data are available concerning the kinetics of diffusion or of absorption of the hormone during treatment. Plasma GH levels were monitered after injection or implantation in eel (Duan and Hirano 1991) and in trout (Le Bail *et al.* unpublished). Significantly higher levels than those of control animals are seen several hours after injection and up to 15 days after implantation. Plasma GH level evolution reveals, in both cases, an extremely high post-administration peak. A similar profile of plasmatic GH is observed in the case of intestinal absorption (Le Bail *et al.* 1989b). It is thus difficult to evaluate the minimal effective GH level as well as the minimal duration of exposure to this level. It is also clear that a significant part of the administered GH does not react. The amounts administered, generally expressed in μg of hormone per gram of live weight per week, cannot be compared between experiments except when using the same method of administration, otherways a large variability would be introduced.

Multifactorial analysis, carried out on the supplement experiments with GH in Coho salmon (Figure 1b), shows that the increase in growth is proportional to the injected dose of bovine GH. Three experiments carried out on Coho salmon (Higgs *et al.* 1978), sockeye salmon (Clarke *et al.* 1977) and carp (Adelman 1977) using 4 or 5 doses of bovine GH, show that the increase in growth is proportional to the logarithm of the dose (Donaldson *et al.* 1979). This dose effect has been confirmed by numerous other experiments which use bovine GH (Down *et al.* 1988, 1989), chicken GH (Gill *et al.* 1985) or fish GH (Noso *et al.* 1988, Rand-Weaver *et al.* 1989, Danzman *et al.* 1990, Moriyama *et al.* 1990, Smal *et al.* unpublished).

In trout, the lowest effective dose (0.01 μg/g/week) was obtained using cod GH (Rand-Weaver *et al.* 1989). Recombinant bovine GH is the most active of the mammalian GH (0.2 μg/g/week, Down et al. 1989). In certain cases, the strongest doses (>10 μg/g bGH/week) do not generate supplementary gain (Higgs *et al.* 1977) which implies that the totality of growth potential is expressed at weaker doses.

High doses may also have a depressive effect on trout growth (Farbridge and Leatherland 1988, Agellon *et al.* 1988) or provoke high mortality in cat fish (*Ictalurus melas*, Kayes 1977). These effects are as yet unexplained and could result from numerous factors, for exemple, toxic contaminants, insulin effect, metabolism rate too high for the environmental conditions etc.
So, the values of effective (or saturating) doses must be chosen, taking into account animal husbandry conditions and the receptiveness of the species for the GH used.

II) RECEPTIVENESS OF THE ANIMAL TO TREATMENT

a) Effect of stress

It is well documented that, in fish, a halt in growth can be provoked by a state of stress (Pickering, in press). Generally, studies have not taken into account the impact of stress during GH treatment even though less growth was seen in control animals which were injected with saline or had an implant without GH, than in controls which were simply anesthetized. It should be noted that control animals which were never handled were not taken into account in this type of experiment.
The experiments of Pickering *et al.* (1991), the most probing, show that the levels of plasma GH decrease several minutes after application of the stress factor (confinement in fish). After several hours, the levels increase (Cook and Peter 1984, Takahashi *et al.* 1991b). This increase probably reflects a decrease in the hepatic receptivity to GH and in consequence a decrease in the secretion of somatomedines (directly responsible for tissue growth) which are no longer able to control, negatively, the pituitary GH secretion. Under these conditions, weight growth would slow down or would even be blocked and exogenous GH may have no effect. However, this hypothesis remains to be proved.

b) Nutritional state

Multifactorial analysis applied to experiments carried out with trout gave no correlation between level of feeding and receptivity to exogenous GH (Figure 1a).
However, in a more specific experiment using Coho salmon, Markert *et al.* (1977) showed that an increase in calorie intake (quantity or quality of feed) brought about an increase in receptiveness to GH.
Taking these data into consideration we suggest that, in experiments where feed is limited inducing differences in food uptake between fish, the variability of the response

to GH treatment might be increased, more so as GH stimulates the food uptake (Donaldson *et al.* 1979, Jalabert *et al.* 1982).

c) Influence of growth rhythms

Multifactorial analysis does not show any effect of animal size on response to growth hormone. Results obtained in brown trout, *Salmo trutta*, show that the animals are able to respond as soon as they hatch (Vandeputte 1990), however, we do not know whether above a certain size, the degree of the GH response diminishes or is abrogated.

Generally, the duration of GH treatment in salmonids varies between 15 days and 2 months and it is not certain that GH treatment carried out for longer periods has the same effect. However, Weatherley and Gill (1987), using a GH treatment for 10 months on American pike (*Esox americanus*), observed a GH response throughout the experiment and obtained animals whose sizes were greater than record sizes of animals captured in the wild.

Fish growth follows annual (Marchand and Peter 1986) and monthly rhythms (Wagner and McKeown 1985, Cotten 1987) which can be seen with growth striations found in the bones and scales. A nycthemeral secretion of GH in the form of pulses (Le Bail *et al.* 1991), which are mainly synchronous (Marchand and Peter 1986, Takahashi *et al.* 1991b), has also been seen in fish. These variations in growth rate and GH secretion which are observed even when the breeding conditions are constant, probably influence the receptiveness to exogenous GH.

c) Influence of external factors

Several studies have demonstrated that long photoperiods stimulate growth in young Atlantic salmon (Stefanson *et al.* 1989a, 1989b, Saunders and Harmon 1990). This acceleration in growth is accompanied by an increase in pituitary GH content (Komourdjian *et al.* 1989). Multifactorial analysis shows that there is a link between low gain of growth rate due to GH treatment (controls had a high growth rate) and the summer photoperiod in Rainbow trout (Figure 1a) and in Coho salmon (Figure 1b). These experiments were carried out in natural conditions, the photoperiod and temperature were correlated and it is thus difficult to draw conclusions about which of these two factors is determinant in response to GH treatment.

The increase in growth rate due to GH is inversely proportional to temperature in Coho salmon and trout (multifactorial analysis). At very high temperatures (19°C), the effects of GH are suppressed (Danzman *et al.* 1990). In an experiment where the effect of GH at different temperatures was studied, Yao *et al.* (unpublished) did not find this effect clearly. They showed however, that in controls, the level of circulating GH increased together with a decrease in hepatic receptiveness, when the temperature increased. This increase in plasma GH with temperature has also been observed in natural conditions (Barett and McKeown 1989). These results imply that salmonids raised at low temperatures are more suited to respond to a growth hormone treatment. When the temperature increase inhibits growth of the controls, GH treatment can once again be effective, as has been observed in carp (Adelman 1977).

The influence of water velocity has never been taken into account in studies on the effects of GH. However, a high water velocity stimulates growth in the arctic char, *Salvelinus alpinus* (Christiansen and Jobling 1990) and in the Rainbow trout (Le Bail *et al.* unpublished). This imposed physical exercise is accompanied by an increase in plasma GH (Barett and McKeown 1988a, 1988b, 1989). Response to exogenous GH of animals submitted to various strengths of current remains to be evaluated. These studies would have to take into account behaviour interactions between individuals and feeding conditions.

Throughout the passage from fresh water to sea water for euryhalin salmonids, the levels of GH (Sweeting *et al.* 1985, Collie *et al.* 1989, Boeuf *et al.* 1989, Rydevik *et al.* 1990) and IGF increase (Lindahl *et al.* 1985). This increase could explain the high growth rate observed in animals during their sea phase. Throughout this phase, the animals are still able to respond to GH treatment, as has been shown in Coho salmon (Down *et al.* 1988). These data are not sufficient to draw conclusions on whether the saline environment modifies receptivity to GH or not.

e) Species and Strain

The only species that have undergone a large number of experiments are Rainbow trout (39) and Coho salmon (43). Thus, a comparison can be made on their receptiveness to GH (Table 2), but the large variability in experimental conditions make it difficult to draw meaningful conclusions. It seems, however, that Coho salmon is more responsive to GH treatment than is trout, as 51 % of experiments carried out on the first species have a weight gain factor greater than 1.75 as compared to 27 % in trout.

The strains used in fish farms have very different geographical origins and have undergone a great deal of genetic selection pressure for growth. This selection which is unique to each fish culture, may act at different physiological levels in the mecanismes controlling growth (GH secretion, tissue receptiveness etc.). Thus, it would be logical to observe subspecies which are more receptive than others, but no comparative experiments of this type have so far been carried out in fish to disprove this. In our laboratory, GH treatment separately carried out on two strains of trout induce systematically a lower gain of growth rate in one of these strains. This "strain" effect is probably responsible for the differences in response, observed in numerous experiments in fish.

III) ENDOCRINE MECHANISMS IMPLICATED

From the somewhat fragmentary results obtained from fish (see references Ch. II) and taking into account information aquired from mammals (Pell and Bates 1990, Ross and Buchanan 1990, Clemmons and Underwood, 1991), it is possible to draw up a regulation diagram to partially explain the "receptivity" of animals to GH (Figure 2). Stress diminishes the secretion of pituitary GH and probably increases the resistance of target GH tissues (Cook and Peter 1984, Pickering *et al.* 1991, Takahashi *et al.* 1991).

Hypotheses put forward to explain the effects of nutrition are drawn from results obtained from starved fish. We found that starvation decreases the apparent number of GH receptors (Yao and Le Bail, unpublished) in the target tissues, which probably induces GH resistance. Plasma levels of IGF decrease (Komourdjian and Idler 1978, Yao and Le Bail unpublished), which probably has the effect of increasing the plasma levels of GH (Barett and McKeown 1988a, Sumpter *et al.* 1991a, 1991b), as an inhibitory action of IGF on the pituitary GH secretion has been demonstrated (Perez-Sanchez *et al.* 1991c). In these circumstances, exogenous GH is without effect on growth.

Increase in receptivity to GH in salmonids raised at low temperatures is explained by the lower levels of circulating GH. The number of free hepatic GH receptors also increase (Yao and Le Bail, unpublished), which suggests a better potential receptivity of animals to exogenous GH treatment. However, the number of free receptors is inversely proportional to the level of circulating GH (Le Bail *et al.* unpublished), which might demonstrate the amount of receptor occupation and not the total number of receptors (Sakamoto and Hirano, in press). So, we do not know yet if temperature could influence the total number of GH receptors.

The affinity of the GH receptors, which is not modified by variations of the water temperature (Yao and Le Bail, unpublished) or by variations in the salinity (Sakamoto and Hirano, in press), would have no influence on fish receptivity to GH treatment.

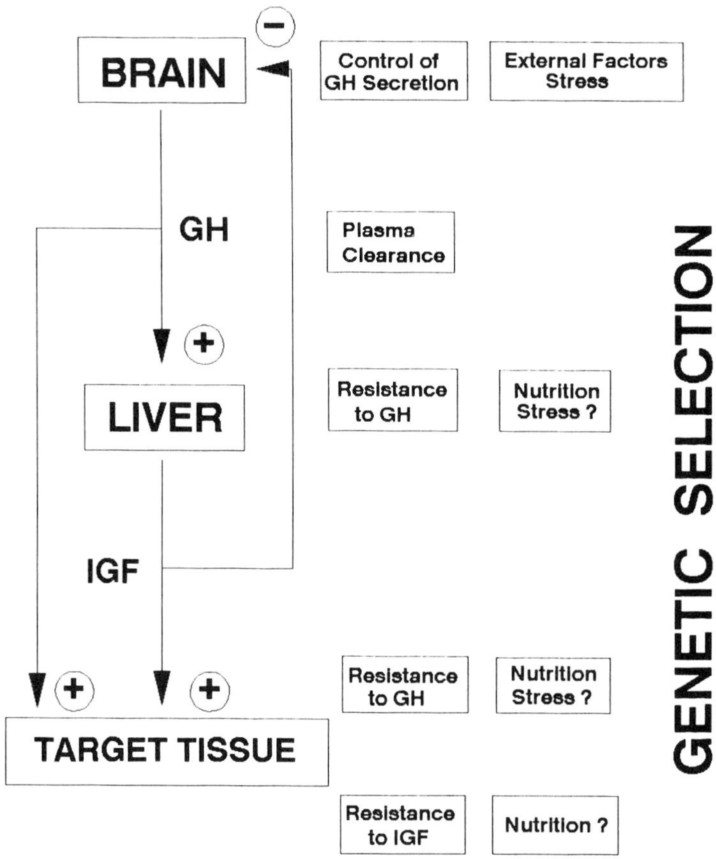

FIGURE 2: DIAGRAM OF ENDOCRINE MECHANISMS IMPLICATED IN THE RECEPTIVITY TO GH TREATMENT.

Modifications in GH plasma clearance rate could also be a regulatory factor. As few studies have been carried out in this field, no conclusions can be drawn (Le Bail et al., 1989; Le Bail and Perez-Sanchez, unpublished). However, Duan and Hirano (1991) did not observe any significant modifications in clearance between eels raised in fresh-water or in salt-water.

From the information available, it is reasonable to assume that temperature, and perhaps other external factors, act principally on pituitary GH secretion.

CONCLUSION

GH from all zoological origins are active in fish. The biological activity of fish GH, determined using fish radioreceptor assays, is higher than that of mammalian GH. However, the quality of fish GH preparations is very variable. The amount of GH injected or implanted influences the plasma levels attained but also the length of time necessary for these to return to normal levels. Thus, the effect measured results of the combination of time effect and dose effect. This lack of control over the methods of administration explains that, when different GH are tested *in vivo* on the acceleration of weight growth, the differences are not as clear as when testing them in a radioreceptor assay. A standardization of *in vivo* and *in vitro* bioassays should be developed in order to make future experimental results coherent and comparable.

The multifactorial analysis (Figure 1) demonstrates a strong negative correlation between the growth increment in treated animals and the growth rate of control animals. This phenomenon is observed in trout and in Coho salmon. The growth rate reflects and integrates the factors which modulate growth physiology, such as environmental factors. This implies that the more the breeding conditions (other than food) are unfavourable to growth, the greater the response is to GH treatment.

The same reasoning could apply concerning the genetic characteristics of animals. If a strain has a high growth rate, it should respond less to GH (especially if this selection increases the endogenous GH secretion). This situation might correspond to that in mammals and in birds where GH treatment has little or no effect on growth rate of farmed subspecies, which have undergone large selection pressures to increase their growth rates. Only dwarf animals are receptive to exogenous GH.

Fish would thus have similar growth control mechanisms to mammals. The spectacular differences in receptiveness to GH treatment observed between mammals and fish might be explained for the greater part by:

- genetic selection which has not yet used all the growth potential linked to GH secretion in fish.

- rearing temperature which is often far from the optimal growth temperature of fish, which, it should be remembered, are poikilothermic animals.

ACKNOWLEDGEMENTS

The results obtained by our laboratory and cited in this review have been financed by "la region Bretagne (Britta)" and by the E.E.C. (SAST). We are grateful to D[r] J. Smal from EUROGENTEC for communicating unpublished data and to D[r] F. Le Gac for critically reading the manuscript.

REFERENCES

Adelman I.R. (1977). Effects of bovine growth hormone on growth of carp (*Cyprinus carpio*) and the influence of temperature and photoperiod. **J. Fish. Res. Board Can.** 34, 509-515.

Adelman I.R. (1982). Enhancement of growth of common carp by injection of homogenate of carp pituitary glands. **Prog. Fish-Cult.** 44 (2), 94-97.

Agellon L.B., Emery C.J., Davies S.L., Dingle A.D. and Chen T.T. (1988). Promotion of rapid growth of rainbow trout (*Salmo gairdneri*) by a recombinant fish growth hormone. **Can. J. Fish Aquat. Sci.** 45, 146-151.

Ash P.J. (1977). Incorporation of (^{35}S)sulfate into mucopolysaccharide by teleost cartilage *in vitro*: The influence of mammalian growth hormone, teleost plasma and mammalian plasma. **Gen. Comp. Endocrinol.** 32, 187-194.

Barrett B.A. and McKeown B.A. (1988a). Sustained exercise augments long-term starvation increases in plasma growth hormone in the steelhead trout, *Salmo gairdneri*. **Can. J. Zool.** 66, 853-855.

Barrett B.A. and McKeown B.A. (1988b). Growth hormone response to sustained swimming in exercise-acclimated steelhead trout, *Salmo gairdneri*. **J. Fish Biol.** 32, 799-800.

Barrett B.A. and McKeown B.A. (1989). Plasma growth hormone levels in *Salmo gairdneri*: studies on temperature and the exercise intensity/duration relationship. **Comp. Biochem. Physiol.** 94A, 791-794.

Björnsson B. T., Yamauchi K., Nishioka R. S., Deftos L. J. and Bern H. (1987). Effects of hypophysectomy and subsequent hormonal replacement therapy on hormonal and osmoregulatory status of coho salmon, *Oncorhynchus kisutch*. **Gen. Comp. Endocrinol.** 68, 421-430.

Boeuf G., Le Bail P.-Y. and Prunet P. (1989). Growth hormone and thyroid hormones during atlantic salmon, *Salmo salar* l., smolting, and after transfer to seawater. **Aquaculture** 82, 257-268.

Boeuf G., Le Roux A., Severe A., Prunet P. and Le Bail P.-Y. (1990). The role of growth hormone in the adaptability of atlantic salmon (*Salmo salar*) to seawater. **Can. Tech. Rep. Fish. Aquat. Sci.** 1761, 125-131.

Bolton J.P., Takahashi A., Kawauchi H., Kubota J. and Hirano T. (1986). Development and validation of a salmon growth hormone radioimmunoassay. **Gen. Comp. Endocrinol.** 62, 230-238.

Bolton J.P., Young G., Nishioka R.S., Hirano T. and Bern H. (1987). Plasma growth hormone levels in normal and stunted yearling coho salmon, *Oncorhynchus kisutch*. **J. Exp. Zool.** 242, 379-382.

Charrier J. and Martal J. (1988). Growth hormone I. Polymorphism. **Reprod. Nutr. Dev.** 28, 857-886.

Chartier-Baraduc M. M. (1959). Influence de l'hormone somatotrope sur les teneurs en eau et en électrolytes du plasma et du muscle de la truite arc-en-ciel (*Salmo gairdneri*). **C. R. Soc. Biol.** 153, 1757-1761.

Cheema I. R. and Matty A. J. (1978). Increased uptake of L-leucine-14C in the skeletal muscle of rainbow trout, *Salmo gairdneri*, after administration of growth hormone. **Pakistan J. Zool.** 10 (2), 119-123.
Christiansen J.S. and Jobling M. (1990). The behaviour and relationship betwen food intake and growth of juvenile arctic charr, *Salvelinus alpinus*, subjected to sustained exercise. **Can. J. Zool.** 68, 2185-2191.
Clarke W.C., Farmer S.W. and Hartwell K.M. (1977). Effect of teleost pituitary growth hormone on growth of *Tilapia mossambica* and on growth and sea water adaptation of sockeye salmon (*Oncorhynchus nerka*). **Gen. Comp. Endocrinol.** 33, 174-178.
Clemmons D.R. and Underwood L.E. (1991). Nutritional regulation of IGF-I and IGF binding proteins. **Annu. Rev. Nutr.** 11, 393-412.
Collie N.L., Bolton J.P., Kawauchi H. and Hirano T. (1989). Survival of salmonids in seawater and the time-frame of growth hormone action. **Fish Physiol. Biochem.** 7, 315-321.
Cook A.F. and Peter R.E. (1984). The effects of somatostatin on serum growth hormone levels in the goldfish, *Carassius auratus*. **Gen. Comp. Endocrinol.** 54, 109-113.
Cook A.F., Wilson S.W. and Peter R.E. (1983). Development and validation of a carp growth hormone radioimmunoassay. **Gen. Comp. Endocrinol.** 50, 335-347.
Cotten C. (1987). Effet de l'hormone de croissance humaine (h-GH) et de la testostérone sur la croissance de la truite arc-en-ciel (*Salmo gairdneri*). **Mémoire fin d'étude de DEA d'Université de Rennes I.**
Danzmann R.G., Van der Kraak G.L., Chen T.T., and Powers D.A. (1990). Metabolic effects of bovine growth hormone and genetically engineering rainbow trout growth hormone in rainbow trout (*Oncorhynchus mykiss*) reared at a high temperature. **Can. J. Fish. Aquat. Sci.** 47, 1292-1301.
De Luze A. and Leloup J. (1984). Fish growth hormone enhances peripheral conversion of thyroxine to triiodothyronine in the eel (*Anguilla anguilla* L.). **Gen. Comp. Endocrinol.** 56, 308-312.
De Luze A., Leloup J., Papkoff H., Kikuyama S. and Kawauchi H. (1987). Régulation hypophysaire de la 5'-désiodation périphérique de la thyroxine chez l'anguille. **J. Physiol.** 82, 26a.
Degani G., and Gallaguer M, L. (1985). Effects of dietary 17a-Methyltestosterone and bovine growth hormone on growth and food conversion of slow- and normally- growing american elvers (*Anguilla rostrata*). **Can. J. Fish. Aquat. Sci.** 42, 185-189.
Donaldson E.M., Fargerlund U.H.M., Higg D.A. and McBride J.R. (1979). Hormonal enhancement of growth. **Academic Press New York San Francisco London. In "Fish Physiology" (W.S. Hoar D.J. Randall and J.R. Brett Eds.)**, VIII, 455-492.
Doneen B.A. (1976). Biological activities of mammalian and teleostean prolactins and growth hormone on mouse mammary gland and teleost urinary bladder. **Gen. Comp. Endocrinol.** 30, 34-42.
Down N.E.T., Donaldson E.M., and Dye H.M. (1989). A potent analog of recombinant bovine somatotropin accelerates growth in juvenile coho salmon (*Oncorhynchus kisutch*). **Can. J. Fish. Aquat. Sci.** 46, 178-183.
Down N.E.T., Donaldson E.M., Dye H.M., Langley K., and Souza L.M. (1988). Recombinant bovine somatotropin more than doubles the growth rate of coho salmon (*Oncorhynchus kisutch*) acclimated to seawater at ambient winter conditions. **Aquaculture** 68, 141-155.
Duan C. and Hirano T. (1991). Plasma kinetics of growth hormone in the japanese eel, *Anguilla japonica*. **Aquaculture** 95, 179-188.
Duan C.M. and Inui Y. (1990). Effects of recombinant eel growth hormone on the uptake of(^{35}S)Sulfate by ceratobranchial cartilages of japanese eel, *Anguilla japonica*. **Gen. Comp. Endocrinol.** 79, 320-325.

Farbridje K.J. and Leatherland J.F. (1988). Interaction between ovine growth hormone and triiodo-L-thyronine on metabolic reserves of rainbow trout, *Salmo gairdneri.* **Fish Physiol. Biochem.** 5 (3), 141-151.
Farbridje K.J. and Leatherland J.F. (1991). The development of a noncompetitive enzyme-linked immunosorbent assay for Oncorhynchid growth hormone using monoclonal antibodies. **Gen. Comp. Endocrinol.** 83, 7-17.
Foster A.R., Houlihan D.F., Gray C., Medale F., Fauconneau B., Kaushik S.J. and Le Bail P.-Y. (1991). The effect of ovine growth hormone on protein turnover in rainbow trout. **Gen. Comp. Endocrinol.** 81, 111-120.
Fryer J.N. (1979). A radioreceptor assay for purified teleost growth hormone. **Gen. Comp. Endocrinol.** 39, 123-130.
Fryer J.N. and Bern H.A. (1979). Growth hormone binding to tissues of normal and stunted juvenile coho salmon *Oncorhynchus kisutch.* **J. Fish. Biol.** 15, 527-533.
Furuya A., Ohtomo T.I., Inada T. and Yoshida H. (1987). Generation and application of monoclonal antibodies against salmon somatotropin (salmon growth hormone) and salmon prolactin. **Agric. Biol. Chem.** 51, 2331-2335.
Gill J.A., Sumpter J.P., Donaldson E.M., Dye H.M., Souza L., Berg T., Wypych J. and Langley K. (1985). Recombinant chicken and bovine growth hormones accelerate growth in aquacultured juvenile pacific salmon, *Oncorhynchus kisutch.* **Biotechnology** 3, 643-646.
Grau E.G. and Stetson M.H. (1979). Growth hormone is thyrotropic in *Fundulus heteroclitus.* **Gen. Comp. Endocrinol.** 39, 1-8.
Gray E.S., Young G.G., and Bern H.A. (1990). Radioreceptor assay for growth hormone in coho salmon (*Oncorhynchus kisutch)* and its application to the study of stunting. **J. Exp. Zool.** 256, 290-296.
Hayashida T. (1970). Immunological studies with rat pituitary growth hormone (rGH). II. Comparative immunochemical investigation of GH from representative of various vertebrate classes with monkey antiserum to rGH. **Gen. Comp. Endocrinol.** 15, 432-452.
Higgs D.A. and Eales J.G. (1977). Influence of food deprivation on radioiodothyronine and radioiodine kinetics in yearling brook trout, *Salvelinus fontinalis* (Mitchill), with a consideration of the extent of L-thyroxine conversion to 3,5,3'-triiodo-L-thyroxine. **Gen. Comp. Endocrinol.** 32, 29-40.
Higgs D.A., Donaldson E.M., Dye H.M. and McBride J.R. (1975). A preliminary investigation of the effect of bovine growth hormone on growth and muscle composition of coho salmon, *Oncorhynchus kisutch.* **Gen. Comp. Endocrinol.** 27, 240-253.
Higgs D.A., Donaldson E.M., Dye H.M. and McBride J.R. (1976). Influence of bovine growth hormone and L-thyroxine on growth, muscle composition, and histological structure of gonads, thyroid, pancreas, and pituitary of coho salmon, *Oncorhynchus kisutch.* **J. Fish. Res. Board Can.** 33, 1585-1603.
Higgs D.A., Donaldson E.M., McBride J.R. and Dye H.M. (1978). Evaluation of the potential for using a chinook salmon (*Oncorhynchus tshawytscha*) pituitary extract versus bovine growth hormone to enhance the growth of coho salmon (*Oncorhynchus kisutch*). **Can. J. Zool.** 56, 1226-1231.
Higgs D.A., Fagerlund U.H.M., McBride J.R., Dye H.M. and Donaldson E.M. (1977). Influence of combinaisons of bovine growth hormone, 17a-methyltestosterone, and L-thyroxine on growth of yearling coho salmon, *Oncorhynchus kisutch.* **Can. J. Zool.** 55, 1048-1056.
Hirano T. 1991. Hepatic receptors for homologous growth hormone in eel. **Gen. Comp. Endocrinol.** 81, 383-390.
Inui Y. and Ishioka H. (1985a). *In vivo* and *in vitro* effects of growth hormone on the incorporation of ^{14}C-leucine into protein of liver and muscle of the eel. **Gen. Comp. Endocrinol.** 59, 295-300.
Inui Y. Miwa S. and Ishioka H. (1985b). Effects of mammalian growth hormone on amino nitrogen mobilisation in the eel. **Gen. Comp. Endocrinol.** 59, 287-294.

Jalabert B., Fostier A. et Breton B. (1982). Contrôle endocrinien de la croissance chez les poissons téléostéens: données majeures et applications. **Oceanis** 8, 551-577.
Kawauchi H., Moriyama S., Yasuda A., Yamaguchi K., Shirahata K., Kubota J. and Hirano T. (1986). Isolation and characterization of chum salmon growth hormone. **Arch. Biochem. Biophys.** 244, 542-552.
Kawauchi H., Yasuda A. and Rand-Weaver M. (1990). Evolution of prolactin and growth hormone family. **In Progress in Comparative Endocrinoloy, Epple A., Scanes C.G. and Stetson M.H. (eds). 1990 Wiley-Liss**, Inc, 47-53.
Kawazoe I., Noso T., Kuriyama S.I., Akasaka A. and Kawauchi H. (1988). Isolation and characterization of growth hormone from yellowtail *Seriola quineqeradiata.* **Nippon Suisan Gakkaishi.** 54, 393-399.
Kayes T. (1977). Effects of hypophysectomy, beef growth hormone replacement therapy, autotransplantation , and environmental salinity on growth, in black bullhead (*Ictalurus melas*). **Gen. Comp. Endocrinol.** 33, 371-381.
Kayes T. (1978). Effects of hypophysectomy, beef growth hormone replacement therapy on morphometric and biochemical indicators of growth in the fed versus starved black bullhead (*Ictalurus melas*). **Gen. Comp. Endocrinol.** 35, 419-431.
Kishida M. and Hirano T. (1988). Development of radioimmunoassay for eel growth hormone. **Nippon Suisan Gakkaishi** 54,1321-1327.
Kishida M., Hirano T., Kubota J., Hasegawa S., Kawauchi H., Yamaguchi K. and Shirahata K. (1987). Isolation of two forms of growth hormone secreted from eel pituitaries *in vitro.* **Gen. Comp. Endocrinol.** 65, 478-488.
Komourdjian M.P. and Idler D.R. (1978). Hepatic mediation of hormonal and nutritional factors influencing the *in vitro* sulfur uptake by rainbow trout bone. **Gen. Comp. Endocrinol.** 36, 33-39.
Komourdjian M.P. and Idler D.R. (1979). Chum salmon pitutary fractions: somatotropic activity and cytoimmunofluorescence studies. **Gen. Comp. Endocrinol.** 37, 343-349.
Komourdjian M.P., Burton M.P. and Idler D.R. (1978). Growth of rainbow trout, *Salmo gairdneri*, after hypophysectomy and somatotropin therapy. **Gen. Comp. Endocrinol.** 34, 158-162.
Komourdjian M.P., Fenwick J.C. and Saunders R.L. (1989). Endocrine-mediated photostimulation of growth in atlantic salmon. **Can. J. Zool.** 67, 1505-1509.
Komourdjian M.P., Saunders R.L. and Fenwick J.C. (1976). The effect of porcine somatotropin on growth, and survival in seawater of atlantic salmon (*Salmo salar*) parr. **Can. J. Zool.** 54, 531-535.
Le Bail P.-Y., Boulard G., Barenton B. and Zygmunt M. (1989a). Purification of chinook salmon (*Oncorhynchus tshawytscha*) GH for receptor study. **Fish Physiol. Biochem.** 7, 243-251.
Le Bail P.-Y., Sire M.F. and Vernier J.F. (1989b). Intestinal transfer of growth hormone into the circulatory system of the rainbow trout, *Salmo gairdneri*: interference by granule cells. **J. Exp. Zool.** 251, 101-107.
Le Bail P.-Y., Sumpter J.P., Carragher J., Mourot B., Niu P.D. and Weil C. (1991). Development and validation of high sensitive radioimmunoassay to chinook salmon (*Oncorhynchus tshawytscha*) growth hormone. **Gen. Comp. Endocrinol.** 83, 75-85.
Le Gac F., Ollitrault M., Loir M., and Le Bail P.-Y. (1992). Evidence for binding and action of growth hormone (GH) in trout testis. **Biol. Reprod.** (in press).
Leatherland J.F. and Nuti R.N. (1981). Effects of bovine growth hormone on plasma FFA concentrations and liver, muscle and carcass lipid content in rainbow trout, *Salmo gairdneri* R. **J. Fish Biol.** 19, 487-498.
Lewis U.J., Singh R.N.P., Seavey B.K., Lasker R., and Pickford G.E. (1972). Growth hormone- and prolactin-like proteins of the blue shark (*Prionace glauca*). **Fish. bull.** 70, 933-939.

Lindahl K.I., Sara V., Fridberg G. and Nishimiya T. (1985). The presence of somatomedin in the baltic salmon, *Salmo salar*, with special reference to smoltification. **Aquaculture** 45, 177-183.

Madsen S.S. (1990). The role of cortisol and growth hormone in sea water adaptation and development of hypoosmoregulatory mechanisms in sea trout parr (*Salmo trutta trutta*). **Gen. Comp. Endocrinol.** 79, 1-11.

Marchant T.A. and Peter R.E. (1986). Seasonal variations in body growth rates and circulating levels of growth hormone in the goldfish, *Carassius auratus*. **J. Exp. Zool.** 237, 231-239.

Markert J.R., Higgs D.A., Dye H.M. and MacQuarrie D.W. (1977). Influence of bovine growth hormone on growth rate, appetite, and food convertion of yearling coho salmon (*Oncorhynchus kisutch*) fed two diets of different composition. **Can. J. Zool.** 55, 74-83.

McLachy D.L. and Eales J.G. (1990). Growth hormone stimulates hepatic thyroxine 5'-monodeiodinase activity and 3,5,3'-triiodothyronine levels in rainbow trout, *Salmo gairdneri*. **Gen. Comp. Endocrinol.** 78, 164-172.

McLean E., Donaldson E.M., Dye H. and Souza L. M. (1990). Growth acceleration of coho salmon (*Oncorhynchus kisutch*) following oral administration of recombinant bovine somatotropin. **Aquaculture** 91, 197-203.

Miwa S. and Inui Y. (1985). Effects of L-thyroxine and ovine growth hormone on smoltification of amago salmon (*Oncorhynchus rhodorus*). **Gen. Comp. Endocrinol.** 58, 436-442.

Mori I., Sakamoto T. and Hirano T. in press. Growth hormone (GH) dependent hepatic GH receptors in the japanese eel, *Anguilla japonica*: effects of hypophysectomy and GH injection. **Gen. Comp. Endocrinol.**

Moriyama S. and Kawauchi H. (1990). Growth stimulation of juvenile salmonids by immersion in recombinant salmon growth hormone. **Nippon Suisan Gakkaishi** 56, 31-34.

Niu P.-D. (1990). Etude du polymorphisme de l'hormone de croissance (GH) chez la truite arc-en-ciel (*Oncorhynchus mykiss*). **These de Docteur d'Université de Rennes I.** 125 p.

Noso T., Yasuda A., Kawazoe I., Takehara H., Takahasi A., Sakai K., and Kawauchi H. (1988). Isolation and characterization of growth hormone from a marine fish, bonito (*Katsuwonus pelamis*). **Int. J. Peptide Protein Res.** 32, 579-589.

Pector R., Zamal H., Ollevier F., Drot S., Lamproye A., Lebecque S., Lieffrig F., Poncin A. and Smal J. (1990). Influence of injected recombinant trout growth hormone on growth and body composition of rainbow trout, *Oncorhynchus mykiss*. **15th Conference of European Comparative Endocrinologists, 9-14 September 1990 Leuven. Abstr.**

Pell J.M. and Bates P.C. (1990). The nutritional regulation of growth hormone action. **Nutr. Res. Rev.** 3, 163-192.

Perez-Sanchez J., Le Bail P.-Y. and Zohar Y. (1991a). Development and validation of homologous radioimmunoassay and radioreceptor assay for sea bream (*Sparus aurata*) growth hormone. **13th Conference of European Society for Compative physiology and Biochemistry, Antibes-Juan les Pins 6-10 October 1991**, 142.

Pérez-Sanchez J., Smal J. and Le Bail P.-Y. (1991b). Location and characterization of growth hormone binding sites in the central nervous system of a teleost fish (*Oncorhynchus mykiss*). **Growth Regulation** (in press).

Pérez-Sanchez J., Weil C. and Le Bail P.-Y. (1991c). Effects of a human insulin-like growth factor-I on the release of growth hormone by rainbow trout (Oncorhynchus mykiss) pituitary cells. **J. Exp. Zool.** (in press).

Peter R.E., Monckton E.A. and McKeown B.A. (1976). The effect of gold thioglucose on food intake, growth and forebrain histology in goldfish, *Carassius auratus*. **Physiol. Behav.** 17, 303-312.

Pickering A.D. (in press). Growth and stress in fish production. **Aquaculture**.

Pickering A.D., Pottinger T.G., Sumpter J.P., Carragher J.F. and Le Bail P.Y. (1991). Effects of acute and chronic stress on the levels of circulating growth hormone in the rainbow trout, *Oncorhynchus mykiss*. **Gen. Comp. Endocrinol.** 83, 86-93.

Pickford G.E and Thompson E.F. (1948). The effect of purified mammalian growth hormone on the killifish (*Fundulus heteroclitus* L.). **J. Exp. Zool.** 109, 367-383.

Pickford G.E. (1954). The response of hypophysectomized male killifish to purified fish growth hormone, as compared with the response to purified beef growth hormone. **Endocrinology** 55, 274-287.

Pickford G.E. (1973). Introductory remarks. **Amer. Zool.** 13, 711-717.

Pickford G.E. (1957). Part IV. the growth hormone. **In Pickford G.E. and Atz J.W. The physiology of the pituitary gland of fishes. N. Y. Zool. Soc. 1957**, 84-99.

Pickford G.E., Lofts B., Bara G., and Atz J.W. (1972). Testis stimulation in hypophysectomized male killifish *Fundulus heteroclitus*, treated with mammalian growth hormone and/or luteinizing hormone. **Biol. Reprod.** 7, 370-386.

Pickford G.E., Wilhelmi A.E. and Nussbaum N. (1959). Comparative studies of the response of hypophysectomized killifish, *Fundulus heteroclitus*, to growth hormone preparation. **Anat. Rec.** 134, 624-625.

Prack M., Caiati M.A., Roskowski M., Treacy T., Vodicnik M.J. and De Vlaming V.L. (1980). The effects of mammalian prolactin and growth hormone on goldfish (*Carassius auratus*) growth, plasma amino acid levels and liver amino acid uptake. **Comp. Biochem. Physiol.** 67a, 307-310.

Rand-Weaver M., Walther B.T., and Kawauchi H. (1989). Isolation and characterization of growth hormone from atlantic cod (Gadus morha). **Gen. Comp. Endocrinol.** 73, 260-269.

Richman III N.H. and Zaugg W.S. (1987). Effects of cortisol and growth hormone on osmoregulation in pre- and desmoltified coho salmon (*Oncorhynchus kisutch*). **Gen. Comp. Endocrinol.** 65, 189-198.

Ross R.J.M. and Buchanan C.R. (1990). Growth hormone secretion: its regulation and the influence of nutritional factors. **Nutr. Res. Rev.** 3, 143-162.

Rydevik M., Borg B., Haux C., Kawauchi H. and Björnsson T. (1990). Plasma growth hormone levels increase during seawater exposure of sexually mature atlantique salmon parr (*Salmo salar*). **Gen. Comp. Endocrinol.** 80, 9-15.

Sakamoto T. and Hirano T. (1991). Growth hormone receptors in the liver and osmoregulatory organs of rainbow trout: characterization and dynamics during adaptation to seawater. **J. Endocrinol.** (in press).

Saunders R.L. and Harmon P.R. (1990). Influence of photoperiod on growth of juvenile atlantic salmon and development of salinity tolerance during winter-spring. **Trans. Am. Fish. Soc.** 119, 689-697.

Schulte P.M., Down N.E.(Ted), Donaldson E.M. and Souza L.M. (1989). Experimental administration of recombinant bovine growth hormone to juvenile rainbow trout (*Salmo gairdneri*) by injection or by immersion. **Aquaculture** 76, 145-156.

Sekine S., Mizukami T., Nishi T., Kuwana Y., Saito A., Sato M., Itoh S., and Kawauchi H. (1985). Cloning and expression of cDNA for salmon growth hormone in *Escherichia coli*. **Proc. Natl. Acad.Sci. USA** 82, 4306-4310.

Sheridan M.A. (1986). Effects of thyroxin, cortisol, growth hormone, and prolactin on lipid metabolism of coho salmon, *Oncorhynchus kisutch*, during smoltification. **Gen. Comp. Endocrinol.** 64, 220-238.

Singh H., Griffith R.W., Takahashi A., Kawauchi H., and Stegeman J.J. (1988). Regulation of gonadal steroidogenesis in *Fundulus heteroclitus* by recombinant salmon growth hormone and purified salmon prolactin. **Gen. Comp. Endocrinol.** 72, 144-153.

Skyrud T., Andersen O., Alestrom P. and Gautvik K.M. (1989). Effects of recombinant human growth hormone and insulin-like growth factors 1 on body growth and blood metabolites in brook trout (*Salvelinus fontinalis*). **Gen. Comp. Endocrinol.** 75, 247-255.

Stefansson S.O., Hansen T. and Taranger G.L. (1989a). The influence of light on growth and smolting of atlantic salmon (*Salmo salar*): effects of spectral composition, intensity and photoperiod., 79-84. **In R.L. Saunders (ed) Proceedings of Canada-Norway finfish aquaculture workshop, September 11-14, 1989. Can. Tech. Rep. Fish Aquat. Sci.** 1761.

Stefansson S.O., Naevdal G. and Hansen T. (1989b). The influence of three unchanging photoperiods on growth and parr-smolt transformation in atlantique salmon, *Salmo salar.* **J. Fish Biol.** 35, 237-247.

Sumpter J.P., Le Bail P.-Y., Pickering A.D., Pottinger T.G. and Garragher F.J. (1991a). The effect of starvation on growth and plasma Growth Hormone concentration of rainbow trout, *Oncorhynchus mykiss.* **Gen. Comp. Endocrinol.** 83, 94-102.

Sumpter J.P., Lincoln R.F., Bye V.J., Carragher J.F. and Le Bail P.Y. (1991b). Plasma growth hormone levels during sexual maturation in diploid and triploid rainbow trout *Oncorhynchus mykiss.* **Gen. Comp. Endocrinol.** 83, 103-110.

Swetting R.M., Wagner G.F. and McKeown B.A. (1985). Changes in plasma glucose, aminoacid, nitrogen and growth hormone during smoltification and seawater adaptation in coho salmon *Oncorhynchus kisutch.* **Aquaculture** 45, 185-197.

Swift D.R. (1954). Influence of mammalian growth hormone on rate of growth of fish. **Nature (London)** 173, 1096.

Takahashi A., Kawazoe I. and Kawauchi H. (1991a). A competitive immunoassay for chum salmon growth hormone. **Nippon Suisan Gakkaishi** 57 (2), 267-272.

Takahashi A., Ogasawara T., Kawauchi H. and Hirano T. (1991b). Effects of stress and fasting on plasma growth hormone levels in the immature rainbow trout. **Nippon Suisan Gakkaishi** 57 (2), 231-235.

Tarpey J.F. and Nicoll C.S. (1985). Characterization of hepatic growth hormone binding sites in two fish species, *Gillichthys mirabilis* (teleostei) and *Acipenser transmontanus* (chondrostei). **Gen. Comp. Endocrinol.** 60, 39-50.

Van der Kraak G., Rosenbloom P. and Peter R.E. (1990). Growth hormone-dependent potentiation of gonadotropin-stimulated steroid production by ovarian follicles of the goldfish. **Gen. Comp. Endocrinol.** 79, 233-239.

Vandeputte M. (1990). Contribution à l'étude de l'endocrinologie de la croissance aux premiers stades du développement chez la truite commune, *Salmo trutta fario.* **Mémoire de DEA de l'Université de Rennes I. 40 p.**

Venugopalan V.K. (1967). Effect of growth hormone injection on the levels of nucleic acids in the liver of intact fish (*Ophicephalus striatus*). **Gen. Comp. Endocrinol.** 8, 332-336.

Wagner G.F. and McKeown B.A. (1985). The purification, partial characterization and bioassay of growth hormone from two species of pacific salmon. **Current Trends in comparative Endocrinology. (eds. B. Lofts & W.H. Holmes). Hong Kong University Press,** 1985, 557-561.

Wagner G.F. and McKeown B.A. (1986). Development of a salmon growth hormone radioimmunoassay. **Gen. Comp. Endocrinol.** 62, 452-458.

Weatherley A.H. and Gill H.S. (1982). Influence of bovine growth hormone on the growth of mosaic muscle in relation to somatic growth of rainbow trout, *Salmo gairdneri* R. **J. Fish Biol.** 20, 165-172.

Weatherley A.H. and Gill H.S. (1987). Growth increases produced by bovine growth hormone in grass pickrel, *Esox americanus vermiculatus*, and the underlying dynamics of muscle fiber growth. **Aquaculture** 65, 55-56.

Weatherley A.H., Gill H.S. and Rogers S.C. (1980). Growth dynamics of mosaic muscle fibers in fingerling rainbow trout (*Salmo gairdneri*) in relation to somatic growth rate. **Can. J. Zool.** 58, 1532-1541.

Yao K., Niu P.D., Le Gac F. and Le Bail P.-Y. (1991). Presence of specific growth hormone binding sites in rainbow trout (*Oncorhynchus mykiss*) tissues: characterization of the hepatic receptor. **Gen. Comp. Endocrinol.** 81, 72-82.

Young G. (1988). Enhanced response of the interrenal of coho salmon (*Oncorhynchus kisutch*) to ACTH after growth hormone treatment *in vivo* and *in vitro*. **Gen. Comp. Endocrinol.** 71, 85-92.

12

The Neuroendocrine Control of the Gonadotropin (GTH2) Secretion in Teleost Fish

B. Breton, T. Mikolajczyk, and W. Popek

In fish the control of the pituitary maturational gonadotropin (GTH2) secretion is mainly under the influence of the gonadotropin releasing hormone (GnRH). This paper will deal with recent advances in the knowledge of the GnRH system in teleost fish, and on the other neuromediators which can modulate positively or negatively this system or act directly on the GTH2 secretion, hormone mainly involved in the control of the last phases of the reproductive cycle, from the end of the vitellogenesis to maturation and ovulation on which have been done most of the works to date.

THE GONADOTROPIN RELEASING SYSTEM

Characterization and biological activity

Until now 6 different forms of GnRH have been characterized from the vertebrate brain. At least 2 forms are generally present in teleost fish. The salmon GnRH form (sGnRH = Trp^7 Leu^8 GnRH), has been found in most species as well as the chicken GnRH II form (cGnRH II = His^5 Trp^7 Tyr^8 GnRH). Other forms have also been demonstrated in fish, such as the chicken GnRH I form (Gln^8 GnRH) in tilapia, hake and winter flounder, and the mammalian form in some species such as the eel and the sturgeon (reviewed by King and Millar 1987, Sherwood and Coe 1991). In the goldfish, sGnRH, cGnRH I and II are all biologically active on the *in vivo* stimulation of the GTH2 secretion (Peter et al. 1987), cGnRH II being even more active than sGnRH, may be demonstrating a possible physiological significance of the presence of these different GnRH forms in fish (Chang et al. 1990).

The biological effects of GnRH are dose dependent and related to the pituitary responsivness. In salmonids this responsiveness increases with the advancement of the gametogenesis and becomes maximum during the spawning period (Weil and Marcuzzi 1990a and b). In the goldfish, the pituitary responsivenes to GnRH also varied with the gonadal stage (Sokolowska et al. 1985), but cannot be demonstrated without the removal of a dopamine inhibition over GnRH action on the pituitary (for review Peter et al. 1986). This problem will be discuss in more detail in another part of this paper. The biological action of the peptide can also depends upon the GnRH catabolism which can varied according to species. In the sea bream *Sparus aurata*, sGnRH is rapidly cleaved at the 5-6 and 9-10 bonds by pituitary endopeptidases. The modification of these two bonds blocked the cleavage sites, sGnRHa being only intensively cleaved at the unmodified bond (Zohar et al. 1990). In the rainbow trout, the 5-6 bond cleavage is on the contrary very weak and slow in the pituitary, GnRHa being not degradated *in vitro* by pituitary cells in culture (Weil et al. 1991). Thus there would be species differences in the endopeptidase GnRH cleaving activity according to species. The highest resistance of fish GnRH compared to mammals, could also be linked to a GnRH serum binding protein with a low affinity site (Ka= 2,2 $10^{-6}M^{-1}$) and a high capacity (Ro=3,56 10^{-10}M/mg of serum protein), as demonstrated in the goldfish, which would prevent clearance of GnRH by the kidney and protect GnRH against degradation by serum enzymes (Huang and Peter 1988).

Numerous works have been done using mammalian GnRH analogs (GnRHa) generally substituted in position 6 and 9 in order to investigated the structure activity relationships in fish. GnRHa generally induced a long lasting stimulation of the GTH2 secretion (Peter et al. 1987), in addition treatment by a sustained release form of GnRHa stimulated the GtH2 secretion for several weeks in the rainbow trout (Breton et al 1990b), indicating that contrary to mammals there is no rapid down regulation of the GnRH action at the pituitary level.(Breton et al. 1990). Because of their prolonged effect on the stimulation of GTH2 secretion, GnRHa are now widely used as potent maturation inducers in numerous cultured species.

Localization

Immunoreactive sGnRH structures are localized along a column extending from the rostral olfactory bulbs to the pituitary. Cells bodies are observed within the olfactory nerves and bulbs, along the ventromedian telencephalon, the ventrolateral preoptic area and laterobasal hypothalamus. There are two main pathways reaching the pituitary, where GnRH has been localized in the neural tissue of the neurointermediated lobe (Kah et al. 1986). This has been confirmed using quantitative RIA determinations of both sGnRH and cGnRH II. In the goldfish, (Yu et al. 1988) and in the rainbow trout (Okusawa et al.1990), cGnRH II is found in the same areas as sGnRH, but in addition it is preminent in the caudal brain (cerebellum, medula, spinal cord and optic tectum) in which no specific pathways leading to the pituitary have been so far identified, making questionnable the physiological role of GnRH contained in these structures. However, this suggests that cGnRH II and sGnRH immunocytochemical functions are anatomically separated, and that these peptides could have partially different physiological functions, at the pituitary level and also on other structures.

Involvement of GnRH in the control of gonadotropin secretion during the annual reproductive cycle

One of the main problem in studying GnRH secretion in teleost, is the lack of an hypothalamo-hypophysial system avoiding the possibility to measure GnRH blood portal levels in relation with GTH secretion. In mammals there is a correlation between luteinizing hormone (LH) and GnRH rythms of secretion. In fish, it is thus questionnable to know if the GnRH brain and pituitary contents could reflect a GnRH secretory activity, and be representative of a pituitary gonadotrophic activity in the absence of knowledge on such rythms. However, seasonal variations of sGnRH brain and pituitary contents have been demonstrated in the rainbow trout (Breton et al. 1986) and the goldfish (Yu et al. 1987), but generally not correlated with GTH2 blood plasma levels, excepted the periovulatory period during which GTH2 secretion follows a circadian rythm in both species (Hontela and Peter 1980, Zohar et al. 1986b). In the goldfish the GTH2 ovulatory surge is accompanied by a dramatic decrease of the pituitary, telencephalon and olfactory bulbs sGnRH contents, may be corresponding to a sGnRH release inducing the gonadotropin surge. In a natural environment, in male and in female roach *Rutilus rutilus*, sGnRH contents were the highest, correlated with the lowest plasma GTH2 levels the day preceding fish gathering on the spawning grounds and the first ovulations. The following day, GTH2 blood plasma concentrations increased correlated with a decrease of brain and pituitary sGnRH contents, That occured in all fish including male , ovulated and unovulated females exhibiting a spawning behavior (Breton et al. 1988a and b). This let suppose that, in natural populations, it would exist a synchronizing factor activating the hypothalamo-pituitary axis. In the goldfish, GnRH contents decreased in all parts of the rostral brain including the olfactory bulbs, providing evidence of functionnal relations between these structures and other brain areas (Yu and Peter 1990b). This could be an important link between environmental cues and the neuroendocrine system. Among the others, pheromonal stimulations certainly participate, because in the roach, in a natural environment, GnRH contents decreased in all fish independently of the maturationnal stage (Breton et al. 1988a and b), and in the goldfish exposure of male to a female exibiting a prostaglandin induced spawning behavior, modified GnRH contents in the brain including in the olfactory bulbs (Yu and Peter 1990b). This situation appears original compared to mammals in which the olfactory bulbs tonically inhibited the gonadotropin secretion , by a mechanism which would not depends upon chemosensitory signals as in the golden hamster (Pieper et al. 1990).

GnRH mechanism of action

The first step of GnRH action is the binding to GnRH pituitary receptors. They have been demonstrated in the winter flounder (Crim et al. 1988), the African catfish (de Leeuw et al. 1988) and the stickleback (Anderson et al.1989) in which there are a single class of low affinity and high capacity binding sites. In the goldfish two classes of binding sites have been identified, a low affinity high capacity site, and a high affinity low capacity site (Habibi et al. 1987) which would be respectively 51000 M_1 and 71000 M_1 proteins (Habibi et al. 1990). There are seasonal variations in pituitary GnRH receptor activity in the catfish and the goldfish (Habibi et al. 1989), highest binding capacities occuring during the late stages of gonadal recrudescence when the pituitary responsivness to exogenous GnRH stimulation becomes maximum. There is also a correlation between GnRHa binding affinity and their functionnal ability to stimulate pituitary GtH secretion (Habibi et al. 1989).

Recent studies demonstrated a desensitizing effect of GnRH which depends upon the dose and the frequency of application of the decapeptide. This process is partly the result of a down regulation of the pituitary binding capacity (Habibi et al. 1991). A pulsatile GtH secretion exists in the rainbow trout (Zohar et al. 1986a), although GnRH pulsatility cannot be demonstrated in fish, it could be supposed that the GtH pulsatility would be the result of a GnRH pulsatile pituitary delivery, may be essential for the maintenance of a normal pituitary GtH release in fish as in mammals.

The post membrane action of GnRH in fish is mainly mediated by extra cellular Ca^{++} as second messenger, in catfish (Van Asselt et al. 1989), tilapia (Levavi-Shimoni and Yaron 1989) and carp (Mikolajczyk et al. 1990a). The entrance of Ca^{++} into gonadotropes would be the result of the activation of voltage dependent calcium channels (VDCC), but in carp, contrary to mammals in which LHRH activated L types of VDCC, sGnRH activated N or T types of VDCC. On another hand, in carp the specific blockade of L type VDCC by nifedipine also induced an *in vitro* GtH2 release in absence of GnRH and their specific activation by Bay K8644, induced a dose dependent inhibition of GtH2 release (Mikolajczyk et al. 1990b), may be indicating that other mediators have a direct trophic effect on gonadotropes through the modification of L type VDCC. Other relays than Ca^{++}, have also been demonstrated in fish. This is the case of the protein kinase C (PKC) in tilapia (Levavi-Shimoni and Yaron 1989), which is also involved in LHRH action in mammals resulting of PKC intracellular redistribution following Ca^{++} mobilization; and arachidonic acid (AA) in the goldfish (Chang et al. 1989). In this last species sGnRH and cGnRH II bind to the same receptors, but they could have different second intra cellular messengers, AA mediating more especially the action of cGnRH II (Chang et al. 1991).

OTHER NEUROMEDIATORS AND THEIR INTERACTION WITH THE GNRH SYSTEM

The dopaminergic sytem

The role of dopamine (DA) as a gonadotropin release inhibiting factor (GRIF) is well established in goldfish, in which its main action is the inbition of GnRH stimulates GtH release as well *in vivo* as *in vitro*, depending on the gonadal stage of development. The inhibitory tone exerts by dopamine increased with the advancement of the gametogenesis, reaching a maximum level at the time of ovulation. As a consequence, in this species the artificial induction of ovulation absolutely requires a combined treatment using GnRHa and a DA antagonist such as pimozide (for review Peter et al. 1986). The DA inhibitory tone also exists in other teleost species (de Leeuw et al. 1986), but its intensity appears to depend on the species as indicated by the ovulatory response to GnRHa treatments. In salmonids for instance, 100% ovulations can be obtained using GnRHa alone (Breton et al. 1990), although it had been mentionned that pimozide potentiated GnRH stimulated GtH release in the coho salmon. Other species such as the chinese loach (*Paramisgurnus dabryanus*) present intermediary situation, in which DA antagonist also potentiated GnRH induced GtH release, but only increase the frequency of induced ovulation (Lin et al. 1986). This variability of the DA inhibitory tone may be representative of evolution throughout teleost fish, but could also depend on other factors, and among the more or less important dependance of final maturation process to strict environmental cues as in the goldfish, where the spontaneous GtH ovulatory surge cannot occur in absence of aquatic vegetation (Stacey et al. 1979). These include pheromonal interactions which could activate or inhibit neuronal circuits leading to dopaminergic inhibition or its removal as we already hypothetized in the roach (Breton et al.1988a and b).see social factors.

There are also differences in the DA metabolic pathways according to species. In teleost, the main pathway for DA degradation is linked to the monoamine oxydase activity; in addition, in some species such as the rainbow trout and the catfish, a second pathway through catechol o methyl transferase (COMT) has also been demonstrated (Van Asselt et al. 1989, Saligault et al. 1990). By the way,there will be a competition between catechol estrogens and DA for COMT, leading to an additionnal modulation of DA degradation. This process depends upon estradiol levels and aromatase activity (Van Asselt et al. 1989). Research have failed to demonstrate such a mechanism in the goldfish (Trudeau et al. 1991). Thus it can be speculated that the metabolic pathway for DA degradation could be also part of the mechanisms which determine DA inhibitory tone on GnRH action, in relation with steroids.

In the goldfish and the catfish, DA action is exerted through D_2 dopamine receptors (Olmejaniuk et al. 1986, Van Asselt et al.1990), and results in several effects such as a down regulation of the pituitary GnRH receptors (de Leeuw et al. 1988, 1989), and the modification of GnRH contents in the brain structures excepted the pituitary in the normal male goldfish. In male induced to spawning behavior following exposure to prostaglandin treated female, the increase in GnRH brain contents is abolished after DA antagonist treatment (Yu and Peter 1990b). All these data demonstrate the important role of the central dopaminergic system in fish as an important inhibitory component of the regulatory circuits of brain GnRH levels, GnRH release and GnRH action at the pituitary level.

Neuropeptide Y (NPY)

This 36 aminoacids peptide which, in the mammals, had very pleiotropic actions, is involved in the central regulation of the luteinizing hormone (LH) secretion. It had also been characterized in the goldfish (Pontet et al. 1989), and the rainbow trout (Danger et al. 1991). It had a widespread brain distribution. In particular, NPY immunoreactive neurons are located in the nucleus enteropeduncularis and the preoptic nucleus; but there is no colocalization with sGnRH . On the contrary in the pituitary, NPY immunoreactive fibers are in direct contact with gonadotrophs (Kah et al. 1989), may be indicative of direct action of the peptide at the pituitary levels and indirect effects at the upper level. In the rainbow trout, it modulated the *in vitro* GtH2 secretion from perifused pituitary fragments, being either inhibitory during vitellogenesis or stimulatory during the periovulatory period. This modulation of NPY action depends upon the *in vivo* estradiol levels. Indeed, the *in vitro* response to NPY of pituitary taken from vitellogenic females treated with ATD, an inhibitor of the aromatase activity, is similar to that of pituitary taken from ovulating females in which estradiol levels are very low (Breton et al. 1989). The NPY action at the pituitary level is certainly complex, because its action is inhibited in presence of a GnRH antagonist in pituitary fragments, may be indicating interaction between the two peptides. On another hand, it also stimulated GtH2 secretion from perifused dispersed pituitary cells indicating a direct effect on the gonadotrops (Breton et al. 1990). It equally stimulates the *in vivo* GtH2 secretion either after intraperitoneal treatment in the rainbow trout or intraventricular infusion in the carp. In both species it potentiates the *in vivo* and *in vitro* action of GnRHa (Breton et al. 1991). Thus besides direct effects of NPY on the gonadotrops, there are clear evidence of interaction between GnRH and this peptide at the pituitary level. The mechanisms by which NPY enhances GnRH action;, and GnRH antagonist blocks its action is unsolved in fish. In mammals, it had been demonstrated that NPY increased GnRH binding to pituitary membranes, may be explaining in part the action of this neuropeptide to potentiate GnRH induced LH release (Parker et al. 1991), such a mechanism must be investigated in fish. In the rat, one of the action of NPY at the pituitary level is mediated through the activation of VDCC (Crowley et al. 1990). In the rainbow trout, NPY as well as GnRH do not activited the pituitary adenyl cyclase activity (unpublished data), that let suppose that NPY action is mediated by another second messenger. The *in vivo* stimulation of the GtH2 secretion by NPY could be linked to its effects on the GnRH release at the central level as well in fish as in mammals (Khorram et al. 1987). The *in vivo* demonstration of such a mechanism in fish is nearly impossible because the lack of hypothalamo-pituitary portal vessel system. However, in the rainbow trout, NPY stimulates the *in vitro* sGnRH release from slices of telencephalon but not from the pituitary (Breton et al. unpublished data). Thus one of the effects of NPY would be the stimulation of GnRH from telencephalon perykaryons, which in turn would stimulated the pituitary GtH2 secretion. We have recently completed these experiments and demonstrated that the *in vitro* action is blocked in presence of propanonol (unpublished data), a noradrenergic antagonist, indicating that NPY action could be mediated by norepinephrine as already demonstrated in the rabbit (Berria et al.1991).

All these results show that NPY is an important component of an excitatory neural circuit involoved in the control of the pituitary gonadotropin secretion.

Other neuromediators

There are scattered informations on the effects of other neurotransmitters on the control of GtH secretion, mainly obtained in the rainbow trout and the goldfish.

Norepinephrine has been demonstrated to stimulate the GtH2 secretion in the goldfish (Chang and Peter 1984). We have already mentionned the possible involvement of the adrenergic system in the mediation of NPY stimulated GnRH release from telencephalon perykaryons in trout. Supporting this hypothesis, a direct effect of norepinephrine on the *in vitro* sGnRH release from slices of the preoptic anterior hypothalamic region had been demonstrated in the goldfish (Yu et al.1991), in which , in addition it directly stimulated GtH2 secretion independently of sGnRH receptors (Chang et al. 1991). Thus this mediator could exert a dual control on the regulation of GtH2 secretion, having both a direct effect on the pituitary and indirectly modulating GnRH release from GnRH cell bodies.

Serotonin also stimulated *in vivo* and *in vitro* GtH2 secretion in the goldfish, probably through a direct action at the pituitary level (Somoza et al. 1988, Somoza and Peter 1991).

In the male goldfish naloxone (NAL), an opioid receptor antagonist caused an *in vivo* decrease in serum GtH2, it also blocked the stimulatory effect of GnRH during late gonadal recrudescence, whereas it stimulated GnRH induced GtH2 secretion during early recrudescence. Other experiments have shown that there is interaction between NAL and the dopaminergic system (Rosenblum and Peter 1989). These data suggest that opioids may also play an important role in the regulation of GtH2 secretion.

Direct treatment of goldfish with gamma amino butyric acid (GABA), induced an increase of GtH2 secretion depending upon the reproductive stage and the steroid environment (Kah et al. 1991). This action may be mediated by direct or indirect GABA influence on sGnRH release from pituitary peptidergic endings. In sexually mature carp, modifications of the GABA metabolism leading either to increase or decrease the endogenous GABA concentrations, had no effect on the GtH2 basal levels, but modified the response to GnRH treatment used alone or in combination with the DA antagonist pimozide. Thus in the female and male carp, 3 mercapto propionic acid (MPA), an inhibitor of GABA synthesis, enhances the response to GnRH-pimozide treatment, whereas the GABA mimetic compoud valproïc acid decreases the GnRH stimulted GtH2 levels in the male (Roelants et al. 1989). In mammals, GABA interacted with DA , especially in the control of LH secretion (Jason et al. 1980), on which GABA and DA had opposite effects (Elias et al. 1983). Our results suggest that GABA also interacts with DA in the control of gonadotropin secretion in fish, through an inhibitory tone. In the rainbow trout in which DA inhibition on GnRH action is low, the same kind of experiments did not modified either basal or GnRH stimulated GtH2 levels (unpublished data). This again point out the existence of species differences in the control of GtH2 secretion.

There are numerous works on the study of the role of the pineal on the control of the reproductive cycle in fish, but there is not clear conclusion on the relation between pineal and GtH secretion. Although pinealectomy could affected rythm of GtH2 secretion in the goldfish (Hontela and Peter 1980), until now there is no data on the effects of pineal secretions as elements involved in the regulation of gonadotropin secretion in fish. We have recently investigated the effects of melatonin on GtH2 secretion in the female carp. Melatonin intraperitonally injected or infused into the third ventricule induced a stimulation of the GtH2 blood plasma levels only if the treatments are done at the middle of the dark phase, treatments realized either at the beginning of the dark phase or during the light phase being ineffective. During the dark phase, these treatments also increase the response to GnRH injection, and their effects are suppressed in presence of apomorphine and enhanced with pimozide (unpublished data). These results clearly demonstrated that melatonin is involved in the control of GtH2 secretion in the carp, may be partially through the modulation of the dopaminergic inhibitory tone on GnRH action. The fact that melatonin is only effective during a short period of the nycthemer lets suppose that this mediator could have a permissive effect on the gonadotropin release and could play a role in the induction of the GtH2 ovulatory surge occuring during the last phase of the dark period in the goldfish (Stacey et al. 1979).

STEROID HORMONES

Feedback effects of steroids on the GtH2 secretion in fish was first demonstrated after castration experiments in the male (Billard et al. 19776) and the femelle rainbow trout (Bommelaer et al. 1981). Steroid hormones do not seem to have similar effects on the hypothalamo-pituitary axis in salmonids and cyprinids. A positive feedback of steroids on GtH2 synthesis had been demonstrated in immature fish in both species (Crim and Evans 1979, Crim et al 1981), but in goldfish only aromatisable androgens were active, whereas non aromatisable steroids were also effective in the rainbow trout (Goverdina et al. 1983), may be suggesting that testosterone acted differently or its action was differently modulated in these two species (see Peter et al. 1991 for review). This was confirmed by the long term effects of testosterone implantation in immature fish, which stimulated GtH2 secretion in trout (Crim et al 1982, Gielen and Goos 1984) but not in goldfish, may be resulting of a DA inhibition on spontaneous GtH2 secretion in the goldfish (see Peter et al. 1986 for review). We have reinvestigated this problem using a triploïd rainbow trout model, in which testosterone and estradiol blood plasma levels are still undetectable in two years old fish , when normal fish of the same strain undergo gonadal recrudescence (unpublished results). As in normal juvenile fish, testosterone and estradiol simulate the GtH2 synthesis, testosterone being more potent than estradiol, but only testosterone specifically stimulates GtH2 secretion in a dose dependent manner. In addition we have determined the effects of steroids on sGnRH synthesis and accumulation In the brain and the pituitary. testosterone and estradiol also increase sGnRH contents in areas containing sGnRH perykaryons (telencephalon and preoptic area), Testosterone being anew more potent than estradiol. There is no change in the hypothalamus sGnRH contents, whereas testosterone induces a rapid pituitary sGnRH accumulation, occuring less than a month after steroid implantation, and a month earlier than the stimulation of the GtH2 secretion, although the pituitary GtH2 synthesis was already stimulated (Breton et al. unpublished data). That might be indicative of other mechanisms, depending or not upon steroids, occuring progressively during the brain sexualization, and which could be involved in the determinism of GnRH release from pituitary peptidergic endings or in that of the pituitary sensitivity to GnRH for instance.

This last process had been demonstrated to be influenced by steroids in the goldfish, where testosterone implants increased the pituitary GnRH responsiveness at all stages of gonadal development, whereas estradiol exerted a similar effect only in sexually regressed fish (Trudeau et al. 1991a). Although this action could be mediated through the modulation of DA turnover in the goldfish (Trudeau et al. 1991b), in the rainbow trout, using pituitary cells cultures, Weil and Marcuzzi (1990a andb), demonstrated direct effects of steroids on the pituitary responsiveness to GnRH. These direct effects could be partly mediated by the effects of steroids on the regulation of GnRH pituitary receptors as shown in the goldfish and the catfish.

From all these results, it is clear that steroids can act at multiple hypothalamic pituitary sites, and differently according to species. this is equally true for DA, and moreover they can interact together. This point out the necessity to develop comparative studies on different fish models.

SOCIAL FACTORS

In the male goldfish, exposure to a sexual stimuli resulted in a stimulation of the GtH2 secretion and an increase of the milt volume (Kyle et al. 1985). In the rainbow trout GtH2 plasma levels were also higher in males in contact with ovulated females exhibiting a spawning behavior (Liley et al. 1986). In the goldfish, it has been further demonstrated that these effects were mediated by a sex hormone originated from maturating females (Dulka et al. 1986, Kobayashi et al 1986), which has been identified as the maturation inducing steroid (MIS), 17 hydroxy 20ß dihydro progesterone released in the water by ovulating females (Stacey and Sorensen 1986, Sorensen et al. 1987). The effect of the MIS on the GtH2 secretion is not a direct feedback on the pituitary, but results from the activation of neuronal circuits leading to GtH release,in which the olfactory system might play a key role, MIS activating the electrical activity of the olfactory epithelium (Sorensen et al. 1987). On another hand, olfactory tract lesions inhibited MIS induced - GtH2 surge (Dulka and Stacey 1991). As we already mentionned, in spawning male goldfish, GnRH contents decreased in all part of the brain including the olfactory bulbs (Yu and Peter 1990b), may be as the results of MIS action on these structures.

In conclusion the neuroendocrine control of the gonadotropin secretion in fish is a multicomponent system in which each factor interact with others. Besides GnRH, according to species other neuromediators do not seem to have the same importance. This is especially the case of dopamine which strongly inhibits GnRH action in some species such as the cyprinids , but has only a weak influence in others such as the salmonids. This raise the problem of the choice of the experimental model according to the factors to study, and the type and the strengh of their interactions. For that reason it is now nearly impossible to draw a general model of the neuroendocrine regulation on gonadotropin secretion in which all components would have the same importance in all species. In our opinion comparative studies must be developped in other species than cyprinids and salmonids, in which most of the results have been obtained to date. Fish diversity offers such an opportunity.

LITTERATURE REFERENCES

Anderson E, Borg B and De Leeuw R (1989) Characterization of gonadotropin-releasing hormone binding sites in the pituitary of the three-spine stickleback *Gasterosteus aculeatus*. Gen. Comp. Endocrinol. 76:41-45

Berria M, Pau K.Y and Spies H.G (1991) Evidence for ${}_1$-adrenergic involvement in neuropeptide Y-stimulated GnRH release in female rabbit. Neuroendocrinology 53:480-486

Billard R, Richard M and Breton B (1976) Stimulation de la sécrétion gonadotrope hypophysaire après castration chez la truite arc-en-ciel; variation de la réponse au cours du cycle reproducteur. C.R. Acad. Sci. Paris Ser. D 283:171-174

Bommelaer M.C, Billard R and Breton B (1981) Changes in plasma gonadotropin after ovariectomy and estradiol supplementation at different stages at the end of the reproductive cycle in the rainbow trout (*Samo gairdneri* R). Reprod. Nutr. Dévelop. 21:989-997

Breton B, Motin A, Billard R, Kah O, Geoffre S and Precigoux G (1986) Immunoreactive gonadotropin-releasing hormone like material in the brain and the pituitary gland during the periovulatory period in the brown trout (*Salmo trutta*): relationships with the plasma and pituitary gonadotropin. Gen. Comp. Endocrinol. 61:109-119

Breton B, Sambroni E and Gillet C (1988a) Gonadotropin releasing hormone (GnRH) and gonadotropin (GtH) variations around the spawning period in a wild population of roach (*Rutilus rutilus*) from Leman lake. I- The female. Aquat. Living Resour. 1:93-99

Breton B, Sambroni E and Gillet C (1988b) Gonadotropin releasing hormone (GnRH) and gonadotropin (GtH) variations around the spawning period in a wild population of roach (*Rutilus rutilus*) from Leman lake. II The male. Aquat. Living Resour. 1:101-106

Breton B, Mikolajczyk T, Danger J.M, Gonnet F, Saint Pierre S and Vaudry H (1989) Neuropeptide Y (NPY) modulates *in vitro* gonadotropin release from rainbow trout perifused pituitary glands. Fish Physiol. Biochem. 7:77-83

Breton B, Mikolajczyk T, Weil C, Danger J.M and Vaudry H (1990a) Studies on the mode of action of neuropeptide Y (NPY) on maturational gonadotropin (GtH) secretion from perifused rainbow trout pituitary glands. Fish Physiol. Biochem. 8:339-346

Breton B, Weil C, Sambroni E and Zohar Y (1990b) Effects of acute versus sustained administration of GnRHa on GtH release and ovulation in the rainbow trout *Oncorhynchus* mykiss. Aquaculture 91:373-383

Breton B, Mikolajczyk T, Popek W, Bieniarz K and Epler P (1991) Neuropeptide Y stimulates *in vitro* gonadotropin secretion in teleost fish. Gen. Comp. Endocrinol. 84:277-283

Chang J.P and Peter R.E (1984) Influences of norepinephrine and -adrenergic mechanisms on gonadotropin secretion in female goldfish *Carassius auratus*. Gen. Comp. Endocrinol. 55:89-95

Chang J.P, Freedman G.L and De Leeuw R (1989) Participation of arachidonic acid metabolism in gonadotropin-releasing hormone stimulation of goldfish gonadotropin release. Gen. Comp. Endocinol. 76:2-11

Chang J.P, Yu K.L, Wong A.O.L and Peter R.E (1990) Differential actions of dopamine receptors subtypes on gonadotropin and growth hormone release in vitro in goldfish. Neuroendocrinology 51:664-674

Chang J.P, Cook H, Freedman G.L, Wiggs A.J, Somoza G.M, De Leeuw R and Peter R.E (1990) Use of pituitary cell dispersion method and primary culture system for the study of gonadotropin-releasing hormone action in the goldfish, *Carassius auratus*. Gen. Comp. Endocrinol. 77:256-273

Chang J.P and Jobin R.M (1991a) Two native GnRH peptides stimulate gonadotropin release in the goldfish by different second messenger components. Abstracts 4th International Symposium on Reproductive Physiology of Fish. Norwich 7-12 July U.K.

Chang J.P, Van Goor F and Acharya S (1991b) Influences of norepinephrine, and adrenergic agonist and antagonist on gonadotropin secretion from dispersed pituitary cells of goldfish *Carassius auratus*. Neuroendocrinology 54:202-210

Crim L.W and Evans D.M (1979) Stimulation of pituitary gonadotropin by testosterone in the juvenile rainbow trout (*Salmo gairdnei*). Gen. Comp. Endocrinol. 37:192-196

Crim L.W, Peter R.E and Billard R (1981) Onset of gonadotropic hormone accumulation in the immature trout pituitary gland in response to estrogen or aromatisable androgen steroid hormones. Gen. Comp. Endocrinol. 44:374-381

Crim L.W and Evans D.M (1982) Positive testosterone feedback on gonadotropic hormone in the rainbow trout. p23. Proc. 2nd International Symposium on Reproductive Physiology in Fish. Richter C.J.J and Goos H.J.Th edts. Pudoc Press Wageningen, The Netherlands.

Crim L.W, Wilson C, St Arnaud R, Evans D.M and Harim S.A (1987) Characteristics of fish pituitary LHRH receptors. Proc. 3rd International Symposium on Reproductive Physiology in Fish. pp 21-23. Idler D.R, Crim L.W and Walsh J.M Edts. St John's Nfld Canada

Crowley W.R, Shah G.V, Caroll B.L, Kennedy D, Dockter M.E and Kalra S.P (1990) Neuropeptide-Y enhances luteinizing hormone (LH) releasing hormone-induced LH release and elevations in cytosolic Ca^{2+} in rat anterior pituitary cells: evidence for involvement of extracellular Ca^{2+} influx through voltage-sensitive channels. Endocrinology 127:1487-1494

Danger J.M, Breton B, Vallarino M, Fournier A, Pelletier G and Vaudry H (1991) Neuropeptide-Y in the trout brain and pituitary: localization, characterization, and action on gonadotropin release. Endocrinolgy 128:2360-2368

De Leeuw R, Goos H.J.Th and Van Oordt P.G.W.J (1986) The dopaminergic inhibition of the gonadotropin-releasing hormone-induced gonadotropin release: an *in vitro* study with fragments and cells suspensions from pituitaries of the African catfish *Clarias gariepinus* (Burchell). Gen. Comp. Endocrinol. 63:171-177

De Leeuw R, Van't Veer C, Goos H.J.Th and Van Oordt P.G.W.J (1988) The dopaminergic regulation of gonadotropin-releasing hormone receptor binding in the pituitary of the African catfish, *Clarias gariepinus*. Gen. Comp. Endocrinol. 72:408-415

De Leeuw R, Conn P.M, Van't Veer C, Goos H.J.Th and Van Oordt P.G.W.J (1988) Characterization of the receptor for gonadotropin-releasing hormone in the pituitary of the African catfish *Clarias gariepinus.* Fish Physiol. Biochem. 5:99-107

De Leeuw R, Habibi H.R, Nahorniak C.S and Peter R.E (1989) Dopaminergic regulation of pituitary gonadotrophin releasing hormone receptor activity in the goldfish (*Carassius auratus*). J. Endocr. 121:239-247

Dulka J.G, Stacey N.E, Sorensen P.W and Van Der Kraak G.J (1987) A steroid sex pheromone synchronizes male-female spawning readiness in goldfish. Nature 325: 251-253

Dulka J.G and Stacey N.E (1991) Effects of olfactory tract lesions on gonadotropin and milt responses to the female sex pheromone 17 ,20ß-dihydroxy-4-pregnen-3-one, in male goldfish. J. Experiment. Zool. 257:223-229

Farhaenus-van Ree G.E, van Vlaardingen M and Gielen J.Th (1983) Effect of 17 -methyl testosterone, estradiol-17ß and synthetic LHRH on production of gonadotropic hormone in pituitaries of rainbow trout in culture. Cell Tissue Res. 232:157-176

Gielen J.Th and Goos H.J.Th (1984) The brain-pituitary-gonadal axis in the rainbow trout *Salmo gairdneri*. III. Absence of an inhibiting action of teststerone on gonadotrophin release in juveniles. Gen.Comp. Endocrinol. 56:457-465

Habibi H.R, Peter R.E, Sokolowska M, Rivier J.E and Vale W.W (1987) Characterization of gonadotropin-releasing hormone (GnRH) binding to pituitary receptors in goldfish (*Carassius auratus*). Biol. Reprod. 36:844-853

Habibi H.R, De Leeuw R, Nahorniak C.S, Goos H.J.Th and Peter R.E (1989) Pituitary gonadotropin-releasing hormone (GnRH) receptor activity in goldfish and catfish : seasonal and gonadal effects. Fish Physiol. Biochem. 7:109-118

Habibi H.R, Marchant T.A, Nahorniak C.S, Van der Loo H, Peter R.E, Rivier J.E and Vale W.W (1989) Functional relationship between receptor binding and biological activity for analogs of mammalian and salmon gonadotropin-releasing hormones in the pituitary of goldfish (*Carassius auratus*). Biol. Reprod. 40:1152-1161

Habibi H.R, Peter R.E and Hazum E (1990) Photoaffinity labelling of pituitary gonadotropin-releasing hormone receptors in goldfish (*Carassius auratus*). Biol. Reprod. 43:1006-1011

Habibi H.R (1991) Desensitization to native molecular forms of gonadotropin-releasing hormone in the goldfish pituitary: dependence on pulse frequency and concentration. Gen. Comp. Endocrinol. 84:199-214

Hontela A and Peter R.E (1980) Effects of pinealectomy, blinding and sexual condition on serum gonadotropin levels in the goldfish. Gen. Comp. Endocrinol. 40:168-179

Huang Y.P and Peter R.E (1988) Evidence for a gonadotropin-releasing hormone binding protein in the goldfish (*Carassius auratus*). Gen. Comp. Endocrinol. 69:308-316

Kah O, Breton B, Dulka J.G, Nunez-Rodriguez J, Peter R.E, Corrigan A, Rivier J.E and Vale W.W (1986) A reinvestigation of the Gn-RH (gonadotrophin-releasing hormone) systems in the goldfish brain using antibodies to salmon Gn-RH. Cell Tissue Res. 244:327-337

Kah O, Trudeau V.L, Sloley B.d, Martinoli M.G, Chang J.P, Yu K.L and Peter R.E (1991) Implication of GABA in the neuroendocrine regulation of gonadotropin secretion in the goldfish (*Carassius auratus*). Abstracts 4th international Symposium on Reproductive Physiology in Fish. Norwich 7-12 July UK.

Khorram O, Pau K.Y.F and Spies H.G (1987) Bimodal effects of neuropeptide Y on hypothalamic release of gonadotropin-releasing hormone in concious rabbits. Neuroendocrinology 45:290-297

King J.A, Millar R.P (1987) Phylogenic diversity of LHRH. In : LHRH and its analogs. Contraceptive and therapeutic applications. Book 2. Vickery B.H, Nestor J.J Eds. MTP press limited, Lancaster.

Kobayashi M, Aida K and Hanyu I (1986) Pheromone from ovulatory female goldfish induces gonadotropin surge in male. Gen. Comp. Endocrinol. 63:451-455

Kyle A.L, Stacey N.E, Peter R.E and Billard R (1985) Elevations in gonadotropin concentrations and milt volumes as a result of spawning behavior in the goldfish. Gen. Comp. Endocrinol. 57:10-22

Levavi-Shimoni B and Yaron Z (1989) Gonadotropin secretion from perifused tilapia pituitary in relation to gonadotropin-releasing hormone, extracellular calcium, and activation of protein kinase C. Gen. Comp. Endocrinol. 75:187-194

Liley N.R, Breton B, Fostier A and Tan E.S.P (1986) Endocrine changes associated with spawning behavior and social stimuli in a wild population of rainbow trout (*Salmo gairdneri*). I. Males. Gen. Comp. Endocrinol. 62:145-156

Lin H.R, Van Der Kraak G, Peter R.E and Breton B (1986) Effects of (DAla^6, Pro^9-NEt)-LHRH and catecholaminergic drugs on gonadotropin secretion and ovulation in the chinese loach (*Paramisgurnus dabryanus*). Gen. Comp. Endocrinol. 64:389-395

Mikolajczyk T, Weil C, Epler P and Breton B (1990a) Calcium ions as mediator in GnRH action on gonadotropin release in the common carp (*Cyprinus carpio* L). Reprod. Nutr. Dev. 30:483-492

Mikolajczyk T, Weil C, Epler P and Breton B (1990b) Involvement of voltage-dependent calcium channels (VDCC) in the action of GnRH on GtH release in the common carp (*Cyprinus carpio* L): comparison with K^+ action. Reprod. Nutr. Dev. 30:619-628

Olmeljaniuk R.J, Shih S.H and Peter R.E (1987) In vivo evaluation of dopamine receptor-mediated inhibition of gonadotropin secretion from the pituitary gland of the goldfish *Carassius auratus*. J. Endocr. 114:449-458

Parker S.L, Kalra S.P and Crowley W.R (1991) Neuropeptide Y modulates the binding of a gonadotropin-releasing hormone (GnRH) analog to anterior pituitary GnRH receptor sites. Endocrinology 128:2309-2316

Peter R.E, Chang J.P, Nahorniak C.S, Omeljaniuk R.J, Sokolowska M, Shih S.H and Billard R (1986) Interactions of catecholamines and GnRH in regulation of gonadotropin secretion in teleost fish. Recent Progress in Hormone Research 42: 513-546

Peter R.E, Nahorniak C.S, Shih S.H, King J.A and Millar R.P (1987) Activity of position-8-subsituted analogs of mammalian gonadotropin-releasing hormone (mGnRH) and chicken and lamprey gonadotropin-releasing hormones in goldfish. Gen. Comp. Endocrinol. 65:385-393

Peter R.E, Trudeau V.L, Sloley B.D, Peng C and Nahorniak C.S (1991) Actions of catecholamines, peptides and sex steroids in regulation of gonadotropin-II in the goldfish. Abstracts 4th International Symposium on Reproductive Physiology in Fish. Norwich 7-12 July UK.

Pieper D.R, Lobocki C.A, Thompson M and Subramanian M.G (1990) The olfactory bulbs tonically inhibit serum gonadotropin and prolactin in male hamster on long or short photoperiod. J. Neuroendocr. 2:707-715

Pontet J.A, Danger J.M, Dubourg P, Pelletier G, Vaudry H, Calas A and Kah O (1989) Distribution and characterization of neuropeptide Y-like immunoreactivity in the brain and pituitary of the goldfish. Cell Tissue Res. 255:529-538

Roelants Y, Epler P, Mikolajczyk T, Breton B, Bieniarz K and Ollevier F (1990) A presumptive role for GABA in the stimulatory effects of Des- Gly^{10}, (D-Ala^6)-LHRH-ethylamide and pimozide on the gonadotropin release in carp. Life Sci. 47:1801-1812

Rosenblum P.M and Peter R.E (1989) Evidence for the involvement of endogenous opiods in the regulation of gonadotropin secretion in male goldfish *Carassius auratus*. Gen. Comp. Endocrinol. 73:21-27

Saligault C, Bailhache T, Salbert G, Breton and Jego P (1990) Dynamic characteristics of serotonin and dopamine metabolism in the rainbow trout brain: a regional study using liquid chromatography with electrochemical detection. Fish Physiol. Biochem. 8:199-205

Sherwood N.M and Coe I.R (1991) Neuropeptides and their genes in fish. Abstracts 4th International Symposium on Reproductive Physiology in Fish. Norwich 7-12 July U.K.

Sokolowska M, Peter R.E, Nahorniak C.S and Chang J.P (1985) Seasonal effects of pimozide and Des Gly^{10} (D-Ala^{6}) LHRH ethylamide on gonadotropin secretion in the goldfish. Gen. Comp. Endocrinol. 57:472-479

Somoza G.M, Yu K.L and Peter R.E (1988) Serotonin stimulates gonadotropin release in female and male goldfish *Carassius auratus*. Gen. Comp. Endocrinol. 72:374-382

Somoza G.M and Peter R.E (1991) Effects of serotonin on gonadotropin and growth hormone release from *in vitro* perifused goldfish pituitary fragments. Gen. Comp. Endocrinol. 82:103-110

Sorensen P.W, Hara T.J and Stacey N.E (1987) Extreme olfactory sensitivity of mature and gonadally - regressed goldfish to a potent steroidal pheromone, 17 ,20ß-dihydroxy-4-pregnen-3-one. J.Comp. Physiol A 160:305-313

Stacey N.E, Cook H.F and Peter R. E (1979) Spontaneous and gonadotropin-induced ovulation in the goldfish *Carassius auratus* L: effects of external factors. J.Fish Biol. 15:349-361

Stacey N.E, Cook A.F and Peter R.E (1979) Ovulatory surge of gonadotropin in the goldfish *Carassius auratus*. Gen. Comp. Endocrinol. 37:246-249

Stacey N.E and Sorensen P. W (1986) 17 ,20ß-dihydroxy-4-pregnen-3-one: a steroidal primer pheromone increasing milt volume in the goldfish *Carassius auratus*. Can. J. Zool. 64:2412-2417

Trudeau V.L, Peter R.E and Sloley B.D (1991) Testosterone and estradiol potentiate the serum gonadotropin response to gonadotropin-releasing hormone in goldfish. Biol. Reprod. 44:951-960

Trudeau V.L, Sloley B.D, Wong A.O.L and Peter R.E (1991) Interaction of sex steroids with dopamine (DA) and gonadotropin releasing hormone (GnRH) in the control of gonadotropin (GtH) secretion in the goldfish. Abstracts 4th International Symposium on Reproductive Physiology in Fish. Norwich 7-12 July U.K.

Van Asselt L.A.C, Timmers R.J.M, Goos H.J.Th, Lambert J.G.D and Van Oordt P.G.W.J (1989a) Catecholestrogens inhibit dopamine methylation in the gonadotrops of the african catfish *Clarias gariepinus*. Fish Physiol. Biochem. 6:207-212

Van Asselt L.A.C, Goos H.J.Th, van Dijk W and Braas J (1989b) Role of calcium ions in action of gonadotropin-releasing hormone on gonadotropin secretion in the African catfish *Clarias gariepinus*. Gen. Comp. Endocrinol. 76:46-52

Van Asselt V.A.C, Goos H.J.Th, De Leeuw R, Peter R.E, Hol E.M, Wassenberg F.P and Van Oordt P.G.W.J (1990) Characterization of dopamine D_2 receptors in the pituitary of the african catfish *Clarias gariepinus*. Gen. Comp. Endocrinol. 80:107-115

Weil C and Marcuzzi O (1990a) Cultured pituitary cell response to GnRH at diffrent stages of rainbow trout oogenesis and influence of steroid hormones. Gen. Comp. Endocrinol. 79:483-491

Weil C and Marcuzzi O (1990b) Cultured pituitary cell response to GnRH at different stages of rainbow trout spermatogenesis and influence of steroid hormones. Gen. Comp. Endocrinol. 79:492-498

Weil C, Breton B, Sambroni E, Zmora N and Zohar Y (1991) Bioactivity of various forms of GnRH in relation to their resistance to degradation at the pituitary level in the rainbow trout *Oncorhyncus mykiss*. Abstracts 4th International Symposium on Reproductive Physiology in Fish. Norwich 7-12 July U.K.

Yu K.L, Nahorniak C.S, Peter R.E, Corrigan A, Rivier J.E and Vale W.W (1987) Brain distribution of radioimmunoassayable gonadotropin-releasing hormone in female goldfish: seasonal variation and periovulatory changes. Gen. Comp. Endocrinol. 67:234-246

Yu K.L, Sherwood N.M and Peter R.E (1988) Differential distribution of two molecular forms of gonadotropin-releasing hormone in discrete brain areas of goldfish (*Carassius auratus*). Peptides 9:625-630

Yu K.L and Peter R.E (1990a) Alterations in gonadotropin-releasing hormone immunoactivities in discrete brain areas of male goldfish during spawning behavior. Brain Res. 512:89-94

Yu K.L and Peter R.E (1990b) Dopaminergic regulation of brain gonadotropin-releasing hormone in male goldfish during spawning behavior. Neuroendocrinoilogy 52:276-283

Yu K.L, Rosenblum P.M and Peter R.E (1991) *In vitro* release of gonadotropin-releasing hormone from the brain preoptic-anterior region and pituitary of female goldfish. Gen. Comp. Endocrinol. 81:256-267

Zohar Y, Breton B and Fostier A (1986a) Short term profiles of plasma gonadotroppin and estradiol 17beta levels in the female rainbow trout from early recrudescence and throughout vitellogenesis. Gen. Comp. Endocrinol. 64:172-1888

Zohar Y, Breton B and Fostier A (1986b) Short term profiles of gonadotropin and 17 alpha-hydroxy 20 beta dihydroprogesterone levels in the female rainbow trout at the periovulatory period. Gen. Comp. Endocrinol. 64:189-198

Zohar Y, Goren A, Fridkin M, Elhanati E and Koch Y (1990) Degradation of gonadotropin-releasing hormones in the gilthead seabream *Sparus aurata*. II. Cleavage of native salmon GnRH, mammalian LHRH, and their analogs in the pituitary, kidney, and liver. Gen. Comp. Endocrinol. 79:306-319

13

Estrogen Receptor Gene Expression and Regulation in the Liver of Rainbow Trout

G. Flouriot, G. Salbert, F. Le Menn, C. Pelissero and Y. Valotaire

I - INTRODUCTION.

Estrogens, in common with other steroids, regulate gene expression in target cells by specifically binding to their cognate receptor. The hormone-receptor complex is a transcriptional factor which modulates the activity of responsive genes by binding to hormone responsive element of the promoters. The gene and cDNA of these receptors have been cloned and characterized in various species and the biological activities of the receptor domains have been extensively studied in the last few years. For a review, see Nuclear Hormone Receptors (PARKER M.G., 1991).

As regards Fish species, knowledge of estrogen receptor (ER) gene structure and expression is mainly restricted to salmonids. We have previously described the cloning sequencing and *in vitro* expression of a full-length rainbow trout ER cDNA (PAKDEL *et al.*, 1991). This full cDNA randomly labelled was used to study the regulation of expression of hepatic rainbow trout ER mRNA *in vivo* or *in vitro* using trout hepatocyte cell culture.

II - EXPRESION IN THE LIVER AND ITS REGULATION.

1. *Basal expression level and estradiol induction.*

The adult male rainbow trout liver express two ER mRNA species (3.5 and 4.5 kb). In avian liver, only one ER messenger has been shown (KRUST *et al.*, 1986 - MAXWELL *et al.*, 1987). However, in xenopus liver multiple receptor messengers were detected (WEILER *et al.*, 1987). In rainbow trout, the smaller mRNA is the more abundant and represents more than 95 % of the total ER mRNA. The basal concentration in the liver of these animals was estimated to 2 pg ER mRNA/µg total RNA but the fluctuation of this basal level is quite important from one animal to another. According to this number and to the mean value of 5 pg of total RNA hépatocyte, there are 5 copies of the 3.5 kb ER mRNA per cell. This number of copies is quite important when compared to the value given by BARTON and SHAPIRO (1988) in xenopus liver (less than 1 copy/cell). In the immature male or female, there is no significant variation of ER mRNA levels compare to those of adult male. On the other hand, mature females at the end of vitellogenesis exhibit a 5 to 10 fold higher ER mRNA level (> 50 copies/cell) indicating that like vitellogenin mRNA, ER mRNA accumulation should be regulated by estrogen. This hypothesis has been confirmed by estradiol *in vivo* injection into mature males and by direct estradiol stimulation of hepatocytes in primary culture. Figure 1 shows

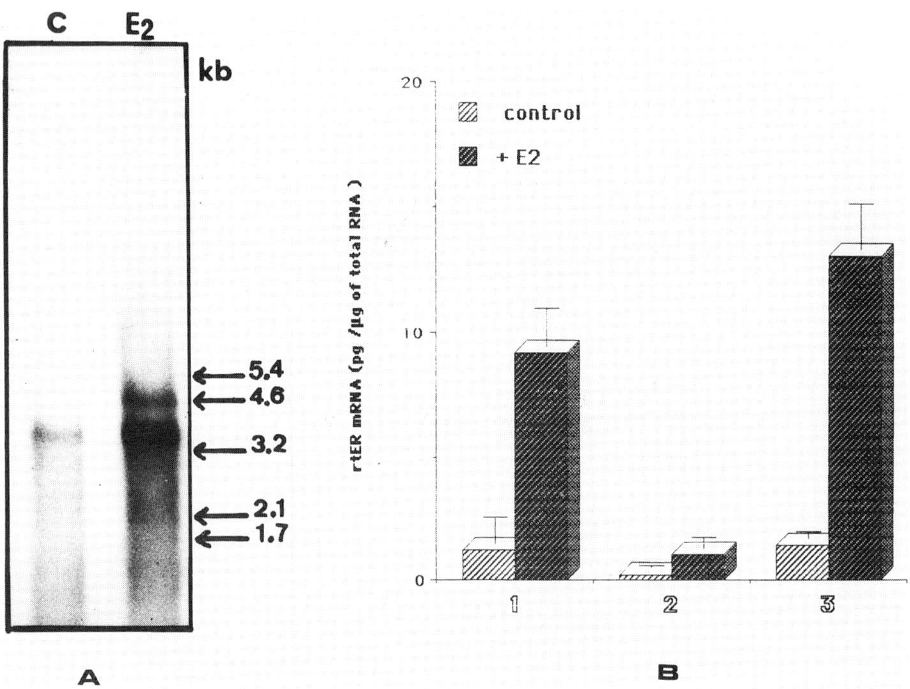

Figure 1 - (A) Total RNA was isolated from 24 h E_2 stimulated or control hepatocytes 1 : *In vivo* 2 : Monolayer culture (MEM medium). 3 : Aggregat culture (DME/F12 medium complemented with Ultroser SF 2 %).
The amounts of rtER mRNA were estimated after slot blot hybridization and autoradiography . Values are given as mean ± standard deviation.

(B) Northern blot analysis of poly A^+ RNA from livers.
(C) Control and (E_2) stimulated animals.

that 24 h after E2 injection, the level of ER mRNA is increased by a factor of 5. The maximum level is obtained 1 to 2 days post-injection and at this time the concentration of ER mRNA in the hepatocyte is very similar to that of the female at the end of vitellogenesis. Figure 1 also shows that the level of the two ER mRNA species are together increased by estradiol treatment. We already showed that Vg mRNA was inducible by estrogens in monolayer-cultured rainbow trout hepatocytes (VAILLANT *et al.*, 1988) ; however the level of induction was quite low. Recently, the response of hepatocytes to estrogens was greatly improved by using aggregate-cultured cells in DME/F12 medium complemented with Ultroser SF. As shows in figure 1, the induction level obtained in this system reaches the observed *in vivo* value. After E2 addition in the culture medium of this primary cultured hepatocytes, the accumulation rate of ER mRNA increases very quickly up until 2 hours then decreases (Fig. 2B). The accumulation of the messenger reaches a plateau at 24 hours (fig. 2A). This result shows that estradiol increases the transcription rate or/and the half-life of the ER mRNA.

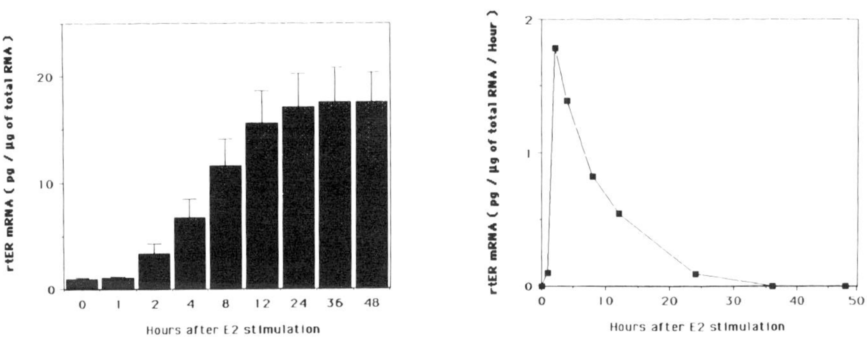

A B

Figure 2 - (A) rtER mRNA accumulation after E2 stimulation. Male rainbow trout hepatocytes cultured in aggregats (2.0 x 10^6 cells / plate) were stimulated by E2 10^{-6}M after one week of preculture. At the indicated time, the cells were collected and total RNA prepared. Slot blot hybridizations were performed with 5 µg of total RNA using the rtER probe. The accumulation was calculated after densitometry of the autoradiogram.
(B) Accumulation rate was calculated from the results of fig. 2A.

The first hypothesis, which is in concordance with the presence of a potential ERE in the ER promoter, needs to be confirmed by run on experiments. After *in vivo* induction with E2 in male Fish, ER mRNA and nuclear receptor levels decrease and return to the basal value 2 or 3 weeks after the treatment (PAKDEL *et al.*, 1991). This decrease follows the disappearance of E_2 in the plasma of the animals, so that E_2 level in the hepatocytes become too low to saturate the receptors. So, it is possible that the free receptors are not able to bind the ERE and that the transcription is no longer activated. However, negative feedback, by binding and competition of free receptors to ERE may also be involved as it has been shown that free receptors can bind ERE without gene transcription activation (KUMAR *et al.*, 1987). Nevertheless, our results are different from those described in xenopus (BARTON and SHAPIRO, 1988). In that species, it has been shown that ER mRNA induction in the liver is permanent and stable for at least 4 months after E_2 stimulation, the xER gene becomes more sensitive to the hormone and the low level of endogenous estrogen is sufficient to maintain activation of the gene. In rainbow trout, the difference in the results could be explained by the lower affinity of the receptor for estrogen (2-5 nM) or by a different mechanism of gene activation during the first stimulation.

These last few years, the role of estradiol in the regulation of ER mRNA levels has been studied in various tissues. Up regulations have been reported for xenopus liver and oviduct or rat liver and pituitary (BARTON and SHAPIRO, 1988 - VARRIALE and TATA, 1990 - SHUPNIC *et al.*, 1989) and down regulations for pig and rat uterus, rat hypothalamus and MCF7 cells (SAHLIN *et al.*, 1990 - SHUPNIC *et al.*, 1989 - SIMERLY and YOUNG, 1991 - CHO *et al.*, 1991).

If these regulations are generally explained by transcription modifications there is however no report of a direct involvement of some transcriptional factors in the mechanism. Recently, using co-injection of rtER promoter construct and rtER cDNA in xenopus oocytes, we have detected and increase of CAT activity in the presence of estradiol (N. THEZE, personal communication). This preliminary result which has to be confirmed would definitively demonstrate that the estrogen receptor gene is autoregulated.

2. *Other regulations of expression.*

Regulation of ER mRNA by other factors than estradiol has been reported only for cancer cell lines : progestin in T47D breast cancer cell line (ALEXANDER *et al.*, 1990), phorbol ester and calcium in MCF7 cells (REE *et al.*, 1991). Using the aggregate-cultured hepatocytes, we have recently tested the influence of various endogenous and exogenous factors on rtER gene expression. We have found that a wide variety of natural or synthetic compounds are more or less able to increase rtER mRNA levels. This is the case for all natural and synthetic estrogens (estriol, estrone, diethylstilbestrol (DES)) and for phytoestrogens (Equol, Biochanin A, Genistein, Coumestrol) which have been shown to stimulate vitellogenesis in sturgeon (PELISSERO *et al.* 1991). rtER mRNA levels can also been increased by a wide variety of herbicides and pesticides (polychlorinated hydroxybiphenyls (PCB), lindan and chlordecone (Fig. 3), most ot these compounds have been previously shown to be competitors of estradiol on the ER of various target tissues (KORACH *et al.*, 1987 - WILLIAM *et al.*, 1988). All these molecules are also able to induce Vg mRNA in the hepatocyte cell culture, indicating that an ER dependent mechanism is involved in the regulation of mRNA levels by these pesticides.

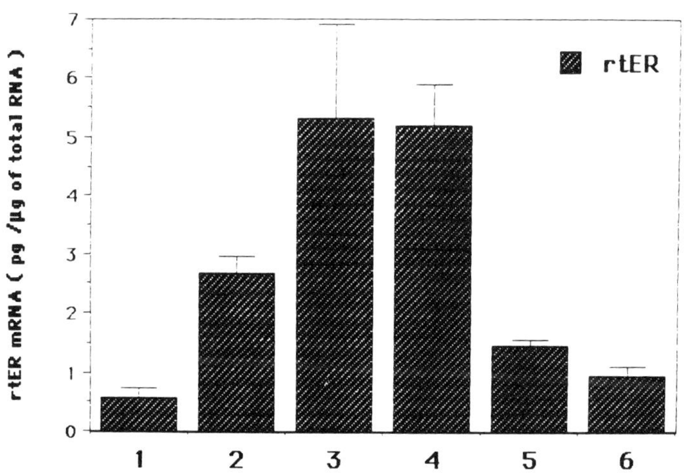

Figure 3 - Xenobiotic induction of rtER mRNA.
The xenobiotics were added in the culture medium. After 24 hours the cells were collected and total RNA prepared. The amount of rtER mRNA was estimated by slot blot hybridization.
1 : Control. 2 : E_2 10^{-8}M. 3 : E_2 10^{-5}M. 4 : chloredecone 10^{-5}M.
5 : Lindane 10^{-5}M. 6 : PCB 10^{-5}M.

We also found that other steroids (testosterone, 17α,20β-diOH progesterone, androsterone) and pituitary peptide hormones (prolactine and growth hormone) have no effect, but cortisol was shown to down regulate rtER mRNA. Negative effects of cortisol on vitellogenesis have been

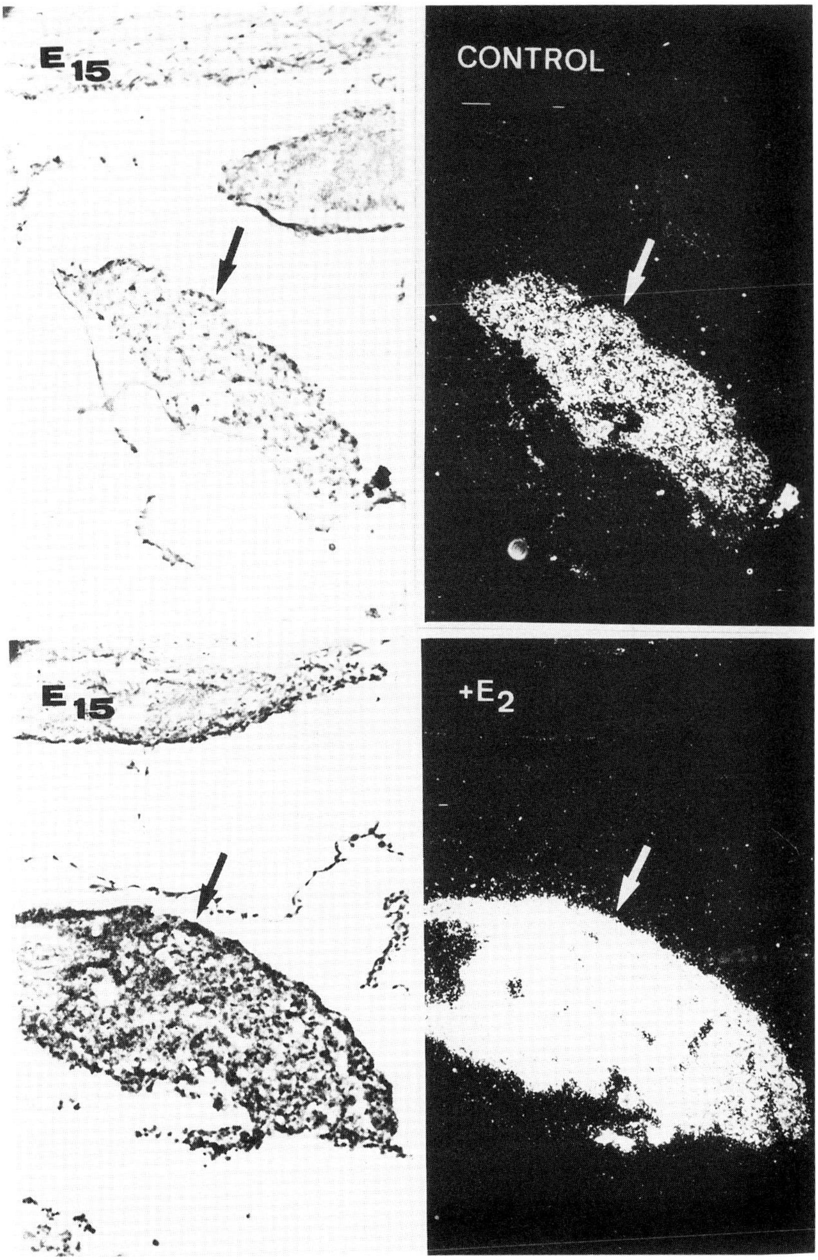

Figure 4 - 15 days old (E15) trout embryos were treated by estradiol (+E2) or not (control) and analysed by *in situ* hybridization. The signal was only detectable in the liver (arrows) and was very high after estradiol stimulation.

previously reported in rainbow trout (CARRAGHER *et al.*, 1989) and as high levels of cortisol have been detected before ovulation (JALABERT, 1976), it is possible that glucocorticoids play an important role in stopping vitellogenesis at the ovulation time.

3. *Liver ontogenesis of rtER mRNA.*

Slot and Northern hybridizations of hepatic RNA show that detectable levels of rtER mRNA are present in the fry as soon as they hatch. There is no quantitative or qualitative difference with adult Fish, as the levels and the ratio between the 4.5 and 3.5 kb mRNA are not significantly different. Furthermore, rtER and Vg mRNA are already E_2 inducible. In the embryos, *in situ* hybridization shows that rtER mRNA are also present and inducible by estrogen 15 days after fecondation (Fig. 4) but we do not know yet if Vg mRNA is also inducible. These results show that liver sensitivity to estrogen appears early during embryogenesis. In xenopus laevis, the presence of ER have also been shown very early in the todd liver (stage 54) but estrogen inducibility of ER and Vg are only observed after metamorphosis (stage 62). In the chicken, estrogen inducibility of Vg and apolipoprotein is observed during embryogenesis (ELBRECHT *et al.*, 1984 - EVANS *et al.*, 1987).

4. *ER mRNA in the liver of other Fishes.*

Using rtER cDNA probe and northern blot analysis in stringent conditions (THOMAS, 1980), we have investigated the possibility of cross hybridization with ER mRNA in the liver of various Fish species after E_2 stimulation. The 3.5 kb mRNA is very well conserved in all salmonids tested (pacific and atlantic salmons, brown and brook trouts). For other species, the size is more variable, (Fig. 5) for instance, corregone shows a 5 kb mRNA and sole a 3 kb one. In addition, there is no direct relationship between Fish classification and intensity of cross hybridization as sole which is quite far from salmonids give a very strong signal and pike which is quite near gave no detectable one.

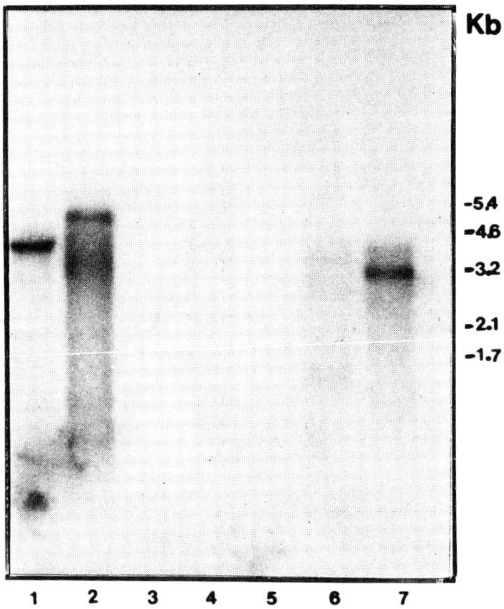

Figure 5 - Northern blot analysis of E_2 stimulated hepatic RNA from different fish species. The fishes were injected by E_2 (500 µg / kg BW). 48 hours after injection, the animals were killed and the livers removed. Poly A^+ RNA was prepared from 500 µg total RNA separated on a denaturing formaldehyde agarose gel, transferred to a nylon membrane, hybridized in Thomas's conditions with rtER probe and autoradiographed.
1 : rainbow trout (5 µg total RNA). 2 : corregone. 3 : pike. 4 : eel. 5 : carp. 6 : lamprey. 7 : sole (5 µg Poly A^+ RNA).

References.

Barton M.C. and Shapiro D.J. (1988) Transient administration of estradiol-17β establishes and autoregulatory loop permanently inducing estrogen receptor mRNA. Proc. Natl. Acad. Sci. U.S.A. **85**:7119-7123.

Carragher J.F., Sumpter J.P., Pottinger T.G. and Pickering A.D. (1989) The deleterious effects of cortisol implantation on reproductive function in two species of trout *Salmo trutta* and *Salmo gairdneri* Richardson. Gen. Comp.. Endocrinol. **76**:310-321.

Cho H., Ng P.A. and Katzenellenbogen B.S. (1991) Differential regulation of gene expression by estrogen in estrogen growth independant and dependent MCF.7 human cancer cell sublines. Mol. Endocrinol. **5**:1323-1330.

Elbrecht A., Lazier C.B., Protter A. and Williams D.L. (1984) Independant developmental programs for two estrogen regulated genes. Science **225**: 639-641.

Evans R.M., O'Malley P.J., Krust A. and Burch J.B.E. (1987) Developmental regulation of the estrogen receptor and the estrogen responsiveness of five yolk protein genes in avian liver. Proc. Natl. Acad. Sci. U.S.A. **84**: 8493-8497.

Jalabert B. (1976) *In vitro* oocyte maturation and ovulation in Rainbow trout (*Salmo gairdneri*), Northern pike (*Essox Lucius*) and Gold fish (*Carassius auratus*). J. Fish Res. Board. Can. **33**:974-988.

Korach K.S., Sarver P., Chae K., McLachlan J.A. and McKinney J.D. (1987) Estrogen receptor-binding activity of polychorinated hydroxybiphenyls : conformational restricted structural probes. Mol. Pharmacol. **33**:120-126.

Krust A., Green S., Argos P., Kumar V., Walter P., Bornert J.M. and Chambon P. (1986) The chicken estrogen receptor sequence homology with v-erbA and the human estrogen and glucocorticoid receptors. E.M.B.O. J. **5**:891-897

Kumar V., Green S., Stack G., Berry M., Ruin Jin J. and Chambon P. (1987) Functional domains of the human estrogen receptors. Cell **51**:941-951.

Maxwell B.L., McDonnell D.P., Conneely O.M., Schultz T.Z., Greene G.L. and O'Malley B.W. (1987) Structural organization and regulation of the chicken estrogen receptor. Mol. Endocrinol. **1**:25-35.

Pakdel F., Féon S., Le Gac F., Le Menn F. and Valotaire Y. (1991) *In vivo* estrogen induction of hepatic estrogen receptor mRNA and correlation with vitellogenin mRNA in rainbow trout. Mol. Cell. Endocrinol. **75**: 205-212.

Parker M.G. (1991) In : Nuclear Hormone Receptors. Molecular mechanisms cellular functions, clinical abnormalities. M.G. Parker, editor.

Pelissero C., Bennetau B., Babin P., Le Menn F. and Dunogues J. (1991) The estrogenic activity of certain phytoestrogens in the siberian sturgeon (*Acipenser Baeri*) . J. Steroid Biochem. **38**:293-299.

Ree A.H., Landmark B.F., Walaas S.I., Lahooti H., Eikvar L., Eskild W. and Hansson V. (1991) Down regulation of messenger ribonucleic acid and protein levels for estrogen receptors by phorbol ester and calcium in MCF-7 cells. Endocrinology **124**:2577-2583.

Sahlin L., Rodriguez-Martinez H., Stanchev P., Dalin A.M., Norstedt G. and Eriksson H. (1990) Regulation of the uterine expression of messenger RNAs encoding the estrogen receptor and IGF-I peptides in pig uterus. J. Vet. Med. Ser. **37**:795--800.

Shupnik M.A., Gharib S.D. and Chin W.W. (1989) Tissue specific regulation of rat estrogen receptor mRNAs. Mol. Endocrinol. **3**:660-665.

Simerly R.B. and Young B.J. (1991) Regulation of estrogen receptor messenger ribonucleic acid in rat hypothalamus by sex steroid hormones. Mol. Endocrinol. **5**:424-432.

Thomas P.S.. (1980) Hybridization of denatured RNA and small DNA fragments transferred to nitrocellulose. Proc. Natl. Acad. Sci. U.S.A **77**: 5201- 5205.

Vaillant C., Le Guellec C., Pakdel F. and Valotaire Y. (1988) Vitellogenin gene expression in primary culture of male rainbow trout hepatocytes. Gen. Com. Endocrinol. **70**284-290.

Varriale B. and Tata J.R. (1990) Auto-induction of estrogen receptor is associated with FOSP1 messenger RNA induction by estrogen in primary culture of Xenopus oviduct cells. Mol. Cell. Endocrinol. **71**:25-32.

Weiler I.J., Lew D. and Shapiro D.J. (1987) The Xenopus laevis estrogen receptor : sequence homology with human and avian receptors and identification of multiple estrogen receptor messenger ribonucleic acids. Mol. Endocrinol. **1**:355-361.

William J., Eckols K. and Uphoose L. (1988) Estradiol and chlordecone interactions with the estradiol receptors. Toxicol. and Applied Pharmacol. **98**:413-422.

14

Hormonal Pheromones: Recent Developments and Potential Applications in Aquaculture

N.E. Stacey, P.W. Sorensen, and J.R. Cardwell

Introduction

Although it has been known for many years that pheromones play fundamental roles in reproductive physiology and behavior of teleost fish (Liley 1982; Stacey *et al.* 1986), the first clear evidence that some sex pheromones might be released hormones and/or their metabolites is less than a decade old (Colombo *et al.* 1982). More recent studies have shown that hormonal pheromones are widely used by teleosts, and have diverse chemical structures and functions. These exciting findings have greatly increased our understanding of pheromone function and stimulated speculation on pheromone evolution (Sorensen 1991; Sorensen and Stacey 1990; Stacey and Sorensen 1991). Here, we briefly summarize (referring to reviews where possible) some of the latest developments in hormonal pheromone research, and consider several issues related to the application of hormonal pheromones to fish culture. Throughout, we use the term *pheromone* to refer to chemical(s) released by an individual that elicit reproductive response(s) in conspecifics, recognizing that the term need not imply communication and can be used when referring to component(s) of a complex odor mixture (see Stacey and Sorensen 1991).

---

[1]Department of Fisheries and Wildlife, University of Minnesota
St. Paul, Minnesota U.S.A. 55108

The evidence for hormonal pheromones in fish

The best characterized hormonal pheromones are those of the goldfish (*Carassius auratus*), for which there is information on production, release, and response mechanisms. Initial studies indicated females sequentially release two distinct pheromones during the periovulatory period: a preovulatory *primer* (inducing only a male endocrine response) and a postovulatory *releaser* (eliciting male behavior) (Sorensen and Stacey 1990).

The preovulatory primer is steroidal, the principal component believed to be 17α,20ß-dihydroxy-4-pregnen-3-one (17,20ß-P), the proposed oocyte maturation-inducing steroid of goldfish and some other teleost species. Pheromone production is initiated by the preovulatory gonadotropin (GtH) surge, which stimulates surges in both plasma and released 17,20ß-P that terminate abruptly at ovulation (Stacey *et al.* 1989).

Males exposed to 17,20ß-P increase milt (sperm and seminal fluid) volume within 6 h (Dulka *et al* 1987). This testis response is initiated by stimulation of specific and sensitive (pM threshold) olfactory receptors as shown by electro-olfactogram (EOG) recording from the surface of the olfactory epithelium (Sorensen *et al* 1990). Afferent pheromonal information is carried by mitral cells of the olfactory bulb which receive input from primary olfactory receptors and project centrally via the medial olfactory tracts (Dulka and Stacey 1990; Sorensen *et al* 1991a); the terminal nerve is not influenced by hormonal pheromones (Fujita *et al* 1991). Olfactory input increases blood GtH within 15 min through a mechanism involving removal of tonic dopaminergic inhibition of gonadotropin releasing hormone (GnRH)-stimulated GtH release (Dulka *et al* 1991). Elevated GtH then stimulates testicular 17,20ß-P synthesis (Dulka *et al.* 1987) which increases milt volume through mechanisms which are poorly understood.

Although a pheromonal role for 17,20ß-P appears to be accepted, important questions remain. In terms of function, there is no information as to the degree individual reproductive success is influenced by releasing or responding to 17,20ß-P. We proposed (Stacey and Sorensen 1990) that 17,20ß-P functions simply in *chemical spying* by the male, not in *chemical communication* by the female, because there is no evidence for signal specialization and no indication that it would be likely to evolve through selection based on variation in 17,20ß-P release. The goldfish mating system involves sperm competition and a lack of mate fidelity; females have little to gain by advertising their impending ovulation, but males have much to gain if they use female 17,20ß-P as a cue to increase their fertility in anticipation of ovulation. Physiologically, the route of 17,20ß-P release is not known, although high urinary levels of 17,20ß-P metabolites are found in other species (Scott and Canario 1991). More importantly, the likelihood that 17,20ß-P is only one component of a complex preovulatory primer pheromone is suggested by two recent findings.

First, during the preovulatory surge of 17,20ß-P release, females also release a 17,20ß-P precursor, 17α-hydroxyprogesterone (17-P), and two metabolites, 17,20ß-P-glucuronide (17,20ß-P-G), and 17,20ß-P-sulphate (17,20ß-P-S); all are olfactory stimulants which increase GtH and milt, although 17,20ß-P is the most potent. 17-P and 17,20ß-P-G appear to be detected by the olfactory receptor mechanism that detects 17,20ß-P, and thus may simply amplify the 17,20ß-P signal. However, 17,20ß-P-S, which is detected by a different olfactory receptor mechanism, could be a supplemental cue (Sorensen et al 1990a, 1991b,c and unpub).

Second, the goldfish olfactory system detects C_{19} steroids, of which the most potent tested is androstenedione (AD) (Sorensen *et al* 1990a,b; 1991b,c). AD-like steroids (for convenience termed *androgens*, although 11-ketotestosterone is not detected) appear to inhibit 17,20ß-P action; males exposed to 17,20ß-P have higher GtH and milt levels than males exposed to 17,20ß-P and androgen (Stacey 1991, unpublished). The inhibitory mechanism is not known. Excess AD or T displaces 17,20ß-P from binding sites on olfactory membrane (Rosenblum *et al* 1991); however, cross-adaptation (Sorensen *et al* 1990a) shows that 17,20ß-P and androgens are detected by separate olfactory receptor mechanisms. The function of pheromonal androgen likely will not be clarified until it is known which specific compounds are released, by which sex, and at what reproductive stage. One possibility is that *17,20β-P pheromone* is a mixture whose activity depends on the relative amounts of stimulatory progestins and inhibitory androgens, and that only ovulatory females have a progestin/androgen ratio capable of stimulating GtH and milt increase.

At ovulation, the surge of 17,20ß-P synthesis and release collapses and levels of immunoreactive prostaglandin F2α (PGF) increase in the blood. PGF appears to have two important functions in spawning. Hormonally, PGF acts within the brain to induce female spawning as long as ovulated eggs are in the tract (Stacey 1987). Pheromonally, PGF undergoes unknown metabolic steps and then is released as a potent trigger for male courtship (Sorensen *et al* 1988). EOG studies (Sorensen *et al* 1988) indicate the two known PGs most stimulatory for the goldfish olfactory system (PGF and a mammalian metabolite, 15-keto-PGF) act via separate olfactory receptors. However, recent HPLC analyses of holding water shows that females injected with radiolabelled PGF release several unknown PGF metabolites, but little if any PGF or 15-keto-PGF (Sorensen *et al* 1991c and unpub).

In addition to these modifications of our initial model of goldfish pheromones, we now believe the roles of steroid pheromones extend beyond male responses to female odors: e.g. ovulation rate increases in response to water-borne 17,20ß-P (Sorensen and Stacey 1987), and male interactions that increase or decrease milt volume are consistent with stimulatory and inhibitory actions of progestins and androgens (Stacey 1991). A final caveat is that *primer* and *releaser* are relative terms: the primer effect of 17,20ß-P involves subtle behavioral change, and the releaser effect of PGF can involve GtH increase (Sorensen *et al* 1989).

Goldfish studies show that fully characterizing the nature and functions of a species' hormonal pheromones will be a complex and multidisciplinary undertaking. Such effort is warranted, because hormonal pheromones regulate important processes (ovulation, sperm production, intersexual attraction) and are potent-

Table 1. EOG studies using putative hormonal pheromones.

Family *species*	most sensitive to[a] C_{21}	C_{19}	PGs	Ref[b]
Cyprinidae				
Carassius auratus	17,20ß-P	AD	15-K-PGF	1
Cyprinus carpio	17,20ß-P	AD	15-K-PGF	2
Carassius carassius	(17,20ß-P)	X	X	3
Nocomis biguttatus	17,20ß-P-G	0	PGF=15-K-PGF	2
Ctenopharyngodon idellus	17,20B-P-G	0	PGF	4
Phoxinus eos	17,20ß-P-G	0	PGF	4
Semotilus margarita	17,20ß-P-G	0	15-K-PGF	4
Notropis hudsonius	17,20ß-P-G	0	15-K-PGF	4
Puntius gonionotus	17,21-P	(T)	15-K-PGF	5
P. schwanenfeldi	17,21-P	(T)	15-K-PGF	5
P. altus	17,21-P	(T)	15-K-PGF	5
Brachydanio rerio	0	0	PGF	4
Catostomidae				
Ictiobus bubalus	0	0	PGF=15-K-PGF	2
Moxostoma macrolepidotum	0	0	PGF	2,4
Catostomus catostomus	0	0	15-K-PGF	4
C. commersoni	0	0	15-K-PGF	4
Clariidae				
Clarias gariepinus	P-G	E-G	X	6
Cobitidae				
Misgurnus anguillicaudatus	X	X	13-K-PGF = 15-K-PGF	7
Salmonidae				
Salmo alpinus	X	X	PGF	8
Salmo salar	X	(T)	X	9

[a] C_{21} and C_{19} steroids and F-prostaglandins which had the lowest detection thresholds as measured by electro-olfactogram (EOG) recording; (T) = only compound tested or reported; 0 = not detected; X = not tested; 17,20ß-P = 17α,20ß-dihydroxy-4-pregnen-3-one; 17,20ß-P-G = 17,20ß-P-glucuronide; 17,21-P = 17α,21-dihydroxy-4-pregnen-3-one; P-G = 5ß-pregnane-3α,17α-diol-20-one-3α-glucuronide; T = testosterone; E-G = etiocholanolone glucuronide; PGF = prostaglandin $F_{2\alpha}$; 15-K-PGF = 15-keto-PGF; 13-K-PGF = 13,14-dihydro-15-K-PGF.

[b] 1. Sorensen *et al* 1988, 1990a,b; 2. Sorensen et al1991c; 3. Bjerselius and Olsen 1991; 4. JR Cardwell and JG Dulka (unpub); 5. JR Cardwell (unpub); 6. Resink *et al* 1989; 7. Kitamura and Ogata 1990; 8. Sveinsson and Hara 1990; 9. Moore 1991, Moore and Scott 1991.

ially ubiquitous. Hormonal pheromones have been proposed in six families (Table 1): Cyprinidae (*Brachydanio rerio*: Van Den Hurk and Lambert 1982, Van Den Hurk *et al* 1987); Gobiidae (*Gobius jozo*: Colombo *et al* 1982); Clariidae; Cottidae (*Cottocomephorus grewingki*: Dmitrieva *et al* 1988); Cobitidae; Salmonidae. Recent work (Sorensen *et al* 1991a; Cardwell and Dulka unpub) using EOG to screen hormones and metabolites for olfactory activity suggests hormonal pheromones are widespread among Cyprinids and Catostomids. Also, extreme EOG sensitivity (fM threshold) of rainbow trout (*Oncorhynchus mykis*) to GnRH raises the possibility of a salmonid peptide pheromone (Anderson and Doving 1991); GnRH does not induce EOG responses in goldfish (Sorensen unpub). We expect continued rapid increase in understanding of fish hormonal pheromones because of the ease with which they can be obtained, administered, and measured.

Applying hormonal pheromones to fish culture

Responses of cultured species to water-borne hormones or their metabolites (Stacey et al 1991; Table 1) demonstrate potential aquacultural applications. But developing this potential in any specific instance will depend on such factors as whether regulations prohibit use of the hormone, whether the reproductive process affected by the pheromone is an economic problem, and whether this process can be regulated more cheaply or effectively by alternative techniques. Here, we focus on the biological issues influencing application of hormonal pheromones.

Hormonal pheromones have several advantages over other chemical methods for controlling reproduction. Being externally active, they can be used without disturbing the fish, thus eliminating some of the labor cost and stress associated with hormone/drug injections (Peter *et al* 1988). In monoculture, hormonal pheromones could selectively stimulate only mature individuals or only one sex; in polyculture, a single species could be regulat-

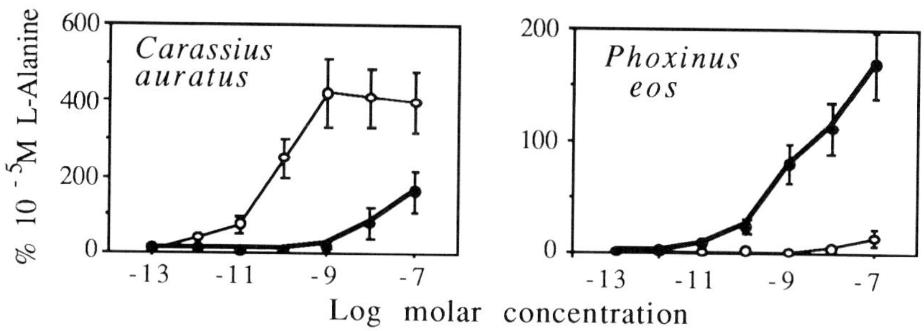

Figure 1. Electro-olfactogram responses to free (open circles) and glucuronated (closed circles) 17,20ß-P. Such olfactory differences between species could enable selective species regulation in polyculture.

ed in cases where the pheromone is species-specific (Sorensen *et al* 1991c) (Fig 1).

Hormonal pheromone application to fish culture is likely to proceed in one of two ways. In the simpler, a known effect could lead immediately to practical application. We will consider the more difficult case, in which a culture problem in a species for which no pheromonal information is available requires first that a pheromone (if one exists) be identified and second that techniques for pheromone delivery be developed.

We believe that a species' reproductive strategy is the primary determinant in the evolution of its hormonal pheromones, and therefore should be an important concern in any research program of pheromonal identification. Knowing the temporal pattern of association among and between the sexes at the time of interest (e.g. ovulation), and how blood hormones of potential pheromone donors change at that time, will provide a solid basis for attempting identification. Unlike their hormonal precursors, which perform a similar set of homologous functions in virtually all teleosts, hormonal pheromones are expected to have evolved only in those specific situations where an individual can benefit by detecting a hormone or metabolite released by conspecifics

(Stacey and Sorensen 1991). Thus, we do not expect that the attractant released by territorial male gobies (Colombo *et al* 1982) and sculpins (Dmitrieva *et al* 1988) occurs in cyprinids which spawn in mixed-sex groups, or that the preovulatory primer released by female goldfish occurs in species where the female is far from the male during oocyte maturation. On the other hand, we expect PGF-based pheromone systems to be widespread, because increased synthesis and release of PGF compounds is expected in ovulated females of many species (Stacey 1987; Sorensen et al 1988), and because (at least for oviparous species) recognizing ovulated females should benefit a male regardless of his mating system.

Presently, the simplest approach to begin characterizing a hormonal pheromone is to use EOG recording to screen hormones and metabolites. EOG recording is inexpensive, sensitive, specific, faster than using a combination of odor fractionation and bioassay (e.g. Sherwood *et al* 1991),.and particularly useful if categories of stimulants are known for related species (e.g. Table 1). However, both negative and positive data must be interpreted cautiously. Thus, failure to detect EOG responses to a variety of C_{21} steroids in Catostomids and *B. rerio* could indicate absence of a 17,20ß-P-like pheromone in these species, or a failure to test the correct compound. This is of particular concern because of our poor understanding of hormone metabolism in fish, and because few metabolites are commercially available for testing. The PGF pheromone appears to be a case in point: goldfish release several unknown PG compounds, but not PGF or 15-keto-PGF, to which they are mosty sensitive (Sorensen *et al* 1991b,c). Although studies of PG metabolism in fish are lacking; studies in mammals indicate species-specific metabolism. If fish are similar, the apparent conservatism of the PGF pheromone (Table 1) may simply result from not having the correct PG compounds for EOG testing. Future research will be obliged to combine EOG recording with studies of hormone metabolism.

Having identified a hormonal pheromone with aquaculture potential, a primary concern will be developing an effective deliv-

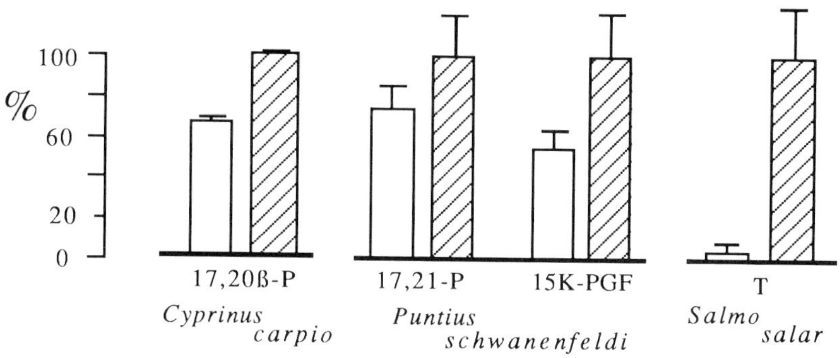

Figure 2. Influence of maturity on electro-olfactogram response (given as a percentage of response of mature fish to 10^{-8}M steroids and 10^{-9}M prostaglandin). *Cyprinus* (Irvine and Sorensen 1991): mature males are more responsive than juveniles; *Puntius* (Cardwell, unpub): spermiated males with tubercles are more responsive than sperm-iated males without tubercles; *Salmo* (Moore and Scott 1991): precocious male parr are most responsive at time of spawning.

ery technique. Success here will require understanding how temporal pattern of exposure affects response, a relationship which likely differs among species. In precocious male Atlantic salmon parr (Moore 1991), a brief pulse of T is sufficient to trigger positive rheotaxis, a response suggestive of upstream movement and of a method for selective removal of precocious males from culture pens. However, in cases where the *active space* (zone of detection) is small (as proposed for goldfish 17,20ß-P; Sorensen and Stacey 1990), the pheromone receiver would encounter repetitive odor pulses, the pattern of which could affect the response. Interactions involving active signalling (communication) (e.g. male gobies) would be the most likely to require a specific exposure pattern, and this might be achieved by pulsatile pheromone release in urine (Scott and Canario 1991).

Successful application of hormonal pheromones also will require information on the degree to which pheromone responsiveness is linked to reproductive state; both peripheral and central mechanisms appear to be involved. EOG studies show that change in olfactory responsiveness is rapid and dramatic at precocious maturity in male Atlantic salmon parr (Moore 1991), but protracted and relatively minor in *Cyprinus carpio* (Irvine and Sorensen 1991) (Fig. 2). In goldfish, EOG response to 17,20ß-P or PGF is not greatly influenced by gonadal maturity (Sorensen *et al* 1987 and unpub), indicating central mechanisms block reproductive responses (GtH increase, male courtship) of gonadally regressed individuals. Similarly, central mechanisms likely account for the common finding (Liley 1982; Stacey et al 1986) that females of many species respond to male odors only if ovulated. Other problems that can be expected with aquaculture application are that pheromonal responses in some species may depend on other sensory cues (visual, auditory) or on complex odor mixtures, both of which could be difficult to duplicate; here, the reflexive responses of goldfish and Atlantic salmon to single compounds may be atypical.

Despite these *potential* problems, the facts that hormonal pheromones are widespread, and known to regulate commercially important reproductive functions, indicate they inevitably will find applications to fish culture.

References

Anderson O, Doving KB (1991) Gonadotropin releasing hormone (GnRH) - a novel olfactory stimulant in fish. Neuroreport 2: 458-460

Bjerselius R, Olsen KH (1991) Olfactory sensitivity to sexual pheromones in crucian carp (Carassius auratus) In: Proc 4th Intl Symp Reprod Physiol Fish. Scott AP, Sumpter JP, Kime D (eds) (in press)

Colombo L, Belvedere PC, Marconato A, Bentivegna F (1982) Endocrinology of teleost reproduction. *In*: Proc 2 nd Intl Symp Reprod Physiol Fish. Richter CJJ, Goos HJTh (eds) Pudoc Netherlands pp 84-94

Dmitrieva, TM, Katsel PL, Baleyev RB, Ostroumov VA, Kozlov UP (1988) The isolation of the sexual pheromone of the male yellowfin Baikal sculpin (*Cottocomephorus grewingki*). Biol Nauki 39-44

Dulka JG, Stacey NE, Sorensen PW, Van Der Kraak GJ (1987a) A sex steroid pheromone synchronizes male-female spawning readiness in the goldfish. Nature 325:251-253

Dulka JG, Stacey NE (1990) Effects of olfactory tract lesions on gonadotropin and milt responses to the female sex pheromone, 17α,20ß-dihydroxy-4-pregnen-3-one. J Exp Zool 257:223-229

Dulka JG, Sloley BD, Stacey NE, Peter RE (1991) A reduction in dopamine turnover in the pituitary is associated with sex pheromone-induced gonadotropin secretion in male goldfish. *In*: Proc 4 th Intl Symp Reprod Physiol Fish. Scott AP, Sumpter JP, Kime D (eds) (in press)

Fujita I, Sorensen PW, Stacey NE, Hara TJ (1991) The olfactory system, not the terminal nerve, functions as the primary chemosensory pathway mediating responses to sex pheromones in goldfish. Brain Behav Evol (in press)

Irvine IAS, Sorensen PW (1991) Seasonal olfactory sensitivity of wild Mississippi river carp to sex steroids with putative pheromonal actions. *In*: Proc 4 th Intl Symp Reprod Physiol Fish. Scott AP, Sumpter JP, Kime D (eds) (in press)

Kitamura S, Ogata H (1990) Olfactory responses of male loach, *Misgurnus anguillicaudatus*, to F-type prostaglandins. *In*: Proc 24 th Jap Symp Taste and Smell pp 163-166

Liley NR (1982) Chemical communication in fish. Can J Fish Aquat Sci 39:22-35

Moore A (1991) Behavioral and physiological responses of precocious male Atlantic salmon (*Salmo salar* L.) parr to testosterone. *In*: Proc 4 th Intl Symp Reprod Physiol Fish. Scott AP, Sumpter JP, Kime D (eds) (in press)

Moore A, Scott AP (1991) Testosterone is a potent odorant in precocious male Atlantic salmon (*Salmo salar* L.). Phil Trans Roy Soc Lond, Ser B 322:241-244

Peter RE, Lin HR, Van Der Kraak GJ (1988) Induced ovulation and spawning of cultured freshwater fish in China: advances in application of GnRH analogues and dopamine antagonists. Aquaculture 74:1-10

Resink JW, Voorthuis PK, Van Den Hurk R, Peters RC, Van Oordt PGWJ (1989) Steroid glucuronides of the seminal vesicle as olfactory stimuli in African catfish, *Clarias gariepinus*. Aquaculture 83:153-166

Rosenblum PM, Sorensen PW, Stacey NE, Peter RE (1991) Binding of the steroidal pheromone 17α,20ß-dihydroxy-4-pregnen-3-one to goldfish, *Carassius auratus*, olfactory epithelium membrane preparations. Chem Senses 16:143-154

Scott AP, Canario AVM (1991) 17α,20ß-dihydroxy-4-pregnen-3-one-20-sulphate: a major new metabolite of the teleost maturation-inducing steroid. Gen Comp Endocrinol (in press)

Sherwood NM, Kyle AL, Kreiberg H, Warby CM, Magnus TH, Carolsfeld J, Price WS (1991) Partial characterization of a spawning pheromone in the herring *Clupea harengus pallasi*. Can J Zool 69:91-103

Sorensen PW (1991)Hormones, pheromones, and chemoreception. *In*: Fish Chemoreception. Hara TJ (ed) Chapman and Hall London (in press)

Sorensen PW, Stacey NE (1987) 17α,20ß-dihydroxy-4-pregnen-3-one functions as a bisexual priming pheromone in goldfish. Amer Zool 27: abst 412

Sorensen PW, Hara TJ, Stacey NE, Goetz FW (1988) F prostaglandins function as potent olfactory stimulants that comprise the postovulatory female sex pheromone in goldfish. Biol Reprod 39:1039-1050

Sorensen PW, Chamberlain KJ, Stacey NE (1989) Differing behavioral and endocrinological effects of two female sex pheromones on male goldfish. Horm Behav 23:317-322

Sorensen PW, Stacey NE (1990) Identified hormonal pheromones in the goldfish: the basis for a model of sex pheromone function in teleost fish. *In*: Chemical Signals in Vertebrates V. MacDonald D, Muller-Schwarze D, Natynczuk SE (eds) Oxford University Press pp 302-311

Sorensen PW, Hara TJ, Stacey NE, Dulka JG (1990a) Extreme olfactory specificity of male goldfish to the preovulatory steroidal pheromone 17α,20ß-dihydroxy-4-pregnen-3-one. J Comp Physiol A 166:373-383

Sorensen PW, Stacey NE, Hara TJ (1990b) Acute olfactory sensitivity and specificity of mature male goldfish to water-borne androgenic steroids: a class of inhibitory pheromones? Chem Senses 15:641

Sorensen PW, Hara TJ, Stacey NE (1991a) Sex pheromones selectively stimulate the medial olfactory tracts of male goldfish. Brain Res (in press)

Sorensen PW, Goetz FW, Scott AP, Stacey NE (1991b) Recent studies indicate that goldfish use mixtures of unmodified hormones and hormonal metabolites as sex pheromones. *In*:

Proc 4 th Intl Symp Reprod Physiol Fish. Scott AP, Sumpter JP, Kime D (eds) (in press)

Sorensen, PW, Irvine IAS, Scott AP, Stacey NE (1991c) Electrophysiological measures of olfactory sensitivity suggest that goldfish and other fish use species-specific mixtures of hormones and their metabolites as sex pheromones. *In*: Chemical Signals in Vertebrates VI. Doty R, Muller-Schwarze D (eds) Plenum Press New York (in press)

Stacey NE (1987) Roles of hormones and pheromones in fish reproduction. *In*: Psychobiology of Reproductive Behavior: An Evolutionary Perspective. Crews D (ed) Prentice Hall New Jersey, pp 28-69

Stacey NE (1991) Hormonal pheromones in fish: status and prospects. *In*: Proc 4 th Intl Symp Reprod Physiol Fish. Scott AP, Sumpter JP, Kime D (eds) (in press)

Stacey NE, Kyle AL, Liley NR (1986) Fish reproductive pheromones. *In*: Chemical Signals in Vertebrates IV. Duvall D, Muller-Schwarze D, Silverstein RM (eds) Plenum Press New York, pp 117-133

Stacey NE, Sorensen PW, Dulka JG, Van Der Kraak GJ (1989) Direct evidence that 17α,20ß-dihydroxy-4-pregnen-3-one functions as a goldfish primer pheromone: preovulatory release is closely associated with male endocrine responses. Gen Comp Endocrinol 75:62-70

Stacey NE, Sorensen PW (1991) Function and evolution of fish hormonal phero-mones. *In*: Biochemistry and Molecular Biology of Fishes Vol I. Hochachka PW, Mommsen TP (eds) Elsevier Amsterdam, pp 109-135

Stacey NE, Sorensen PW, Dulka JG, Cardwell JR, Irvine AS (1991) Fish sex pheromones: current status and potential applications. Bull Zool Academia Sinica (Taiwan) (in press)

Sveinsson T, Hara TJ (1990) Olfactory receptors in Arctic char (*Salvelinus alpinus*) with high sensitivity and specificity for prostaglandin F2α. Chem Senses 15:645-646.

Van Den Hurk R, Lambert JGD (1983) Ovarian steroid glucuronides function as sex pheromones for male zebrafish , *Brachydanio rerio*. Can J Zool 61: 2381-2387

Van Den Hurk R, Schoonen WGEJ, Van Zoelen GA, Lambert JGD (1987) The biosynthesis of steroid glucuronides in the testis of the zebrafish, *Brachydanio rerio*, and their pheromonal function as ovlation inducers. Gen Comp Endocrinol 68:179-188

15

Molecular Biology of Tilapia Prolactins

D. Swennen, B. Sekkali, A.-C. Poncelet, F. Rentier-Delrue, J.A. Martial and A. Belayew

Introduction

Physiological role of prolactin

Prolactin (PRL) is a polypeptide hormone which is mainly secreted by the pituitary gland. In vertebrates, this hormone is implicated in a number of physiological processes such as reproduction, osmoregulation, growth and metabolism. The primary role of teleost PRL is the regulation of water and electrolyte homeostasis in freshwater (Clarke and Bern, 1980; Loretz and Bern, 1982; Hirano, 1986; Wendelaar Bonga and Pang, 1989; Prunet *et al.*, 1990). Tilapia has two forms of PRL (tiPRL-I: 188 amino acids and tiPRL-II: 177 amino acids) which have been well characterized (Specker *et al.*, 1985a, b; Flik *et al.*, 1986; Yamagushi *et al.*, 1988; Young *et al.*, 1988; Prunet *et al.*, 1989; Specker *et al.*, 1989; Dauder *et al.*, 1990a, b). Both PRLs purified from pituitaries are able to promote sodium retention (Specker *et al.*, 1985b). They restore the plasma osmolality and electrolyte concentration to normal levels as well as the integumental transepithelial potential difference of hypophysectomized tilapia (Young *et al.*, 1988). Until now, there have been only limited data in homologous bioassays suggesting differential activities among both kinds of PRLs. The preliminary results of Specker *et al.* (1985b) show that the large PRL (tiPRL-I) promotes growth whereas the small PRL does not. One important prerequesite for investigating possible differences in the roles of the two PRLs is the availability of large amounts of pure active proteins. This is the reason why we have cloned the cDNAs coding for the two tilapia PRLs. We have used these cDNAs to construct bacterial expression vectors and produced biologically active recombinant PRLs in *Escherichia coli*.

Regulation of prolactin production

Tilapia PRLs are secreted by cells located in the rostral pars distalis of the adenohypophysis (Nagahama *et al.*, 1981). The activity of PRL-producing cells is under two types of control mechanisms. One is exerted by plasma osmotic pressure (Helms *et al.*, 1991) and by steroïd hormones such as cortisol which inhibits PRL secretion (Borski *et al.*, 1991) and estradiol which stimulates it (Barry and Grau, 1986). The other kind of control mechanism is by hypothalamic factors such as somatostatin, dopamin, vasoactive intestinal peptide which inhibit PRL secretion (Nagahama *et al.*, 1975; Wigham *et al.*, 1977; Grau *et al.*, 1982, 1987; Kelley *et al.*, 1988; Helms *et al.*, 1991). Calcium and cAMP seem to be secondary messengers involved in activating or maintaining PRL secretion (Richman III *et al.*, 1990, 1991; Borski *et al.*, 1991; Helms *et al.*, 1991).
The only teleost PRL gene to have been cloned and sequenced so far is the carp gene (Chen *et al.*, 1991)

which is organized in five exons and four introns similarly to the mammalian PRL genes. A potential regulatory element, a palindromic sequence, has been identified at positions -79 to -59.
We describe in this paper, the cloning and characterization of the tilapia PRL-I gene.

Results

Cloning and sequencing of tilapia PRL cDNAs

A cDNA library was constructed using poly(A)$^+$ RNA purified from tilapia (*Oreochromis niloticus*) pituitaries. TiPRL cDNA clones were selected by hybridization to a trout PRL cDNA fragment (Mercier *et al.*, 1989) under low stringency. Two types of PRL cDNAs were isolated and their complete nucleotide sequence determined. The larger cDNA (tiPRL-I) codes for a polypeptide of 212 amino acids, including a putative signal sequence of 24 amino acids. The second PRL cDNA (tiPRL-II) encodes a polypeptide of 200 amino acids, including a presumptive signal peptide of 23 amino acids. Both proteins share 69 % amino acid identity at optimal alignment. The comparison of the nucleotide sequences as well as the polypeptide sequences clearly shows that the two forms of PRL are products of distinct genes (Rentier-Delrue *et al.*, 1989).

Production and purification of biologically active recombinant tilapia PRLs

The T7 expression system (Studier *et al.*, 1990) was used to produce recombinant tiPRL-I and tiPRL-II.

Constructions were done in such a way that the cDNAs were expressed as native proteins containing a N-terminal methionine residue (Rentier-Delrue *et al.*, 1989; Swennen *et al.*, 1991). The amount of tiPRL-I and tiPRL-II produced in an aggregated form (inclusion bodies) by this expression system was estimated respectively to 70 and 45 mg/l of *Escherichia coli* culture. Moreover, the identity of the expressed products was confirmed by Western blot analysis using salmon PRL antiserum.

In order to recover active and soluble proteins, the inclusion bodies were isolated by cell lysis followed by low speed centrifugation. Urea 6 M and ß-mercaptoethanol 1 % were used to unfold the polypeptides and conditions were adjusted to allow them to refold correctly (Swennen *et al.*, 1991). The correct refolding was tested by analyzing the bioactivity of these recombinant tiPRLs. The *in vivo* and *in vitro* assays were performed by the laboratory of S.E. Wendelaar Bonga (University of Nijmegen, The Netherlands) and P. Prunet (INRA, Rennes, France). They show that both tiPRLs were able to increase tilapia blood calcium levels, to stimulate the mucocyte proliferation of the skin epithelium and to bind tilapia kidney membranes as do natural tilapia PRLs, suggesting that the recombinant PRLs are biologically active (Swennen *et al.*, 1991).

Cloning and sequencing of tiPRL-I gene

A tilapia (*Oreochromis mossambicus*) genomic bank was constructed by cloning partially Sau3AI-digested sperm DNA into the lambdagem-11 phage vector. The screening of the library with the tiPRL-I cDNA as a probe, resulted in the purification of 8 recombinant clones out of 400,000. One of them comprized a 5,500 bp EcoRI fragment which was subcloned in pGEM. A library of randomly overlapping clones was constructed. DNA of about fourty of these clones was sequenced from both sides on an EMBL automatic sequencer. The nucleotide sequences were computer aligned with the "GCG Fragment Assembly System" (University of Wisconsin). The origin of transcription was determined by the classical primer extension method but the technique was adapted to the automatic sequencing device (Swennen *et al.*, submitted).

Structure of tiPRL-I gene

By comparison with the *Oreochromis niloticus* tiPRL-I coding region, that of *Oreochromis mossambicus* contains four nucleotide substitutions. These modifications do not affect the amino acid sequence.

The tiPRL-I gene has a "TATA box" about 30 nucleotides upstream from the transcription origin similar to most eukaryotic genes (Breathnach and Chambon, 1981). The origin of transcription (CAP site) of the PRL gene is a guanine nucleotide in tilapia and carp (Chen *et al.*, 1991) whereas it is an adenine for mammalian PRL genes. The tiPRL-I gene is 3677 bp long and organized in five exons and four introns (Fig. 1) which is similar to the other known PRL genes. It differs in size only by the length of the introns.

Figure 1: Structure of tilapia PRL-I gene.

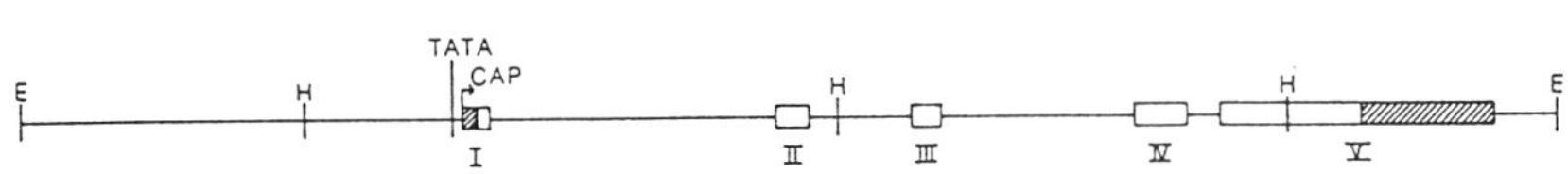

The "TATA box" and CAP site are represented. Open boxes: coding regions (exons); hatched boxes: untranslated regions; lines between boxes: introns. E: EcoRI, H: HindIII. The EcoRI fragment is 5.5 kb long.

The PRL protein belongs to the same hormonal family as growth hormone (GH). Both are secreted by the pituitary gland and share several biological, immunological and structural properties. On the basis of their amino acid sequence (Niall *et al.*, 1971) and later on, when the nucleotide sequences of the cDNAs encoding these hormones became available (Cooke *et al.*, 1980; Martial and Cooke, 1980), their analysis suggested that the members of this gene family have evolved by duplication and divergence from a common ancestral gene (reviewed by Miller and Eberhardt, 1983).

Figure 2: Evolution of growth hormone/prolactin gene family.

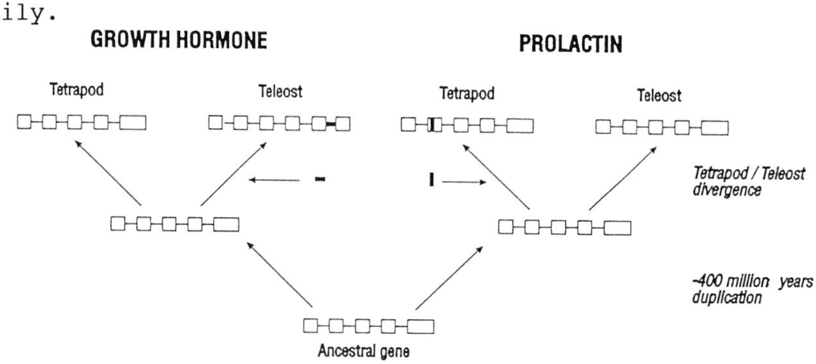

▬ : intron V insertion
I : aminoterminal loop coding fragment insertion

All members of this gene family are organized in five exons and four introns. The first teleost genes of this hormonal family which have been cloned and sequenced are the trout (Agellon *et al.*, 1988) and the salmon (Johansen *et al.*, 1989) GH genes. These two genes differ from the mammalian genes by an extra-intron (intron V) in the last exon. Agellon *et al.*

(1988) have suggested that this intron V has been recently introduced into the fish GH genes. Duplication and divergence of PRL and GH genes occured before the divergence of fish and tetrapods; the PRL and GH genes are present in both groups and the mammalian PRL and GH genes represent similar structure. The absence of intron V in carp and tilapia PRL genes supports this view.

Another evolutionary feature is the introduction in the tetrapod PRL gene of a sequence coding for the amino-terminal loop (Yasuda *et al.*, 1987, 1990) (Fig. 2).

Cell culture and transfections

To identify DNA sequences involved in transcriptional regulation, molecular biologists construct hybrid genes by linking the promoter under study to a reporter gene. Various mutants of this hybrid gene are then introduced in cells grown *in vitro* and the transient expression of the reporter gene is measured, which reflects the transcriptional activity of the linked promoter. We assayed four genes that had been previously described to select a reporter gene that could be used in fish cell transfections. The genes coding for chloramphenicol acetyl transferase (CAT; Gorman *et al*, 1982), ß-galactosidase (ß-gal; Mac Gregor and Caskey, 1989), ß-lactamase (amp^R; Muller *et al*, 1986) and luciferase (LUC; de Wet *et al*, 1987) were used for transient expression studies in fish cells grown *in vitro*. Few promoters that direct efficient expression in fish cells have been characterized to date. Two promoter/enhancer elements have recently been introduced into fish cells grown *in vitro* and found to be very active at expressing a linked CAT reporter gene (Friedenreich and Schartl, 1990). These are the Rous Sarcoma Virus (RSV) long terminal repeat and the human cytomegalovirus (CMV) immediate early region. For each of the four reporter genes, several plasmid constructs were tested: in one of them, no promoter was present (o) whereas in the others, the reporter gene was linked either to the CMV or RSV promoter/enhancer.

The EPC cell line, originating from carp epidermal herpes virus-induced hyperplastic lesions (epithelioma papulosum cyprini) were used as model for fish cells in these experiments. Electroporation (Peers *et al.*, 1990) was used to introduce foreign DNA into EPC cells grown *in vitro* according to Friendenreich and Schartl (1990). The CMV-CAT construct was used in initial experiments in order to set up optimal transfection conditions (Table 1).

Table 1:

	B.G.	(o)	RSV	CMV	CMV/(o)
CAT	0	1,4	22	60	50
ß-gal	0.04	0.04	0.13	0.95	25
LUC	0	4	N.D.	4,700	1,000
AMP^R	0.8	0.8	1.3	N.D.	N.D.

The reporter genes were used either without a promoter (o) or with RSV or CMV promoter/enhancer (see text). The optimal electroporation conditions are: 4 x 10^6 cells in 800 μl volume, 250 volts and 1500 μFarads. The enzymatic activities were measured 48 hours after electroporation with 10 μg test plasmid. CAT: chloramphenicol acetyl transferase enzymatic activity measured as % of acetylated [^{14}C] chloramphenicol (Gorman *et al*, 1982); ß-gal: ß-galactosidase enzymatic activity measured as change in O.D. at 420 nm upon hydrolysis of ONPG (Sambrook *et al*, 19890; LUC: luciferase enzymatic activity measured as light emission upon reaction with luciferin and ATP (de Wet *et al*, 1987); AMP^R: ß-lactamase activity measured as change in O.D. at 482 nm upon hydrolysis of nitrocefin (Muller *et al.*, 1986).

These conditions resulted in about 50% cell survival, and were chosen in all further experiments. The various promoterless reporter genes or the constructs with CMV and RSV promoter/enhancer were assayed in EPC cells under these optimal conditions. Table 1 provides the enzymatic activities measured in these transient expression assays. The highest activities were found for the luciferase (LUC) gene; the CMV promoter/enhancer brings a thousand-fold increase in expression as compared to the promoterless LUC gene. This ratio is only 50 and 25 for the CAT and ß-gal reporter genes, respectively. As for the AMP^R gene, its basal activity is so high that it precludes its use in these experiments.

We have thus chosen to fuse the 1555 bp promoter of the tiPRL-I gene to the LUC reporter gene for mutational analysis in transient expression assays. These experiments are in progress.

Conclusion

We have described in this paper several molecularü biological techniques which have brought important progress to the study of both kinds of tilapia PRL.

The cloning and characterization of the tiPRL cDNA has allowed: 1) unambiguous determination of nucleotide sequence and deduction of the encoded amino acid sequence; 2) construction of expression vector to produce large amounts of recombinant PRLs in *Escherichia coli*; 3) cloning of the corresponding genomic sequence.

Production of large amounts of highly purified biologically active recombinant PRLs provides the possibility to study tilapia PRLs independently without limitation of quantity.

Moreover, the cloning and characterization of tiPRL-I gene will help to understand at a molecular level the regulation of its expression. Further cloning and analysis of tiPRL-II gene will allow the comparison of both regulation mechanisms.

References

Agellon, L.B., Davies, S.L., Chen, T.T. and Powers, D.A. (1988). Structure of a fish (rainbow trout) growth hormone gene and its evolutionary implications. *Proc. Natl. Acad. Sci. USA 85*, 5136-5140.

Barry, T.P. and Grau, E.G. (1986). Estradiol-17ß and thyrotropin-releasing hormone stimulate prolactin release from the pituitary gland of a teleost fish *in vitro*. *Gen. Comp. Endocrinol. 62*, 306-314.

Borski, R.J., Helms, L.M.H., Richman III, N.H. and Grau, E.G. (1991). Cortisol rapidly reduces prolactin release and cAMP and $^{45}Ca^{2+}$ accumulation in the cichlid fish pituitary *in vitro*. *Proc. Natl. Acad. Sci. USA 88*, 2758-2762.

Breathnach, R. and Chambon, P. (1981). Organization and expression of eukaryotic split genes coding for proteins. *Annu. Rev. Biochem. 50*, 349-383.

Chen, H.-T., Chiou, C.-S. and Chang, W.-C. (1991). Cloning and characterization of carp prolactin gene. *Biochim. Biophys. Acta 1088*, 315-318.

Clarke, W.C. and Bern, H.A. (1980). Comparative endocrinology of prolactin. In *Hormonal Proteins and Peptides*, vol.8, pp 105-197. Ed. C.H.Li. New-York: Academic Press.

Cooke, N.E., Coit, D., Weiner, R.I., Baxter, J.D. and Martial, J.A. (1980). Structure of cloned DNA complementary to rat prolactin messenger RNA. *J. Biol. Chem. 255*, 6502-6510.

Dauder, S., Young, G. and Bern, H.A. (1990a). Effect of hypophysectomy, replacement therapy with ovine prolactin, and cortisol and triiodothyronin treatment on prolactin receptors of the tilapia (*Oreochromis mossambicus*). *Gen. Comp. Endocrinol. 77*, 378-385.

Dauder, S., Young, G., Hass, L. and Bern, H.A. (1990b). Prolactin receptors in liver, kidney, and gill of tilapia (*Oreochromis mossambicus*): characterization and effects of salinity on specific binding of iodinated ovine prolactin. *Gen. Comp. Endocrinol.* *77*, 368-377.

de Wet, J.R., Wood, K.V., De Luca, M., Helsinki, D.R. and Subramani, S. (1987). Firely luciferase gene: structure and expression in mammals cells. *Mol. Cell. Biol.* *7*, 725.

Flik, G., Fenwick, J.C., Kolar, Z., Mayer-Gostan, N. and Wendelaar Bonga, S.E. (1986). Effects of ovine prolactin on calcium uptake and distribution in *Oreochromis mossambicus*. *Am. J. Physiol.* *250*, R161-R166.

Friedenreich, H. and Shartl, M. (1990). Transient expression directed by homologous promoter and enhancer sequences in fish cells. *Nucleic Acids Res.* *18*, 3299-3305.

Gorman, C.M., Moffat, L.F. and Howard, B.H. (1982). Recombinant genomes which express chloramphenicol acetyl-transferase in mammalian cells. *Mol. Cell. Biol.* *2*, 1044-1051.

Grau, E.G., Ford, C.A., Helms, L.M.H., Shimoda, S.K. and Cooke, I.M. (1987). Somatostatin and altered medium osmotic pressure elicit rapid changes in prolactin release from the rostral pars distalis of the tilapia, *Oreochromis mossambicus*, *in vitro*. *Gen. Comp. Endocrinol.* *65*, 12-18.

Grau, E.G., Nishioka, R.S. and Bern, H.A. (1982). Effects of somatostatin and urotensin II on tilapia pituitary prolactin release and interactions between somatostatin, osmotic pressure, Ca^{++}, and adenosin 3', 5'-monophosphate in prolactin release *in vitro*. *Endocrinol.* *110*, 910-915.

Helms, L.M.H., Grau, E.G. and Borski, R.J. (1991). Effects of osmotic pressure and somatostatin on the cAMP messenger system of the osmosensitive prolactin cell of a teleost fish, the tilapia (*Oreochromis mossambicus*). *Gen. Comp. Endocrinol.* *83*, 111-117.

Hirano, T. (1986). The spectrum of prolactin action in teleosts. In *Comparative Endocrinology: Developments and Directions*, C.L.Ralph, pp. 53-74. Ed. Alan R. Liss Inc., New-York.

Johansen, B., Johnsen, O.C. and Valla, S. (1989). The complete nucleotide sequence of the growth hormone gene from atlantic salmon (*Salmo salar*). *Gene* *77*, 317-324.

Kelley, K.M., Nishioka, R.S. and Bern, H.A. (1988). Novel effect of vasoactive intestinal polypeptide and peptide histidine isoleucine: inhibition of *in vitro* secretion of prolactin in tilapia, *Oreochromis mossambicus*. *Gen. Comp. Endocrinol.* *72*, 97-106.

Loretz, C.A. and Bern, H.A. (1982). Prolactin and osmoregulation in vertebrates. *Neuroendocrinol.* *35*, 292-304.

Mac Gregor, G.R. and Caskey, C.T. (1989). Construction of plasmids that express *Escherichia coli* ß-galactosidase in mammalian cells. *Nucleic Acids Res. 17*, 2365.

Martial, J.A. and Cooke, N.E. (1980). Isolation and sequence analysis of rat prolactin structural gene. In Control and peripheral regulation of prolactin function. R.M. McLeod and U. Scapagnini Eds. Raven Press, N.-Y. pp 115-123.

Mercier, L., Rentier-Delrue, F., Swennen, D., Lion, M., Prunet, P. and Martial, J.A. (1989). Rainbow trout prolactin cDNA cloning in *Escherichia coli*. *DNA 8*, 119-125.

Miller, W.L. and Eberhardt, N.L. (1983). Structure and evolution of the growth hormone family. *Endocrin. Rev. 4*, 97-130.

Muller, M., Bellefroid, E., Lambert, C., Belayew, A. and Martial, J.A. (1986). Use of a bacterial, ß-lactamase gene for studies of transcription regulation in eukariotic cells. *Arch. Int. Physiol. Biochem. 94*, B35.

Nagahama, Y., Nishioka, R.S., Bern, H.A. and Gunther, R.L. (1975). Control of prolactin secretion in teleosts with special reference to *Gillichthys mirabilis* and *Tilapia mossambica*. *Gen. Comp. Endo-crinol. 25*, 166-188.

Nagahama, Y., Olivereau, M., Farmer, S.W., Nishioka, R.S. and Bern, H.A. (1981). Immunocytochemical identification of the prolactin- and growth hormone-secreting cells in the teleost pituitary with antisera to tilapia prolactin and growth hormone. *Gen. Comp. Endocrinol. 44*, 389-395.

Niall, H.D., Hogan, M.L., Sauer, R., Rosenblum, I.Y. and Greenwood, F.C. (1971). Sequences of pituitary and placental lactogenic and growth hormones: evolution from a primordial peptide by gene reduplication. *Proc. Natl. Acad. Sci. USA 68*, 866-869.

Peers, B., Voz, M.L., Monget, P., Mathy-Hartert, M., Berwaer, M., Belayew, M. and Martial, J.A; (1990). Regulatory elements controlling pituitary-specific expression of the human prolactin gene. *Mol. Cell. Biol. 10*, 1690-4700.

Prunet, P. and Bornancin, M. (1989).Physiology of salinity tolerance in tilapia: an update and basic applied aspect. *Aquat. Living Resour. 2*, 91-97.

Prunet, P., Avella, M., Fostier, A., Björnsson, B., Boeuf, G. and Haux, C. (1990). Role of prolactin in salmonids. In Progress in *Comparative Endocrinology*, pp 547-552. Eds. A. Eple, C.G. Scanes and M.H. Stetson. N.-Y.: Wiley-Liss.

Rentier-Delrue, F., Swennen, D., Prunet, P., Lion, M. and Martial, J.A. (1989). Tilapia prolactin: molecular cloning of two cDNAs and expression in *Escherichia coli*. *DNA 8*, 261-270

Richman III, N.H., Ford, C.-A., Helms, L.M.H., Cooke, I.A., Pang, P.K.T. and Grau, G. (1991). The loss of $^{45}Ca^{2+}$ associated with prolactin release from the tilapia (*Oreochromis mossambicus*) rostral pars distalis. *Gen. Comp. Endocrinol. 83*, 56-67.
Richman III, N.H., Helms, L.M.H., Ford, C.-A., Benishin, C., Pang, P.K.T., Cooke, I.M. and Grau, E.G. (1990). Effects of depolarising concentrations of K+ and reduced osmotic pressure on $^{45}Ca^{2+}$ accumulation by the rostral pars distalis of tilapia (*Oreochromis mossambicus*). *Gen. Comp. Endocrinol. 77*, 292-297.
Sambrook, J., Fritsch, E.F. and Maniatis, T. (1989). Molecular cloning, a laboratory manual. Ed. C. Nolan. New-York, U.S.A.: Cold Spring Harbour Laboratory Press.
Specker, J.L., Brown, P.S. and Brown, S.C. (1989). Unequal activities of the two tilapia prolactins in the whole-animal transepithelial potential bioassay using the red eft. *Fish Physiol. Biochem.* 7, 119-124.
Specker, J.L., King, D.S., Nishioka, R.S., Shirahata, K., Yamagushi, K. and Bern, H.A. (1985a). Isolation and partial characterization of a pair of prolactins released *in vitro* by the pituitary of a cichlid fish, *Oreochromis mossambicus*. *Proc. Natl. Acad. Sci. USA 82*, 7490-7494.
Specker, J.L., King, D.S., Rivas, R.J. and Young, B.K. (1985b). Partial characterization of two prolactins from a cichlid fish. In *Prolactin Basic and Clinical Correlates*, vol.1, pp427-436. Eds R.M. MacLeod, M.O. Thorner and U. Scapagnini. Padua, Italy: Fidia Research Series, Liviana Press.
Studier, F.W., Rosenberg, A.H., Dunn, J.J. and Dubendorff, J.W. (1990). Use of T7 RNA polymerase to direct expression of cloned genes. *Methods in Enzymology 185*, 60
Swennen, D., Rentier-Delrue, F., Auperin, B., Prunet, P., Flik, G., Wendelaar Bonga, S.E., Lion, M. and Martial, J.A. (1991). Production and purification of biologically active recombinant tilapia (*Oreochromis niloticus*) prolactins. *J. Endocrinol. 131*, 219-227.
Wendelaar Bonga, S.E. and Pang, P.K.T. (1989). Pituitary hormones. In *Vertebrate Endocrinology: Fundamentals and Biochemical Implications*, vol.3, pp. 105-136. Eds. P.K.T. Pang and M.P. Schreibman. New-York: Academic Press.
Wigham, T., Nishioka, R.S. and Bern, H.A. (1977). Factors affecting in vitro activity of prolactin cells in the euryhaline teleost *Sarotherodon mossambicus* (*Tilapia mossambica*). *Gen. Comp. Endo-crinol. 32*, 120-131.
Yamagushi, K., Specker, J.L., King, D.S., Yokoo, Y., Nishioka, R.S., Hirano, T. and Bern, H.A. (1988).

Complete amino acid sequences of a pair of fish (tilapia) prolactins, $tPRL_{177}$ and $tPRL_{188}$. *J. Biol. Chem. 263*, 9113-9121.
Yasuda, A., Kawauchi, H. and Papkoff, H. (1990). The complete amino acid sequence of prolactin from sea turtle (*Chelonia mydas*). *Gen. Comp. Endocrinol. 80*, 363-371.
Yasuda, A., Miyajima, K., Kawaushi, H., Peter, R.E., Lin, H.R., Yamagushi, K. and Sano, H. (1987). Primary structure of common carp prolactins. *Gen. Comp. Endocrinol. 66*, 280-290.
Young, P.S., McCormick, S.D., Demarest, J.R., Lin, R.J., Nishioka, R.S. and Bern, H.A. (1988). Effects of salinity, hypophysectomy, and prolactin on whole-animal transepithelial potential in the tilapia *Oreochromis mossambicus*. *Gen. Comp. Endocrinol. 71*, 389-397.

16

The Isolation and Structure of Liver and Globin Genes from Atlantic Salmon

A. Wagner, F. Deryckere, G. Hardiman, L. Byrnes, and F. Gannon

INTRODUCTION

Aquaculture has been a developing activity throughout the world for the past few decades. About 12% of the world fish production in 1990 has been obtained by farming and this percentage is growing every year (Sattaur 1989, MacKenzie *et al* 1989). Having established a base line and a growing level of success in the production of fish from aquaculture, the problems of productivity and quality become more and more important in the overall economic equations. Consequently the problem of desease control is a key parameter since the outbreak of viral, bacterial or other pathogens can have dramatical consequences on the annual production of a farm.

Generally the ability to minimise the impact of disease is directly related to the knowledge of the infecting organisms and of the manner in which the farmed species develops and protects itself from the infecting agent. On the other hand, the ability to increase the fish production depends on the optimisation of all the different aspects of growth, food conversion and development of the animals. This also demands a knowledge of the underlying metabolism and physiology of fish. Over the years major investments have been made in the description of the biochemistry and development of animals of economical importance. It has led, with the development of molecular biology and recombinant DNA technology to the improvment of the farmed species, especially mammals. A similar effort has not been made yiet with fish (Powers, 1989). Moreover, the extrapolation to fish of data collected from other organisms seems hazardous since fish have life cycles which can include migrations between fresh and sea water and consequently implicate adaptation to extreme changes of the environmental conditions. Moreover since their body temperature is very low compared to mammals, important differences in metabolic activity can be expected.

Our aim is to establish basic understanding of the molecular biology of fish and more specifically of the Atlantic Salmon (*Salmo salar*). The interest in this area derives primarily from the close proximity of the major salmon fish farms in the west coast of Ireland. In this paper we report on our progress in two unrelated systems: the salmon globin genes and liver expressed genes.
The analysis of the salmon globin gene family permits the comparison of gene structures and regulations between salmon and higher vertebrates for which these genes have been extensively studied. Moreover, since there is a correlation between the expression of these genes and the fish size, this process may be related to other general growth phenomena which understanding might be relevant to aquaculture. The second type of genes studied are liver expressed. The liver has been the focus of our analysis because of its importance and the body of data which is available from other systems. To date, seven highly expressed liver genes have been isolated of which three have been identified as serum albumin, hemopexin and apolipoprotein-A1. Moreover, the gene for the liver transcription factor HNF1 which regulates the expression of numerous genes in this organ, like the serum albumin, has been cloned. The last type of liver genes studied in our laboratory are the genes involved in the smoltification process that the young salmon undergo during their first migration from fresh water to the sea. This process is probably the most important step in the salmon's life cycle and its molecular basis is only very poorly understood. Again there is a practical relevance to this study in that the efficiency of this step is crucial to the delivery of smolts to the fish farms for their growth towards full size.

THE GLOBIN GENE FAMILY

Globin genes form probably the best known gene family in human and higher vertebrates. They therefore represent a perfect model system to study both gene expression and regulation, and the phylogenetic evolution of different species. Hemoglobin assures the oxygen blood transport from the respiratory organs to the muscles in all vertebrates. It is composed of four non covalently linked globin peptides- 2 α-like and 2 β-like globins. The heme and the ferric ion responsible for the oxygen transport are located in the central cavity formed by the four globin chains. Today, the globin genes of a wide variety of species have been investigated including rabbit, mouse, rat, horse, monkey, chicken, pigeon, frog and salamander.
In most of the studied species, the globin genes are arranged in two loci defined as the α- and β- gene locus respectively which are located on separate chromosomes. In both clusters the genes are arranged in the order of their developmental expression in an 5' to 3' orientation:

5'-fetal gene(s) - embyonic genes(s) - adult gene(s) -3'

In addition to the gene specific regulating regions, some important regulating elements situated about 25kb upstream of the first gene of the β- locus have been identified in man (Van Assendelft *et al* 1989). Recently the same kind of element has been found far upstream of the human α- globin loci (Higgs *et al* 1990). Those regions added to the β- or α- globin promoters allowed a specific expression of transgenes in mice indicating that in fish too the promoter might be suitable for transgenic purposes (Collis *et al* 1990, Hanscombe *et al* 1989).

Two exceptions to this type of arrangment have been found as far. The first one concerns the chicken β-globins which, while located on a different chromosome than the α-genes, are not arranged in their temporal order of expression (Dodgson *et al* 1979). The second exception is found in the Xenopus family

where the α- and β- genes are linked, in the following order (Jeffreys *et al* 1980, Hosbach *et al* 1983):
5'- α larval genes- α adult gene- β adult gene- β larval genes-3'

From the evolutionary point of view, fish as one of the oldest vertebrates represent a very interesting species for the study of globin genes since, phylogenetically, the ancestor of amphibians and reptiles diverged from the bony fish between 425 and 500 million years ago while the amphibians diverged from the reptiles and the ancestor of birds and mammals more recently (see figure 1 below).
According to this tree, the characterisation of the globin gene chromosomal arrangment in fish such as salmon could give insight to the evolution of the globins from the initial duplication of the ancestral gene about 500-570 million years ago to the chromosomal separation of the α- and β- gene clusters existing in the reptilian ancestor of the birds and mammals.

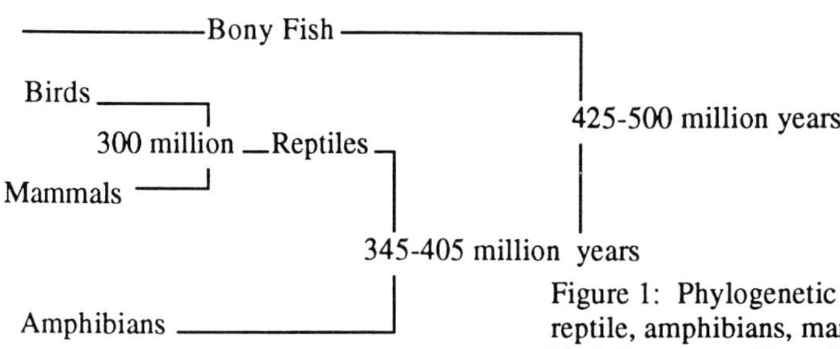

Figure 1: Phylogenetic tree of fish, reptile, amphibians, mammals and bird divergence (from Knochel et al, 1986)

While the salmon hemoglobin structure has been analysed since the early 1960s, the molecular organisation of the corresponding globin genes has been undertaken only recently.

Regarding their electrophoretic migration pattern in non denaturing conditions, the 17 Atlantic salmon hemoglobins have been classified in two groups: the 9 anodal - HgbA- and the 8 cathodal -HgbC- migrating proteins. During the salmon growth there is a progressive shift from the expression of the HgbA -predominant in the young animal - to the expression of HgbC in the adult form. The ratio of total cathodal to anodal proteins in the bloodstream however never exceeds 50%. The proteins of the anodal group are composed of various combination of 4 electrophoretically distinct peptides R, S, T and V. Three other peptides composing the HgbC proteins have been identified: M, P and W. However, in order to justify the existence of the 8 cathodal proteins a fourth peptide must be present which probably migrates to the same position than M, P or W. When the salmon length increases, the R and T peptides tend to be replaced respectively by S and T inside the anodal group, while in each hemoglobin protein the R+S to T+V ratio remains always equal to 1. These observations lead some authors to relate the R and S peptides to α-globins and T-V to β-globins (Wilkins 1967, 1985). Our aim is to clone Atlantic salmon α- and β- globin genes in order to elucidate their chromosomal arrangement as compared to that of the amphibians and the mammals to correlate gene structure with the protein profiles and to study their promoter sequences.

Kidney and spleen are hematopoietic tissues in salmon. In a first step to isolate the globin gene, a complementary DNA (cDNA) library was constructed in the viral vector λgt10 by reverse transcription of messenger RNA (mRNA) extracted from the kidney. The library was then screened with radioactively labelled total cDNA from spleen and kidney. This kind of screening permits the isolation of the abundant cDNAs which correspond to the genes most actively transcribed in each tissue. After characterisation of the positive clones, one 567 base pair long clone was selected. The frequency of occurence in the total library was 2.2%. The deduced amino acid sequence presents a 93% non conservative and 98% conservative homology with the rainbow trout α-globin (Wolff and Gannon 1989).

In a second set of experiments this first cDNA clone - α1 clone- was used to screen the kidney cDNA library as well as a Atlantic salmon genomic DNA library (kindly provided by A. Bannon, University of Arkansas, USA). As a result one other cDNA clone - α2 clone- and one genomic clone - α3- were isolated and sequenced. The isolation of the genomic clone α3 permitted the identification of the corresponding promoter region the activity of which is currently being tested. Table 1 shows the percentages of nucleotidic homology between all three clones in the coding and 3' untranslated regions (UTR) respectively.

Table 1: α-chain Nucleic Acid comparison:Identical matches (%)

	In the coding region:		In the 3'UTR:	
	α2	α3	α2	α3
S.Salar-α1	98.1	98.6	91.1	73.4
S.Salar-α2		96.5		67.1

The homology of the 3 clones in the 5' UTR is equal to 93.3% as compared to α1. In the coding regions, the homology is 96.3% between α2 and α3 and 98.6% between α1 and α2. It is much lower in the 3'UTR where it is equal to 67.1% between α2 and α3. This variability in the 3'UTR has been previously reported for human α-globin and for Xenopus laevis β-globin cDNAs. While the reason of this divergence could be a lower selective pression, it has been shown for the Xenopus genes that itaffects the precise site rather than the efficiency of polyadenylation which could in turn affect the mRNA stability (Mason *et al* 1986, Liebhaber *et al* 1981).

When the deduced amino acid composition of the 3 chains are compared only four amino acid substitutions are observed in α2 and α3 compared to α1. Two of these changes are conservative in α2, whereas only one is conservative in α3. Since the different clones have been identified from libraries constructed from different individuals, it is yet difficult to conclude if the observed divergences are due to polymorphisms between individuals or to different genes.

Table 2 gives the percentage of homology of the three α-globin salmon chains to the α chains from other fish species. The homology between salmon and rainbow trout which are both part of the salmonidae family is about 10% higher than with the carp, the tuna and the torpedo fish (Gorr *et al* 1991). The homology with the adult X. laevis forms is low and equal to 46.5%. Comparisons done with sequences of the adult α-chains from X. tropicalis and X. borealis give a corresponding percentage of homology.

Table 2: α-chain Aminoacid comparison: Identical matches(%)

	S.Salar α1	S.Salar α2	S.Salar α3
S.Salar α1		97.2(99.3)	97.2(97.9)
S.Salar α2		95.1(97.9)	
Other Fish:			
S. Irideus	93.0(95.8)	93.0(95.8)	93.0(95.8)
E. Electricus	69.9(83.1)	69.9(83.1)	69.9(83.1)
C. Carpio	66.9(81.7)	66.9(81.7)	66.9(81.7)
T. Thynnus	60.6(77.5)	62.0(78.2)	60.6(77.5)
Amphibians:			
X.Laevis:			
Tadpole-α3	55.6(68.3)	54.9(67.6)	54.9(66.9)
Tadpole-α5	57.3(73.3)	57.7(74.0)	43.4(73.3)
Adult-α1	46.5(59.9)	46.5(59.9)	46.5(59.9)
Adult-α2	45.8(59.9)	45.8(59.9)	45.8(59.9)
Human:			
Adult-α	57.3(71.1)	57.3(71.1)	57.8(66.9)

Conservative percentages are shown in brackets

The homology with the larval forms of X. laevis is equal to 55.6% which might indicate that the larval genes are closer to the ancestral form common to fish and amphibians than the adult form. It should be noted that the homology with the human adult α-globins is roughly the same as that with the tadpole X. laevis chains. By performing nucleotidic sequence comparisons between β-globins of different species, we were able to identify 2 short well conserved DNA regions. Corresponding consensus oligonucleotides were synthetized and then used to PCR (Polymerase Chain Reactions) amplify the DNA sequence between the two conserved regions using the kidney cDNA library as a template. We obtained a specific product which has been identified as a 217bp fragment of a β-globin cDNA. Southern blot analysis showed that, suprisingly, the β-globin probe obtained as the PCR product hybridizes with some genomic clones containing the α-globin gene. Since, the two cDNAs do not cross-hybridize and as the two probes hybridise to different bands following restriction enzyme digestion, this result suggests that as in Xenopus, the α- and β-globin genes are adjacent in salmon. This result is very important from an evolutionary point of view because it proves that the chromosomal separation of the α- and β- chains occured in the ancestor of birds and mammals after the divergence of reptiles and amphibians about 345-405 million years ago (see figure 1).

LIVER SPECIFIC GENES

The liver, because of its size, its accessibility, its homogeneous architecture, the fact that it is composed primarily of a single cell type and its relatively simple developmental origin has been studied extensively in an attempt to understand the mechanisms underlying tissue specific gene expression. A study of Atlantic salmon liver gene expression was undertaken as part of our effort to establish a molecular biology basis for the study of this species.

Genes involved in smoltification

The Atlantic Salmon is like many salmonids diadromous, its two chief adventures consisting of a feeding migration from the river to the sea and a spawning migration when it returns to the natal river or stream. During the initial migration to the sea the process termed smoltification occurs which is characterised by substantial biochemical, physiological and behavioral changes. The most striking morphological change is the loss of parr marks, a series of darkly pigmented melanin bars due to purine deposition in the scales resulting in the development of silvery pigmentation (Folmar and Dickhoff 1980, Johnston and Eales 1967). A change in the length weight ratio is also observed resulting in a longer more slender fish (Fessler and Wagner 1969). Salinity preference and co-ordinated seaward migration are associated with this process. Gills, kidney and liver are the tissues undergoing the most important changes during the smoltification process.

While, the osmoregulating action of gills and kidney has been reported in other studies, our work focuses on the molecular regulation of liver genes involved in smoltification (Zaug 1972). Plasma proteins, supposingly fron hepatic origin, specific to early, transitional and late smoltification have been isolated. Modification in the population of hepatic mitochondrial proteins during parr-smolt transformation seem to suggest a role for these proteins in seawater adaptation (Bradley 1987, Bradley and Rourke 1988).

In this study differential hybridisation of a salmon liver cDNA library was carried out using cDNA probes generated from parr (pre-smoltification) and salmon (post-smoltification) total liver RNA in order to compare the liver mRNA populations of freshwater and seawater salmon. RNA was prepared from the liver of an adult salmon and a cDNA library was constructed in the λgt10 vector (Byrnes and Gannon, 1990). Clones which exhibited stronger hybridisation with the sea-salmon probe were chosen as candidates for smoltification related expression in the salmon liver. Indeed, if cDNAs and related genes involved in the smoltification process can be isolated, it would not only permit to gain a molecular understanding of this physiologicaly very complex process, but also to isolate promotors inducible at a certain time of the salmons development or in a given set of conditions. An example of differential hybridisation screening of differnt clones with respectively the salmon and parr cDNA are shown in figure 2. It clearly appears on the figure that the hybridisation signal is much higher with the sea salmon probe.

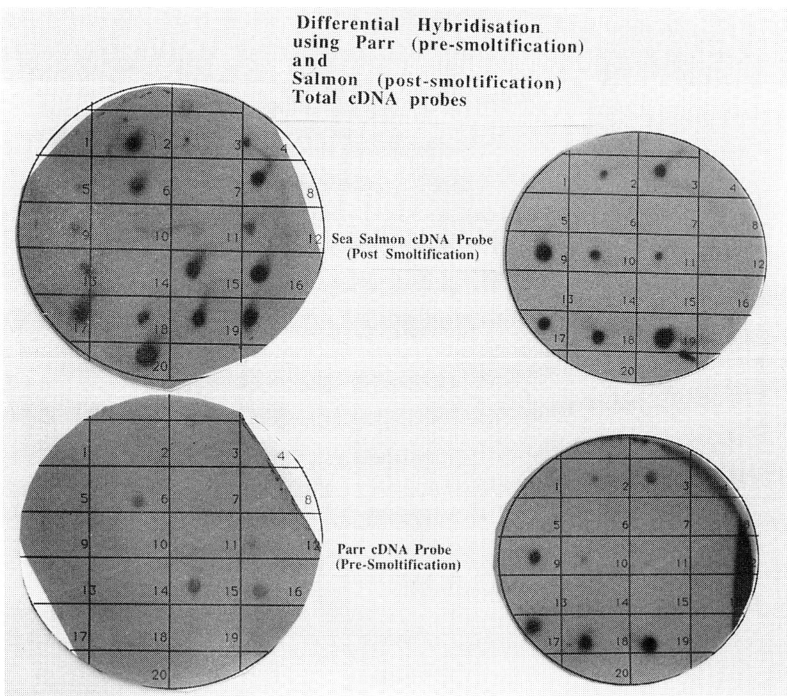

Figure 2: Example of differential screening of liver cDNAs with total parr and sea salmon cDNA probes.

Three clones were selected for further study as they seemed to be the most interesting. DNA of these clones was prepared, labelled and used as probes in Northern blot analysis of liver RNA from salmon undergoing seaward adaptation.

Clone 1 was found to have the greatest expression in sea salmon, Clone 2 increased in expression as the fish entered the final stages whereas Clone 3 increased in expression in the early stages of smoltification. The mRNA sizes were determined

to be 2.7Kb, 0.75Kb and 2.4kb for clones 1,2 and 3 respectively. Back hybridisation to the library revealed them to be abundant clones representing 1.6% ,0.86% and 1.4% of sea salmon liver mRNA. Expression in tissues other than liver was investigated. Clones 3 seems to be strictly liver specific, whereas clone 1 was also found in the intestine and clone 2 in muscle, kidney, spleen and intestine.
DNA sequencing is being carried out on these clones at present and might permit by sequence comparisons to the data bank to identify the 3 clones. The subsequent step will be to clone the corresponding genes and their promoters.

High level expression liver genes

Parallely to the study of smoltification genes, the expression pattern of different highly expressed liver genes is being studied. In this context, the liver λgt10 cDNA library was screened with a total liver cDNA probe in order to select the most abundant clones. Seven cDNA clones, representing abundant salmon liver mRNAs were isolated from the salmon liver cDNA library. Sequence analysis of three of these clones identified them as salmon apolipoprotein A-1, salmon serum albumin and salmon hemopexin.

Salmon apolipoprotein A-1 (apo A-1) is the most abundant cDNA isolated, representing 3% of clones in the unamplified salmon liver library. Apolipoproteins are the protein components of lipoproteins and the major agents for cholesterol and fatty acid transport in the bloodstream. The complete cDNA sequence of this 1115 base pair cDNA was determined (Powell et al, 1991).

The second of the 7 abundant cDNA clones isolated has been identified as serum albumin and represents 1.6% of the unamplified salmon liver library clones. Albumin is the most important serum protein in mammals. It is responsible for the maintenance of the osmotic pressure in the bloodstream and therefore probably plays a major role in the adaptation of the salmon when migrating from the low osmotic fresh water streams to sea water. DNA sequencing and analysis of the deduced amino acid sequence have verified the evolution of the albumin/α-fetoprotein/vitamin D binding protein multigene family based on the triple domain structure of albumin as proposed by Brown, 1976 and Haefliger *et al*, 1989 (Byrnes and Gannon, 1990).

Salmon hemopexin mRNA is the third clone identified and is also abundant in the liver. Partial sequence analysis has revealed 78.3% amino acid homology between human and salmon hemopexin when conservative amino acid substitutions are taken into account .

The tissue expression of six of the seven clones was examined by Northern blot analysis and is shown in figure 3. Figure 3d and 3e correspond respectively to the expression patterns of serum albumin and hemopexin.

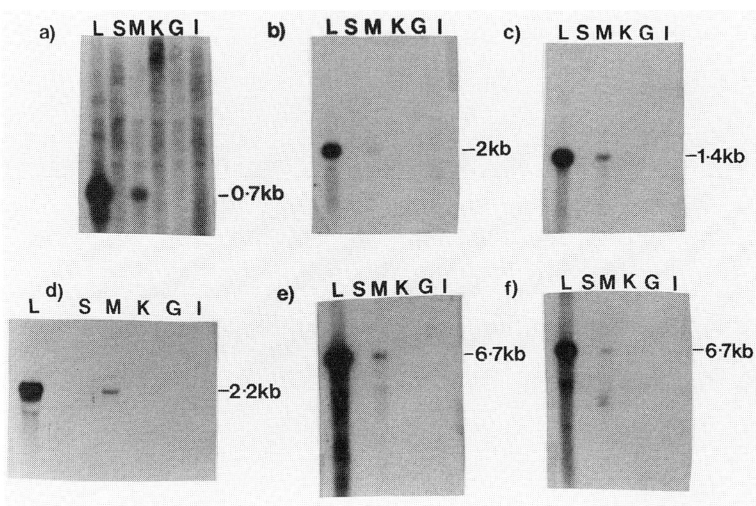

Figure 3: Nothern blot analysis of liver (L), spleen (S), kidney (K), muscle (M), gonad (G) and intestine (I) RNA. a), b), c), d), e), f) were hybridised with labeled DNA from cDNA clones 1, 2, 3, 4, 5, and 6. 5μg of RNA were loaded in a), b), c), e), f) and 10μg in d).

As shown in figure 3, mRNA corresponding to each clone is found in the liver. But, surprisingly all of the abundant salmon liver clones are expressed in muscle too, at approximately 5% of the level of liver expression.

The significance of this result which appears to be particular to salmon, has yet to be evaluated as well as the identity of the 4 remaining clones. While a similar study performed on abundant mouse liver cDNA showed parallel expression in spleen and kidney for some of the clones, no muscle expression could be detected (Derman *et al*, 1981; Powell *et al*, 1984). The fact that salmon exhibits muscle expression of abundant liver messengers may be related to the very special physiology of salmon which allows for the unique characteristics of its life cycle. Salmon fasts from the onset of the spawning migration, from the rich feeding waters of the sea to its natal river. During this time, an appreciable percentage of the body weight is being converted into roe or milt. Thus, the plasma fat transport system is being subjected to important modifications (Nelson and Shore, 1974). These observations may provide a clue to explain the muscle expression of liver sequences but more extensive studies on the mechanism of the control of salmon liver gene expression are required.

Liver specific transactivating factors.

As part of this work, we have been interested in the study of the transactivating factors involved in the liver specific transcription. In mammals, the Hepatocyte Nuclear Factor 1 (HNF1 also known as LF-B1, APF or HP1) has been characterised as a strong activator of the transcription of different liver specific genes including the albumin gene (Cereghini *et al.* 1988), the α1-antitrypsin gene (Monaci *et al.* 1988) and the transthyretin gene (Costa *et al.* 1988). The complete cDNA coding for HNF1 has been isolated and sequenced in rat, man and mouse (Frain *et al.* 1989, Baumhueter *et al.* 1990, Chouard *et al.* 1990, Bach *et al.* 1990, Kuo *et al.* 1991) and contains in its sequence a diverged homeobox (Mendel and Crabtree, 1991). The HNF1 sequence is very well conserved between the three different species and the homology equals 95% at the protein level between each of them.

Our strategy to isolate the salmon equivalent of the HNF1 sequence was based on the hypothesis that the rat and the corresponding salmon sequences, if they actually existed, would be sufficiently conserved to cross hybridise on part of their sequences and most particularly in the homeobox region. The screening of the liver cDNA library with a rat HNF1 homeobox probe (kindly provided by M. Yaniv, Chouard *et al.* 1990) allowed us to isolate several positive clones. Amongst those clones, two reprensent together the entire transcript which has been sequenced.

The overall nucleotidic homology as compared to the sequence of rat HNF1 equals 46% and is quite low compared to the rat-mammalian scores of 95%. The homologies in the homeodomain, POU- and dimerization domains are however much higher and respectively equal to 98%, 85% and 93%. This result highlights the importance of these domains of the protein and confirm quite well results obtained by extensive deletion analysis carried out to determine the important sequences involved in DNA-binding or dimerization functions(Chouard *et al.* 1990, Nicosia *et al.* 1990). The fact that these regions are so well conserved might suggest that the basic mechanims of control of expression in fish are closely related to the ones of higher vertebrates.

Since the important sequence for DNA recognition (third α-helix in the homeodomain) is 100% conserved in the salmon protein, the binding site sequence is probably very well conserved in the albumin promoter.

CONCLUSION

The above studies indicate the current state of progress in our laboratory towards the analysis of gene structure and control of expression in Atlantic salmon. Information has been obtained on awide range of genes that are expressed in liver, kidney and spleen. The detail analysis of each of these genes can be pursued in the expectation that it will provide a full picture on the way in which their expression contributes to the growth and development of salmon. The sequences which have been obtained provide new tools for the study of the salmon biology. For example, the DNA probes which relate to the smoltification genes will allow to identify, monitor and later control the physiological effects of these genes on the animals themselves. In this way, it is expected that molecular biology which we describe will become fully integrated into the strategies which are being put in place to increase the productivity of salmon based aquaculture.

ACKNOWLEDGMENTS

This work has been performed in the context of an EC Bridge contract. A.W. was founded by a postdoctoral EC grant. F.D. was recipient of a postdoctoral EMBO fellowship.

REFERENCES

Bach I, Galcheva-Gargova Z, Mattei MG, Simon-Chazottes D, Guénet JL, Cereghini S and Yaniv M (1990) Cloning of human hepatic nuclear factor 1 (HNF1) and chromosomal localization of its gene in man and mouse. Genomics 8: 155-164

Baumhueter S, Mendel DB, Conley PB, Kuo CJ, Turk C, Graves MK, Edwards CA, Coutois G and Crabtree G (1990) HNF-1 shares three sequence motifs with the POU domain proteins and is identical to LF-B1 and APF. Genes Dev. 4: 372-379.

Bradley TM (1987) Changes in hepatic mitochondrial proteins of steelhead trout during the parr-smolt transformation. AM Fish Soc Sympos 1: 230-235

Bradley TM and Rourke AW(1988) Smoltification in steelhead trout (*Salmo gairdneri*): Developmental aspects of plasma constituents. Fish Physiol Biochem 5: 39-44

Brown RB (1976) Structural origins of mammalian albumin Federation proceedings: OVL 35 JPT 10: 2141-2144

Byrnes L and Gannon F (1990) Atlantic Salmon (*Salmo salar*) Serum Albumin: cDNA Sequence, Evolution and Tissue Expression. DNA and Cell Biol 9: 647-655

Chouard T, Blumenfeld M, Bach I, Vandekerkhove J, Cereghini S and Yaniv M (1990) A distal dimerization domain is essential for DNA-binding by the atypical HNF-1 homeodomain. Nucleic Acids Res. 18: 5853-5863

Cereghini S, Blumenfeld M and Yaniv M (1988) A liver-specific factor essential for albumin transcription differs differs between differentiated and dedifferentiated rat hepatoma cells. Genes Dev. 2: 957-974

Collis P, Antoniou M and Grosveld F (1990) Definition of the minimal requirements within the human β-globin gene and the dominant control region for high level expression. EMBO 9: 223-240

Costa RH, Grayson, DR, Xanthopoulos KG and Darnell JE (1988) A liver-specific DNA-binding protein recognizes multiple nucleotide sites in regulatory regions of transthyretin, α1-antitrypsin, albumin and simian virus 40 genes. Proc. Natl. Acad. Sci. USA 85: 3840-3844

Derman E, Krauter K, Walling L, Weinberger C, Ray M and Darnell JE (1981) Transcriptional control in the Production of Liver-specific mRNAs. Cell 23: 731-739

Dodgson JB, Strommer J and Engel JD (1979) Isolation of the Chicken β-globin gene and a linked Embryonic β-like Globin Gene from a Chicken DNA Recombinant Library. Cell 17: 879-887

Fessler J L and Wagner HH (1969) Some morphological and biochemical changes in steelhead trout during the parr-smolt transformation. J Fish res Board Can 24: 955-964

Frain MG, Swart P, Monaci P, Nicosia A, Stampfli S, Frank R and Cortese R (1989) The liver-specific transcription factor LF-B1 contains a highly diverged homeobox DNA binding domain. Cell 59: 145-157

Folmar LC and Dickhoff W (1980) The parr-smolt transformation (smoltification) and seawater adaptation in salmonids. A review of selected literature. Aquaculture 21: 1-37

Gorr T, Kleinschmidt T and Fricke H (1991) Close tetrapod relationships of the coelacanth Latimeria indicated by haemoglobin sequences. Nature 352: 394-397

Haefliger DN, Moskaitis JE, Schoenberg DR and Wahli W (1989) Amphibian Albumins as Members of the Albumin, Alpha-Fetoprotein, Vitamin D-binding Protein Multigen Family. J Mol Evol 29: 344-354

Hanscombe O, Vidal M, Kaeda J, Luzzatto L, Greaves DR and Grosveld F (1989) High-level, erythroid-specific expression of the human α-globin gene in transgenic mice and the production of human hemoglobin in murine erythrocytes. Genes&Development 3: 1572-1581

Higgs DR, Wood WG, Jarman AP, Lida J, Pretorius IM and Ayyub H (1990) A major positive regulatory region located far upstream of the human α-globin gene locus. Genes & Development 4: 1588-1601

Hosbach HA, Wyler T and Weber R (1983) The *Xenopus laevis* Globin Gene Family: Chromosomal Arrangment and Gene Structure. Cell 32: 45-53

Jeffreys AJ, Wilson V, Wood D and Simons JP (1980) Linkage of Adult α- and β-Globin Genes in *X. laevis* and Gene Duplication by Tetraploidization. Cell 21: 555-564

Johnston CE and Eales JG (1967) Purines in the integumentof the Atlantic salmon (*Salmo salar*) during parr-smolt transformation. J Fish Res Board Can 24: 955-964

Knochel W, Korge E, Basner A and Meyerhof W (1986) Globin Evolution in the Genus Xenopus: Comparative Analysis of cDNAs Coding for Adult Globin Polypeptides of *Xenopus borealis* and *Xenopus Tropicalis*. J Mol Evol 23: 211-223

Kuo CJ, Conley PB, Hsieh CL, Francke U and Crabtree GR (1990) Molecular cloning, functional expression and chromosomal localization of mouse hepatocyte nuclear factor 1. Proc. Natl. Acad. Sci. 87: 9838-9842.

Liebhaber SA, Goossens M and Kan YW (1981) Homology and concerted evolution of the α1 and α2 loci of human α-globin. Nature 290: 26-29

McEvoy T, Stack M, Keane B, Barry T, Sreenan J and Gannon F (1988) The Expression of a Foreign Gene in Salmon embryos. Aquaculture 68: 27-37

MacKenzie D (1989) How to build a better fish. New Scientist 22: 52-56

Mason PJ, Elkington JA, Lloyd MM, Jones MB and Williams J (1986) Mutations Downstream of the Polyadenylation Site of a Xenopus β-globin mRNA Affect the Position but Not the Efficiency of 3' Processing. Cell 46: 263-270

Mendel DB and Crabtree GR (1991) HNF-1, a member of a Novel Class of Dimerizing Homeodomain Proteins. The Journal of Biological Chemistry 266: 667-680
Monaci P, Nicosia A and Cortese R (1988) Two different liver-specific factors stimulate *in vitro* transcription from the human α1-antitrypsin promoter. EMBO J. 7: 2075-2087
Nelson GJ and Shore VG (1974) Chracterization of the Serun High Density Lipoproteins and Apolipoproteins of Pink Salmon. The Journal of Biological Chemistry 249: 536-542
Nicosia A, Monaci P, Tomei L, De Francesco R, Nuzzo M, Stunnenberg H and Cortese R (1990) A myosin-like dimerization helix and an extralarge homeodomain are essential elements of the tripartite DNA binding structure of LFB1. Cell 61: 1225-1236.
Powell DG, Friedman JM, Oulette AJ, Krauter KS and Dranell JE (1984) Transcriptional and Post-transcriptional Control of Specific Messenger RNAs in Adult and Embryonic Liver. J Mol Biol 179: 21-35
Powell R, Higgins DG, Wolff J, Byrnes L, Stack M, Sharp PM and Gannon F (1991) The salmon gene for apolipoprotein A-I: cDNA sequence, tissue expression and evolution. Gene 104: 402-408
Powers DA (1989) Fish as Model Systems. Science 246: 352-358
Sattaur O (1989) The threat of the well-bred salmon. New Scientist 29: 54-60
Sheridan MA, AllenWV and Kersitter TH (1985)
Changes in the fatty acid composition of steelhead trout *Salmo gairdneri* associated with parr-smolt transformation. Comp Biochem Physiol 80B: 671-676
Van Assendelft GB, Hanscombe O, Grosveld F and Greaves DR (1989) The β-globin dominant control region activates homologeous and heterolegeous promoters in a tissue-specific manner. Cell 56: 969-977
Wilkins NP (1968) Multiple Haemoglobins of the Atlantic Salmon (*Salmo salar*). J Fish Res Bd Canada 25: 2651-2653
Wilkins NP (1985) Ontogeny and Evolution of Salmonid Hemoglobins. International Review of Cytology 94: 269-298
Wolff JP and Gannon F (1989) cDNA and deduced amino acid sequence of the *Salmo salar* (Atlantic salmon) adult hemoglobin alpha chain. NAR 17:

Zaugg WS (1982) Some changes in smoltification and seawater adaptability of salmonids resulting from environmental and other factors. Aquaculture 28: 143-151

17

Transgenic Technology in Fish

D. Chourrout

ABSTRACT

Biological experimentation in fish, as well as their genetic improvement for aquaculture could greatly benefit from in vivo gene transfer technologies, if convenient methods of gene introduction permit the integration of the exogenous sequences into their genome. Diverse constraints have determined the choice of cytoplasmic injections to introduce the foreign genes in vivo. This system has a variable efficiency in different studies, but is now clearly validated for the integration. Other methods, such as DNA injections into the oocyte germinal vesicle followed by in vitro maturation, or such as the egg electroporation, have been also tested with more or less success. Satisfactory expression levels of the transgenes have been obtained with reporter genes, and possibilities to manipulate their transcription level and the location of their expression are now permitted. However, the capacity to express any transgene in vivo remains a challenge, especially when heterologous gene constructs are utilized.

INTRODUCTION

After much success in eukaryotic gene isolation and their functional test in transfected cultivated cells, strong efforts have been devoted in the last ten years to establish in vivo gene transfer methodologies. The benefits of the transgenic technology for the advancement of basic biology have been numerous, particularly in the mouse, for which the production of transformed lines is still hard but almost a routine operation in many laboratories.

In the mouse model, the main difficulty is due to the in vitro manipulation of eggs, which have then to be reimplanted in vivo for long term development. Genes are classically transferred as linear DNA fragments and introduced into one of the two egg pronuclei. In very peculiar cases, one may also use vectors derived from retrovirus genomes, in order to infect older embryos, but this operation does not result in animals having the foreign genes in all their cells.

In many fish species, the high fecundity and the external development make much easier any manipulation of embryos. However, research teams interested in gene transfer technology have immediately faced two problems : one is the impossibility to visualize the pronuclei in the fertilized egg, for an eventual gene injection ; the other is the absence of described retroviruses in fish, which could have been used to generate infectious vectors for gene integration.

In this article, we describe the methodological efforts devoted to overcome these problems in order to obtain stable lines of transgenic fishes.

FATE OF FOREIGN DNA INJECTED INTO THE CYTOPLASM

The cytoplasmic injection has been originally tested in the Xenopus for which problems similar to those met in fish were encountered. Very large amounts of circular or linearized conventional plasmids (1 to 1000 pg) were introduced into the egg animal pole, before its first cleavage. The foreign DNA was often considerably amplified during the segmentation, and then almost totally degraded until the end of the gastrulation. However, several copies of the injected sequences have sometimes been detected after the metamorphosis, and in one case in the offsprings of one male (Etkin and Pearman, 1987). The major problem in the Xenopus literature is a lack of informations concerning the fate of foreign DNA in adults resulting from the injections, as well as in the further generations obtained by their reproduction. This is of course fundamental for anyone who wants to know whether the cytoplasmic injection can lead to stable transgenic lines.

In fish as in amphibians, authors who have tested the cytoplasmic injection have followed very diverse protocoles. DNA can be introduced in manually or enzymatically dechorionated eggs, or through the chorion kept around the egg. In the latter case, injections are performed either through a hole previously drilled in the egg shell, or through the egg micropyle. In order to facilitate such injections, a research group has set up a chemical treatment, which softens the chorion and increases its transparency (Yoshizaki et al., 1989). Two other variables may have a particular importance for the subsequent success of the transformation : the amount of

DNA injected which may vary between 1 to 1000 pg in different reports, and the time of injection. Several authors have treated pluricellular embryos, but the majority of them have performed the injections at the first cell stage. However, even in these cases, the position of the injection time in relation to the endogenous DNA replication remains unclear, because basic informations of the first cell cycles are lacking for the fish egg.

Most contributions dealing with this method mention the detection of foreign DNA in embryos or young fishes resulting from the cytoplasmic injection (Zhu et al., 1985 and Yoon et al., 1989, 1990 in goldfish ; Chourrout et al., 1986, Maclean et al., 1987, Mc Evoy et al., 1988 , Fletcher et al., 1988, Rokkones et al., 1989 in salmonids ; Dunham et al., 1987 in channel catfish ; Zhang et al., 1990 in common carp ; Stuart et al., 1988 in zebrafish ; Chong and Vielkind, 1989 in medaka ; Brem et al., 1988 in one tilapia ; Kozlov et al., 1988 et Benyumov et al., 1989 in one loach). These precocious examinations pose the problem of an eventual transient persistence of large amounts of foreign DNA out of the chromosomes, promised to a rapid degradation. At this stage, the yields of positive animals detected in dot blots is usually inferior to 10%, when a single organ is analyzed.

Among the several studies in which older fishes have been examined, ours in the rainbow trout are an exception because we have obtained very high rate of positive fishes (50% and more after two years), which is for example largely superior to those recorded by two other groups working also with salmonids (Maclean et al., 1987 ; Fletcher et al., 1988) ; Penman et al. (1990) have been able to explain partially this major difference of yields, by the very different amounts of DNA injected in different studies.

In our case as well as for other groups, the dot blot analysis of DNA extracted from several organs unequivocally reveals a mosaic distribution of the foreign genes in most animals of the F0 generation (directly issued from the treated eggs). Restriction studies indicate that multiple copies of the injected sequences are generally organized in multimers. Additionnally, the hybridization signals are found in the high molecular weight fractions, when the genomic DNA is analyzed without digestion by restriction enzymes. However, the results of restriction analyses cannot be totally convincing for the question of integration, because the hybridizing bands interpreted as junction fragments (associating the foreign DNA and the fish genome) can be equally explained by foreign DNA rearrangements without subsequent integration. This type of analyses is also complicated by the mosaic distribution of the foreign genes. We must mention one study which has rather clearly suggested the integration from the F0 generation : Penman et al. (1990) have observed hybridization bands at an intermediate level between that of the undigested DNA and that of the introduced plasmid, after cleavage of the fish DNA with restriction enzymes having no site in the injected plasmid.

A more interesting but longer way to address the question of gene integration consists of the examination of F1 and F2 generations of fish, obtained by the reproduction of putatively transgenic F0 adults with wild-type partners. Four research groups have clearly concluded for the transmission of foreign genes through the germline (Stuart et al., 1988 and Culp et al., 1991 in zebrafish ; Chourrout et al., 1988, Guyomard et al., 1989a, 1989b and Tewari et al., unpublished, in rainbow trout ; Shears et al., 1991 in atlantic salmon). The proportions of positive F1 offsprings are usually lower than 50%, indicating that the F0 germline is also mosaic (if the foreign DNA is actually integrated). The integration hypothesis is rather well supported by the analysis of Southern blots, which provide much simpler images than for the F0 generation, containing bands likely to represent junction fragments. In these three species, F1 adults have also been reproduced with wild-type partners, and furnished mendelian proportions of positive offsprings, usually 50% (suggesting a single foreign insertion), but also 75% (suggesting two insertions).

Although all these informations tend to indicate that the cytoplasmic injection into fish eggs leads to a stable integration as does the pronuclear injection in the mouse, we must mention that some of our results in two F2 families (with 50% transmission rate) have raised the possibility that foreign multimers might be stabilized in an extrachromosomal form (Tewari et al., unpublished). As a matter of fact, the images of Southern blots permitted to recognize bands corresponding to the free ends of the injected construct, which should be linked to each other within the multimer and linked to the genome if the multimer is integrated. A detailed examination of these two families with various restriction analyses in diverse conditions of gel electrophoresis (neutral and alkaline migration, PFGE) and with in situ hybridization on chromosomes has permitted to explain these surprising bands without rejecting the integration hypothesis. They result from the originality of the multimer organization after cytoplasmic injection. In mouse, the multiple copies are oriented in the same direction in the multimer, while in Xenopus and in fish, the copies are oriented randomly. This latter organization creates a long palindrom at all the junctions between copies oriented in opposite directions, and it seems that these palindroms induce the formation of secondary structures such as hairpin loops, born by intra-strand pairing. This model permits to interpret the unexpected bands as the result of a cleavage of hairpin loop heads by the restriction enzymes.

FATE OF FOREIGN DNA INTRODUCED BY OTHER WAYS

A Japanese research group has accomplished a very original work on the medaka, in testing two other routes of gene transfer. The first one is the injection of several ten thousand copies of circular plasmid (instead of several ten millions for the cytoplasmic injection), in the germinal vesicle of the immature oocytes (Ozato et al., 1986). This nucleus has the advantage of being much larger than the pronuclei which cannot be visualized in vivo. Copies of the injected plasmids have been subsequently found in the embryos resulting from the oocytes matured in vitro. However, nothing permits yet to know whether this original method leads to stable integration. If it is the case, the plasmids have joined the genome after the first round of replication, because the pictures of in situ hybridization, performed on histological sections of embryos, clearly suggest a mosaic distribution. The second method tested by this group consists of the electroporation of eggs, which has not only led to high rates of foreign DNA persistence in the F0 generation, but also to the germline transmission (Inoue et al., 1990).

Much more intriguing are the results reported on the goldfish by another team, who injected messenger RNAs into the cytoplasm of fertilized eggs. The DNA analyses suggest that the foreign RNA has been retrotranscribed and incorporated as a DNA copy into the host genome (Niu et al., 1989).

Other efforts have been concentrated on attempts to transform the sperm cells before fertilization, by incubating them with linear DNA. The only written report on the subject describes negative results with this method in the rainbow trout (Chourrout and Perrot, in press).

EXPRESSION OF FOREIGN DNA IN TRANSGENIC FISH

The first convincing arguments for the expression of foreign genes has come out of the injection of chicken delta crystallin gene (absent in fish) into the germinal vesicle of medaka oocytes. The protein has been first recognized in seven-days-old embryos by immunohistology and Western blotting (Ozato et al., 1986). A more recent and complementary study (Inoue et al., 1989) has shown an earlier expression in the lens, at the time of its formation. This is in accordance with the informations previously collected with this gene in chicken and in mammalian transfected cells, and therefore illustrates the interest of transformed fish embryos to study the function of heterologous sequences. We have to mention here that this type of expression may be only transient, because the germinal vesicle injection is not proven to lead to stable integration.

The evaluation of several other expression data, obtained after cytoplasmic injection is more difficult (McEvoy et al., 1988 ; Yoon et al., 1990). In addition to technical problems due to obvious "backgrounds" of expression in negative controls, the youth of the animals examined may be a difficulty, because most if not all this expression may result from non-incorporated DNA. This has been clearly suggested in another study of salmonid embryos (Rokkones et al., 1989), after cytoplasmic injection which is known to result in a strong transient amplification of large amounts of foreign DNA.

A large part of the expression studies are based on heterologous gene constructs, because relatively few fish genes or cDNAs had been isolated before the last years. We have for example injected into rainbow trout eggs several genes and cDNAs of mammalian growth hormone coupled with several classical heterologous promoters (Chourrout et al., 1988 ; Guyomard et al., 1989a, 1989b ; Chourrout et al., unpublished results). None of these constructs has been effective in the production of plasmatic hormone. However, two fish only in about one hundred harbouring the bovine growth hormone cDNA (associated with the LTR of RSV and the terminator of the bGH gene) have constitutively produced the bovine growth hormone, showing that this heterologous sequence could be transcribed and that the bGH could be secreted in the trout bloodstream. However, Northern analyses suggest that this expression was accidentally due to favourable position effects in the genome, where endogenous transcription signals have replaced those probably ineffective in the construct. These studies contrast with one on the loach (Benyumov et al., 1989) where the authors mention a growth rate stimulation consecutive to the introduction of similar constructs : however, few animals have been examined here, and it seems that the foreign hormone has not been dosed.

In two in vitro works of fish cell transfections with such constructs associating foreign promoters and mammalian GH coding sequences (Freidereich and Schartl, 1990 ; Bearzotti et al., in press), the presence of foreign hormone has not been detected in significant amounts, neither in the culture medium nor in cellular extracts. However, these studies and another one (Liu et al., 1990) have clearly shown that the expression of the CAT reporter gene can be easily observed in vitro from heterologous and homologous promoters. The difficulty to express mammalian GH genes in fish cells might be due to an inability to process the corresponding immature RNAs (Rokkones et al., 1989 ; Freidereich and Schartl, 1990). But the fact that cDNAs are not expressed either suggests the existence of other obstacles upstream and/or downstream of the RNA maturation.

The CAT gene expression has also been detected several times in vivo, in medakas of F0 generation (Chong and Vielkind, 1989) and more interestingly in F1 transgenic zebrafish (Stuart et al., 1990). Another study reports positive results of CAT expression, but at an early stage of the F0 generation (Liu et al., 1990). In the rainbow trout, a constitutive and ubiquitary expression of the CAT gene has been observed in both F0 and F1 generations with the human CMV immediate-early enhancer/promoter (Tewari et al., unpublished results). Therefore, and at least in the case of an intronless gene coding for a non-secreted protein such as CAT, the problem of foreign gene expression in fish is not always encountered.

In fish gene transfer, there will be a need as for other animals, to manipulate the expression level and localization. Up to now, this has been achieved with heterologous promoters coupled with the CAT gene. The mammalian regulatory elements seem to keep their main properties when they are introduced into fish cells. The promoter of human heat shock protein gene 70 is inducible by heating transfected carp cells (Liu et al., 1990 ; Bearzotti et al., in press). Recently, we have also tested in vivo two constructs associating the CAT gene and mouse immunoglobulin gene enhancer/promoters (Michard-Vanhée et al., unpublished results) : the expression of the reporter gene was restricted to the white cells, and predominant in the B lymphocytes. Therefore, the first attempts to target the expression in certain tissues of transgenic fish are very encouraging.

Problems to express the gene introduced are not expected when homologous fish genes are tested. The expression of a flounder antifreeze gene in the atlantic salmon, although very low up to now (for reasons that are difficult to elucidate), illustrates the interest of this strategy (Davies et al., 1989). A construct coupling the LTR of RSV with an incomplete cDNA of trout GH (without the signal-peptide) has been expressed in the common carp, in the F0 and F1 generations (Zhang et al., 1990). It is not clear whether the hormone was dosed in the plasmas, but the fact that the transgenic carps grew faster than their non-transgenic siblings is a priori surprising.

Finally, one can be extremely surprised by the fact that rabbit globin messengers introduced into goldfish eggs have led to an expression of the protein in the resulting fish (Niu and Tung, 1977 ; Niu et al., 1981 : in Niu et al., 1989). This would mean that endogenous sequences, localized near the DNA copies of the retrotranscribed messengers, have driven the expression.

CONCLUSION

The production of transgenic fish has become a reality and consists in most cases of the cytoplasmic injection of linear DNA. There are no applications yet, that can be seriously foreseen to improve fish genetically with gene transfer, as it is for example the case for cultivated crops. The success, in fish but also in all domestic animals, will depend upon the capacities to control and manipulate the expression levels in vivo. For fish, this may require the isolation and characterization of many homologous genes.

The system of cytoplasmic injection can already be used in many studies of basic biology, as it has been the case in mouse or drosophila with other methods of gene transfer. We have shown how it is possible to study in fish the properties of regulatory elements by associating them with a reporter gene. However, a major difficulty of the system is the management of transgenic fish lines, which cannot be obtained quickly because of the initial mosaicism. Methodological efforts should lead to the development of new tools which will not present this practical setback.

REFERENCES

Bearzotti, M., Perrot, E., Michard-Vanhée, C., Jolivet, G., Attal J., Théron, M.C., Puissant, C., Grabowski, H., Dréano, M., Kopchick, J.J., Powell, R., Gannon, F., Houdebine, L.M. and Chourrout, D. The expression of various gene constructs transfected into fish cells. J. Biotech. in press.

Benyumov, A.O., Yenikopolov, G.N., Barmintsev, V.A., Zelenina, I.A., Sleptsova, L.A., Doronin, Y.K., Golichenkov, V.A., Graschuk, M.A., Georgiev, G.P., Rubtsov, P.M., Skryabin, G., Baev, A.A., Integration and expression of human growth hormone gene in Teleostei. Genetika, 25 (1989) 24-35.

Brem, G., Brenig, B., Hörstgen-Schwark, G., Winnacker, E.L., Gene transfer in Tilapia (Oreochromis niloticus). Aquaculture, 69 (1988) 209-219.

Chong, S.S.C., Vielkind, J.R., Expression and fate of CAT reporter gene microinjected into fertilized medaka (Oryzias latipes) eggs in the form of plasmid DNA, recombinant phage particles and its DNA. Theor.Appl.Genet., 78 (1989) 369-380.

Chourrout, D., Guyomard, R., Houdebine, L.M., High efficiency gene transfer in rainbow trout (Salmo gairdneri Rich) by microinjection into egg cytoplasm. Aquaculture, 51 (1986) 143-150.

Chourrout, D., Guyomard, R., Houdebine, L.M., Techniques for the development of transgenic fish : a review pp. 89-99. In : Transgenic Models in Medecine and Agriculture. R. Church ed., Willey-Liss. Inc. (1990).

Chourrout, D., Guyomard, R., Leroux, C., Pourrain, F., Houdebine, L.M., Integration and germline transmission of foreign genes in trout after injection into the egg cytoplasm. J.Cell.Biochem.Suppl. 12B (1988) 188.

Chourrout, D., Perrot, E. No transgenic rainbow trout produced by fertilization with sperm incubated with linear DNA. Mol. Mar. Biol. Biotech., in press.

Culp, P., Nüsslein-Volhard, C., Hopkins, N. High-frequency germ-line transmission of plasmid DNA sequences injected into fertilized zebrafish eggs. Proc. Natl. Acad. Sci. USA 88 (1991) 7953-7957.

Davies, P.L., Hew, C.L., Shears, M.A., Fletcher, G.L., Antifreeze protein expression in transgenic salmon. J.Cell.Biochem.Suppl. 13B (1989) 169.

Dunham, R.A., Eash, J., Askins, J., Townes, J.M., Transfer of metallothionein human growth hormone fusion gene into channel catfish. Trans.Amer.Fisch.Soc. 116 (1987) 87-91.

Etkin, L.D., Pearman, B., Distribution, expression and germline transmission of exogenous DNA sequences following microinjection into Xenopus laevis eggs. Development 99 (1987) 15-23.

Fletcher, G.L., Shears, M.A., King, M.J., Davies, P.L., Hew, C.L., Evidence for antifreeze protein gene transfer in Atlantic Salmon (Salmo salar). Can.J.Fish Aquat.Sci. 45 (1988) 352-357.

Freidereich, H., Schartl, M., Transient expression directed by homologous and heterologous promoter and enhancer sequences in fish cells. Nucleic Acids Res. 18 (1990) 3299-3305.

Guyomard, R., Chourrout, D., Houdebine, L.M., Production of stable transgenic fish by cytoplasmic injection of purified genes. Gene transfer and Gene therapy pp. 9-18, Eds. I.Verma, R. Mulligan and A. Beauset, Alan R. Liss, Inc. (1989).

Guyomard, R., Chourrout, D., Leroux, C., Houdebine, L.M., Pourrain F., Integration and germline transmission of foreign genes microinjected into fertilized trout eggs. Biochim. 71 (1989) 857-883.

Inoue, K., Ozato, K., Kondoh, H., Iwamatsu, T., Wakamatsu, Y., Fujita, T., Okada, T.S., Stage-dependent expression of the chicken delta-crystallin gene in transgenic fish embryo. Cell.Differ.Dev. 27 (1989) 57-68.

Inoue, K., Yamashita, S., Hata, J.I., Kabeno, S., Asada, S., Nagahisa, E., Fujita, T., Electroporation as a new technique for producing transgenic fish. Cell.Differ.Dev. 29 (1990) 123-128.

Kozlov, A.P., Reshetnikov, V.L., Korzh, V.P., Neyfakh, A.A., The fate of plasmid DNA in the developing embryos of loach, Misgurnus fossilis L. Molecular Biology (Russe) 22 (1988) 1614-1622.

Liu, Z., Moav, B., Faras, A.J., Guise, K.S., Kapuscinski, A.R., Hackett, P.B., Development of expression vectors for transgenic fish. Bio/Technology. 8 (1990) 1268-1272.

Maclean, N., Penman, D., Zhu, Z., Introduction of novel genes into fish. Bio/Technology. 5 (1987) 257-281.

McEvoy, T., Stack, M., Keane, B., Barry, T., Sreenan, J., Gannon, F., The expression of foreign gene in salmon embryo. Aquaculture 68 (1988) 27-37.

Niu, M.C., Xue, G.X., Niu, L.C., Huang, H.Z., Transfer of information from mRNA to chromosomes by reverse transcription in early development of goldfish eggs. Cell.Mol.Biol. 35 (1989) 333-345.

Ozato, K., Kondoh, H., Inohara, H., Iwamatsu, T., Wakamatsu, Y., Okada, T. S., Production of transgenic fish : introduction and expression of chicken delta-crystallin gene in medaka embryos. Cell.Differentiation 19 (1986) 237-244.

Penman, D.J., Beeching, A.J., Penn, S., Maclean, N., Factors affecting survival and integration following microinjection of novel DNA into rainbow trout egg. Aquaculture 85 (1990) 35-50.

Rokkones, E., Allestrom, P., Skjervold, D.H., Gautvik, K.M., Microinjection and expression of a mouse metallothionein human growth hormone gene in fertilized salmon eggs. J.Comp.Physiol. B. 158 (1989) 751-758.

Shears, M.A., Fletcher, G.L., Hew, C.L., Gauthier, S., Davies, P.L. Transfer, expression and stable inheritance of antifreeze protein genes in atlantic salmon (Salmo salar). Mol. Mar. Biol. Biotech. 1 (1991) 58-63.

Stuart, G.W., McMurray, J.V., Westerfield, M., Replication, integration and germline transmission of foreign sequences injected into early zebrafish embryos. Developement 103 (1988) 405-442.

Stuart, G.W., McMurray, J.V., Westerfield, M., Germline transformation of the zebrafish. Gene transfer and gene therapy. pp. 19-28. Eds. I. Verma, R. Mulligan and A. Beauset. Alan R. Liss Inc. 1989.

Stuart, G.W., Vielkind, J.R., McMurray, J.V., Monte, W. Stable lines of transgenic zebrafish exhibit reproducible patterns of transgene expression. Development 109 (1990) 577-584.

Yoon S.J., Liu, Z., Kapuscinski, A.R., Hackett, P.B., Faras, A., Guise, K.S Successful gene transfer in fish. Gene Transfer and Gene Therapy. pp. 29-34, Eds. I. Verma, R. Mulligan and A. Beauset, Alan R. Liss, Inc. 1989.

Yoon S.J., Hallerman, E.M., Gross, M., Liu, Z., Schneider, J.F., Faras, A.J., Hackett, P.B., Kapuscinski, A.R., Guise, K.S. Transfer of the gene for neomycin resistance into goldfish, Carassius auratus. Aquaculture 85 (1990) 21-33.

Yoshisaki, G., Oshiro, T., Takashima, F., Prevention of hardening of chorion and dechorionation for microinjection into fish eggs. Nippon Suisan Gokkaishi 55 (1989) 369.

Zhang, P., Hayat, M., Joyce, C., Gonzalez-Villasenor, L.I., Lin C.M., Dunham, R.A., Chen, T.T., Powers, D.A., Gene transfer expression and inheritance of pRSV-rainbow trout-GH cDNA in the common carp, Cyprinus carpio (Linnaeus) Mol.Reprod.Develop. 25 (1990) 3-13.

Zhu, Z., Li, G., He, L., Chen, S., Novel gene transfer into the fertilized eggs of goldfish. Z. angew. Ichthyol. 1 (1985) 31-34.

18

Cellular and Subcellular Toxicity of Pollutants: Plasma Membrane Transport Systems as Targets

R.K.H. Kinne and E. Kinne-Saffren

INTRODUCTION

Toxicology, similar to its sister science pharmacology, has in recent years moved from observations involving the effect of pollutants on whole animals to the cellular and subcellular level. At these levels its aim is to define the site(s) of action more precisely and to elucidate the mechanism of action of toxins at a molecular level. Once the molecular mechanism is known, it is hoped to develop measures to prevent the influence of pollutants. Ideally, however, - but this is not the topic of this article - similarly measures should be developed from this knowledge to diminish or exclude the formation of pollutants generated by the ever increasing number of people and their demand for decent living conditions. This contribution focusses on the plasma membrane of cells as possible target for pollutants. Such an approach has become possible because techniques have been developed during recent years which allow to study transport systems - and thus the action of pollutants - on individual systems much more precisely [Kinne, 1991a; Kinne and Kinne-Saffran, 1981; Kinne-Saffran and Kinne, 1990]. Thus definite conclusions on the transport system affected and the sidedness of the action of a toxin, i.e. whether it acts intracellularly or extracellularly, can be drawn.

In the years to come, two major tasks can be defined for toxicological research. In light of the progresses made in cloning and determining the sequence and structure of transport systems [Kinne, 1991b; Morrison et al., 1991], one is to define even more precisely at which site of the transport molecules toxins bind and what consequences this interaction has with regard to the structure and function of the molecule. This approach bears strong similarities to the studies on the interaction between DNA and teratogenic compounds conducted so successfully during the last decade [van Zeeland, 1988]. Another equally important task is the integration of the results accumulating at the cellular and molecular level into an understanding of the toxicity at the multicellular, organismic and whole animal level.

PLASMA MEMBRANE TRANSPORT SYSTEMS AND THEIR FUNCTION IN CELLULAR LIFE

The plasma membrane is selectively permeable to inorganic and organic solutes. It is the site where transport systems are located whose functions can in very general terms be subdivided into "*housekeeping*" functions and "*cell-specific*" functions, i.e. functions related to a specific cell type or organ. This categorization is not exclusive but refers to the predominant role of a transporter in the life of a cell. "*Housekeeping*" transport systems are those found in almost every cell investigated and include e.g. Na/K-ATPase, Ca-ATPase, H-ATPases, a Na/H exchange system, and a variety of channels such as potassium channels and probably calcium channels [Kinne, 1991b]. These transporters are mainly involved in maintaining intracellular parameters constant, e.g. the intracellular sodium and potassium concentrations, the intracellular pH, and the intracellular calcium activity. In the context of this contribution "*cell-specific*" transport systems involved in epithelial nutrient transport and salt transport will be discussed. Representatives of this category are the sodium cotransport systems. These systems are characterized by their property to interact simultaneously with their transportate proper and with sodium ions. This interaction makes it possible to use the energy stored in the sodium gradient across the plasma membrane - low intracellular sodium vs. high extracellular sodium and sometimes in addition also the cell inside negative membrane potential - as driving force for transport into or out of a cell. When in addition such cotransport systems are asymmetrically arranged in the plasma membrane areas of a cell, as in epithelial cells, they play a determinant part in transcellular transport of nutrients - such as the amino acid reabsorption in intestine and kidney - and salt transport such as the sodium and chloride transport in gills, salt glands, skin, and again intestine and kidney [Kinne 1991a; 1991b].

Research from a variety of laboratories has shown that such sodium cotransport systems can operate in two different ways: as symport systems - where sodium and the transportate move in the same direction - and antiport systems, where the movements of sodium and the transportate are opposite. Together with their asymmetric distribution in the plasma membrane of polarized cells, symport and antiport systems can be involved either in absorption or secretion, depending on their intracellular localization and their mode of operation. Examples for the role of sodium cotransport systems in epithelial transport are given in figure 1.

SITES OF INHIBITION OF TYROSINE ABSORPTION IN THE FLOUNDER INTESTINE BY MERCURIALS AND ARSENICALS

One example how to unravel the cellular sites of action of a toxin on transepithelial transport is presented in the following. Tyrosine is one of the amino acids that are actively absorbed in the small intestine involving a sodium-tyrosine symport system and the Na/K-ATPase (see figure 1, left panel). The sodium-tyrosine cotransport system is located in the brush border membrane and facilitates the uptake of tyrosine into the cell, using the electrochemical potential difference of sodium across the luminal intestinal membrane as driving force. This driving force is maintained by the operation of the basolaterally

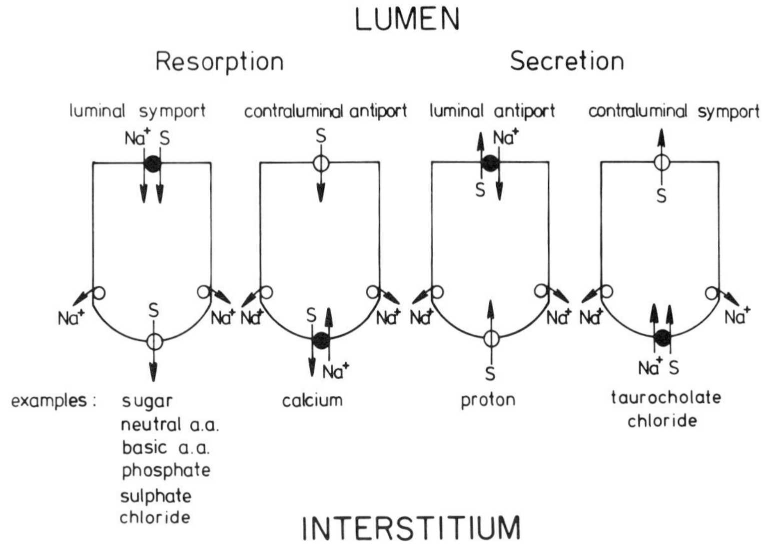

Figure 1: General Scheme for the role of sodium cotransport systems in epithelial absorption and secretion. The sodium pump (O⃗) is located in most epithelial cells in the contraluminal cell side. It generates the driving force for the operation of the sodium cotransport systems (●). The sodium cotransport systems and the sodium-independent transport systems (⊖→) are distributed asymmetrically in the membrane of the epithelial cell. Depending on the kind of sodium cotransport involved (symport or antiport) and on the localization of the cotransport system (luminal or contraluminal) active reabsorption or active secretion of a solute can occur. Symport denotes a cotransport where sodium and solute move in the same direction, during antiport the direction of sodium movement and solute movement is opposite. From [Kinne, 1984] with kind permission.

located Na/K-ATPase, which continuously removes sodium from the cells in exchange for potassium and thus maintains the intracellular sodium concentration low and the transmembranal electrical potential difference high. This membrane potential is mainly caused by the diffusion of potassium ions out of the cell through potassium channels (see also figure 3). The flounder as a marine vertebrate, which continually drinks water to maintain its fluid balance, and as bottom-dweller is particularly poised to be exposed to environmental pollutants, in particular its intestine. In a first series of experiments it was found by Chauncey et al. [1988] that mercurial chloride ($HgCl_2$), p-chloromercuribenzene sulfonate (PCMBS), and arsenical compounds, such as oxyphenylarsine (OPA) inhibit tyrosine absorption in flounder intestine. Thus, at 60 min incubation of the small intestine in vitro, tyrosine transport from the lumen of the intestine to the blood side (mucosal-to-serosal absorption) was inhibited 64% by 100 μM $HgCl_2$, 41% by 1 mM PCMBS, and 48% by 250 μM OPA [Chauncey et al., 1988]. In order to elucidate the cellular site of action of these compounds, the two main transport systems involved either directly (the luminal sodium-tyrosine cotransport system) or indirectly (the Na/K-ATPase) in tyrosine absorption were investigated separately [Musch et al., 1990]. The effects of the toxins on the sodium-tyrosine cotransport system were studied in isolated intestinal brush border vesicles, where the driving forces for the transporter can be kept constant and where metabolic influences on uptake are excluded. The results of these studies are compiled in table 1. If the sodium-dependent uptake of tyrosine is considered, only $HgCl_2$ inhibits the uptake significantly. Under the conditions used in this experiment (preincubation of the brush border vesicles for 1 h with the toxins) the inhibition is about 85% of the control uptake. Neither PCMBS nor OPA alter sodium-tyrosine cotransport in isolated intestinal brush border membranes to an appreciable extent. From these studies it could be conluded that $HgCl_2$ might exert its action on intestinal absorption by inhibiting entry of tyrosine into the cells; the reasons for the inhibition of tyrosine absorption by PCMBS and OPA remained, however, unclear.

Table 1: Mercurial and arsenical effects on brush border tyrosine uptake

Condition	L-tyrosine uptake (pmol/mg protein) 15 sec	90 min
total uptake		
75 mM NaCl medium		
Control	203.7 ± 9.1	153.5 ± 8.2
+ 100 μM $HgCl_2$	186.7 ± 25.0	93.0 ± 15.5*
Control	206.2 ± 10.6	155.0 ± 10.1
+ 500 μM PCMBS	171.8 ± 9.1*	48.6 ± 4.2+
Control	208.7 ± 10.1	143.4 ± 6.9
+ 250 μM OPA	187.6 ± 13.1	161.4 ± 12.1
150 mM mannitol medium		
Control	140.5 ± 10.7	170.9 ± 11.7
+ 100 μM $HgCl_2$	176.3 ± 21.5*	251.3 ± 10.8+
Control	132.0 ± 6.1	173.2 ± 12.6
+ 500 μM PCMBS	100.0 ± 8.8+	96.2 ± 5.8+
Control	144.6 ± 11.7	167.9 ± 11.0
+ 250 μM OPA	133.4 ± 14.2	181.6 ± 12.2
Na^+-dependent uptake		
Control	63.2 ± 7.4	
+ 100 μM $HgCl_2$	10.4 ± 8.2+	
Control	74.2 ± 8.3	
+ 500 μM PCMBS	71.7 ± 8.4	
Control	64.0 ± 7.3	
+ 250 μM OPA	54.2 ± 14.8	

Values shown are means ± SE for six experiments. *$p < 0.05$; +$p < 0.01$ compared with paired control by Students t-test. From [Chauncey et al., 1988] with kind permission.

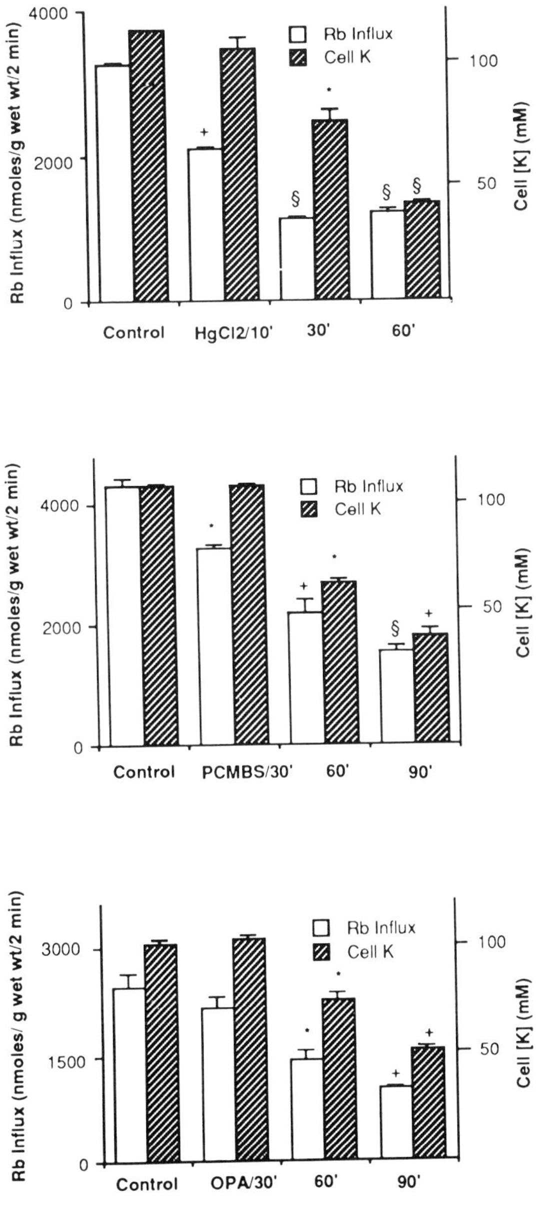

Figure 2: Time course of effects of mucosal 100 μM $HgCl_2$, 1 mM PCBMS, and 250 μM OPA on serosal rubidium influx and cell potassium content. Tissues were prepared and mounted in Ussing chambers as described in [Musch et al., 1990]. After 15 min, additions of metals were begun so that all influxes could be performed at one time 90 min later. Values are means ± SE for three experiments. $^{*}p < 0.05$; $^{+}p < 0.01$; $^{\S}p < 0.001$ compared with paired control by Students t-test. From [Musch et al., 1990] with kind permission.

Investigating the effect of the three pollutants on the cellular activity of the Na/K-ATPase, as reflected by cellular rubidium uptake at the serosal side, two interesting observations were made. First, $HgCl_2$, PCMBS, and OPA inhibited the cellular Na/K-ATPase activity, although as shown in figure 2, after a 60 min incubation period the inhibition by PCMBS and OPA was lower than the inhibition by $HgCl_2$. In addition, the inhibition of the cellular Na/K-ATPase activity was time dependent. Since in the experimental setup used in figure 2 only the mucosal - i.e. the luminal - side of the intestine was exposed to the toxins the compounds must have penetrated the cells to reach the basal-laterally located Na/K-ATPase. $HgCl_2$ had a more rapid onset of action than did PCMBS or OPA, suggesting that permeability of the cells to these agents may be a critical factor in the toxic effect finally exerted on transport.

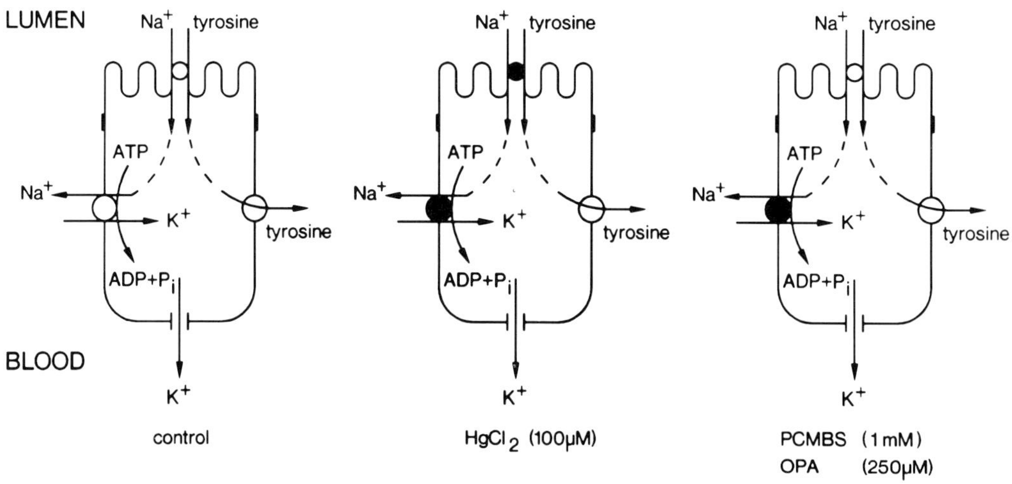

Figure 3: Synposis of the action of $HgCl_2$, PCMBS, and OPA on tyrosine absorption in flounder intestine. Closed circles indicate transport systems that are affected by the toxins. See text for further details.

The results of these studies are summarized in figure 3. $HgCl_2$ has probably two sites of action, one at the luminal cell side by inhibiting sodium-tyrosine cotransport directly and another by inhibiting Na/K-ATPase. Thereby sodium-tyrosine cotransport is affected indirectly via a reduction in driving forces for the transport. Such reduction in driving force can be deduced from the decrease in intracellular potassium content depicted in figure 2. A lower intracellular potassium concentration leads to an increase in intracellular sodium and a decrease in the transmembranal electrical potential difference - both are components of the electrochemical potential difference for sodium across the membrane. PCMBS and OPA inhibit tyrosine transport apparently only indirectly by changing the driving forces via inhibition of the Na/K-ATPase. The dual action of $HgCl_2$ probably explains the stronger inhibition of intestinal tyrosine absorption by $HgCl_2$ than by PCMBS or OPA.

SIDEDNESS OF INHIBITION OF AMINO ACID TRANSPORT BY CADMIUM IN ISOLATED FLOUNDER KIDNEY BRUSH BORDER MEMBRANES

The kidney is one of the main organs affected by chronic poisoning with heavy metals. Usually two stages of intoxication can be discerned. In the proximal tubule first discrete changes in its transport function are observed. After acute exposure of the kidney, solute transport such as reabsorption of amino acids is reduced [Foulkes, 1983]. At a later stage changes in the gross morphology occur which include mitochondrial damage, changes in lysosomes and pinocytic vesicles [Fowler and Nordberg, 1978; Hawkins, 1980]. These morphological changes are associated with proteinuria and enzymuria. Furthermore, as indicated above for intestinal tyrosine transport, it is assumed that there are several cellular sites of interaction of the toxins with plasma membrane transport systems, in particular the renal Na/K-ATPase [Gonick et al., 1980]. The question whether also renal sodium cotransport systems for neutral amino acids can be a target site for the action of heavy metals was investigated recently by Bevan et al. [1989] using brush border membrane vesicles isolated from flounder kidney. It was found that cadmium chloride inhibited sodium-alanine (but not sodium-D-glucose) cotransport in a time- and concentration-dependent manner (figure 4).

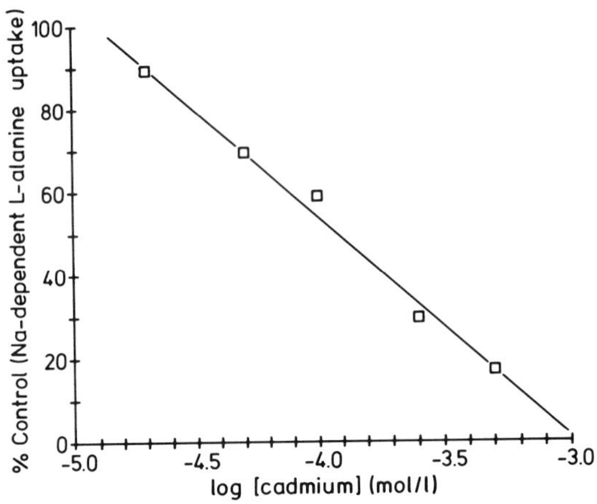

Figure 4: Effect of cadmium on sodium-dependent L-alanine uptake. Vesicles were preincubated for 30 min at 15°C with varying concentrations of $CdCl_2$. Uptake of L-alanine was determined at 15°C as described previously [Bevan et al., 1989]. Sodium-dependent L-alanine uptake was calculated as the difference in L-alanine uptake in the presence of NaCl and in the presence of KCl. Values represent means ± SD from three to five experiments. From [Bevan et al., 1989] with kind permission.

The time dependence of inhibition, as observed at 15°C of sodium-L-alanine cotransport is depicted in figure 5. It is evident that immediately after exposure of the membranes to cadmium no inhibition is observed. The inhibition then increases steadily. It was also found that preincubation at 0°C dramatically decreased the ability of cadmium to inhibit the transporter. Since isolated brush border membrane vesicles are oriented right side out, i.e. the membrane face exposed *in vivo* to the luminal fluid is exposed *in vitro* to the incubation medium, it was hypothesized that the action of cadmium occurred at the cytoplasmic face of the membrane facing the inside of the vesicles *in vitro*. In order to test this hypothesis, cadmium uptake into the brush border membranes was investigated [Bevan et al., 1989; Bevan and Foulkes, 1989]. It could be shown that cadmium uptake into the membrane vesicles was strongly time and temperature dependent; the lowest uptake being observed at short

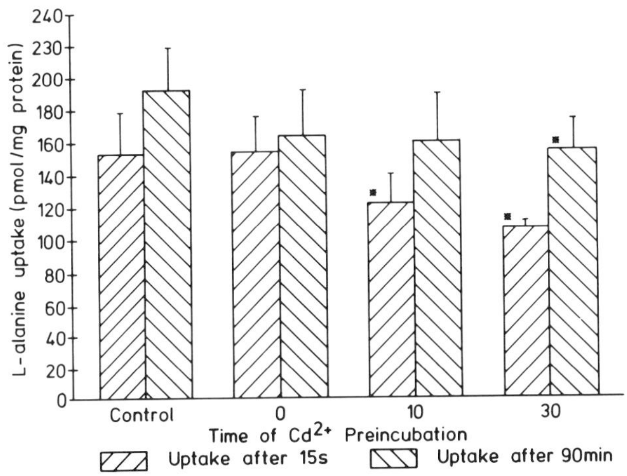

Figure 5: Time dependence of cadmium inhibition of L-alanine uptake in the presence of a sodium gradient. Vesicles were preincubated at 15°C with or without 0.1 mM $CdCl_2$ for the times indicated. L-Alanine uptake was measured as decribed in figure 4. L-Alanine uptake in vesicles preincubated in vesicle buffer in the absence of cadmium was identical over the 30 min period. Values represent means ± SD from three to six experiments. From [Bevan et al., 1989] with kind permission.

incubation times and at 0°C. In addition, exposure of cadmium-pretreated vesicles for 30 min to chelators of cadmium such as EGTA or metallothionein - which reduce the extravesicular cadmium concentration to a level shown to be noninhibitory - did only slightly reverse the inhibition of sodium-alanine cotransport. These data taken together strongly suggest that binding of cadmium to the cytoplasmic face of the brush border membrane is responsible for the inhibition of the sodium-alanine cotransporter.

These studies clearly indicate, that similar to tyrosine transport in the intestine, the renal transport of alanine can be inhibited at several sites by environmental pollutants. They suggest moreover, that cadmium has to enter the renal cells to exert its action at the cytoplasmic face of the brush border membrane. Such a requirement for cell entry could also explain the differences in the time course observed in the inhibition of the cellular Na/K-ATPase activity in the intestinal cells. Indeed, several studies on the mode of inhibition of Na/K-ATPase by mercurials or cadmium indicate that the sites at which these heavy metals interact with the enzyme are located at the cytoplasmic face of the basolateral plasma membrane of epithelial cells [Kinne-Saffran et al., 1991].

CONCLUDING REMARKS

This contribution has focussed on plasma membrane transport systems as potential targets for toxins and environmental pollutants. It first of all has been demonstrated that "*housekeeping*" transport systems (Na/K-ATPase) as well as "*cell-specific*" transport systems (sodium-amino acid cotransport system) can be affected, suggesting that cellular homeostasis and/or specific organ functions can be heavily disturbed by alterations at the plasma membrane level. This suggestion is important as plasma membranes are often the first area of contact of an aquatic animal with environmental pollutants. This holds for the skin, the gills, as well as for the intestine. It was also shown that the agents - at least those studied here - have first to enter the cell in order to exert their action at the cytoplasmic face of the plasma membranes. Such an entry is easily conceivable for lipid soluble compounds, heavy metals may use existing channels for divalent cations for cell entry. Another pathway is the endocytosis of macromolecules to which toxins are tightly adsorbed [Györy and Kinne, 1971]. An example for such a mechanism is the entry of cadmium into renal cells. In the blood, plasma cadmium is tightly bound to metallothionein, a small protein that is filtered in the glomerulum. The cadmium-metallothionein complex is then removed from the primary urine in the proximal tubule via endocytosis. As soon as the cadmium-metallothionein complex reaches the acid intraendosomal compartment, the complex dissociates and cadmium enters the cytoplasma from where it can interact with the cytoplasmic face of the plasma membranes [Squibb et al., 1979; 1984].

Another focus of this article was the demonstration of multiple sites of action of toxins in cells. This multiplicity of interactions together with cellular uptake and intracellular metabolism might explain the irridescent picture of toxicological studies, which often irritate the less versed reader.

ACKNOWLEDGEMENT

Supported by NIEHS grant no. 1-P30-ESO3828 to E.K. and R.K. and a grant from the Deutsche Forschungsgemeinschaft no. Ki 333/2-1 to E.K..

REFERENCES

Bevan C, Foulkes EC (1989) Interaction of cadmium with brush border membrane vesicles from the rat small intestine. Toxicology 54:297-309

Bevan C, Kinne-Saffran E, Foulkes EC, Kinne RKH (1989) Cadmium inhibition of L-alanine transport into renal brush border membrane vesicles isolated from the winter flounder (*Pseudopleuronectes americanus*). Toxicol Appl Physiol 101:461-469

Chauncey B, Schmid ED, Goldstein L (1988) Arsenical and mercurial inhibitions of tyrosine transport by the flounder intestine. J Toxicol Environ Health 23:257-265

Foulkes EC (1983) Nature of inhibition of renal aspartate reabsorption in experimental Fanconi syndrome. Toxicol Appl Pharmacol 71:445-450

Fowler BA, Nordberg GF (1978) The renal toxicity of cadmium-metallothionein: Morphometric and X-ray microanalytical studies. Toxicol Appl Pharmacol 46:609-623

Gonick HC, Indraprasit S, Rosen VJ, Neustein H, van de Velde R, Raghavan SRV (1980) Experimental Fanconi syndrome. III. Effect of cadmium on renal tubular function, the ATP-Na-K-ATPase transporter system and renal tubular ultrastructure. Miner Electrolyte Metabol 3:21-35

Györy AZ, Kinne R (1971) Energy source for transepithelial sodium transport in rat renal proximal tubules. Pflügers Arch 327:234-260

Hawkins WE, Tate LG, Sarphie TG (1980) Acute effects of cadmium on the spot *Leiostomus xanthrurus* (Teleostei): Tissue distribution and renal ultrastructure. J Toxicol Environ Health 6:283-295

Kinne R (1984) Role of sodium cotransport systems in epithelial transport. Ann NY Acad Sci 435:39-47

Kinne RKH (1991a) Selectivity and direction: plasma membranes in renal transport. Am J Physiol 260:F153-F162

Kinne RKH (1991b) From diversity to similarity in biological transport. Issues Biomed 15:69-94

Kinne R, Kinne-Saffran E (1981) Membrane vesicles as tools to elucidate epithelial cell function. Eur J Cell Biol 25:346-352

Kinne-Saffran E, Kinne RKH (1990) Isolation of lumenal and contralumenal plasma membrane vesicles from kidney. Meth Enzymol 191:450-469

Kinne-Saffran E, Pfaff Ch, Maska PS, Kinne RKH (1991) Effect of cadmium ions on partial reactions of the Na-K-ATPase in rectal gland plasma membranes of *Squalus acanthias*. Bull MDIBL 30:38-40

Morrison AI, Panayotova-Heiermann M, Feigl G, Schölermann B, Kinne RKH (1991) Sequence comparison of the sodium D-glucose cotransport systems in rabbit renal and intestine epithelia. Biochim Biophys Acta 1089:121-123

Musch MW, Chauncey B, Schmid EC, Kinne RKH, Goldstein L (1990) Mechanisms of mercurial and arsenical inhibition of tyrosine absorption in intestine of the winter flounder *Pseudopleuronectes americanus*. Toxicol Appl Pharmacol 104:59-66

Squibb KS, Pritchard JB, Fowler BA (1984) Cadmium-metallothionein nephropathy: Relationships between ultrastructural/biochemical alterations and intracellular cadmium binding. J Pharmacol Exp Ther 229:311-321

Squibb KS, Ridlington JW, Carmichael NG, Fowler BA (1979) Early cellular effects of circulating cadmium-thionein on kidney proximal tubules. Environ Health Perspect 28:287-296

van Zeeland AA (1988) Molecular dosimetry of alkylating agents: quantitative comparison of genetic effects on the basis of DNA adduct formation. Mutagenesis 3:179-199

19

Effects of Detergents on Control of Blood Flow Through Fish Gills

D. McKenzie, P. Canciglia, and L. Bolis

The gills of fish are important organs for a number of homeostatic processes, for gas-exchange, ionoregulation and osmoregulation. In order to meet these physiological requirements gills expose a large surface area of well vascularised epithelium to the environment, with very short diffusion distances between water and blood.

The entire cardiac output of fish enters the gills and then divides into two pathways: an arterio-arterial pathway passing across the gas-exchange surface of the gill lamellae and leading to the systemic circulation, and an arterio-venous pathway passing via a central venous sinus and thence back to the heart (Laurent and Dunel, 1976; Laurent, 1984). Effective control of blood flow through these pathways is necessary to match perfusion to ventilation efficiently and thus meet gas-exchange requirements. Control of blood flow through the gills is effected by both humoral and nervous mechanisms.

The circulating catecholamines noradrenaline (NA) and adrenaline (A) constrict the arterio-venous pathway and dilate the arterio-arterial pathway, primarily through the action of ß-receptors (Rankin and Maetz, 1971; Wood, 1974; Payan and Girard, 1977; Bolis and Rankin, 1978; Wahlquist, 1981; Rankin and Bolis, 1984). Autonomic nerves also contribute to control of gill blood flow; adrenergic fibres constricting the arterio-venous pathway and dilating the arterio-arterial pathway, and cholinergic fibres constricting the latter (Nilsson and Peterson, 1981). Under conditions of hypoxia or exercise, when gas-exchange requirements change, a release of circulating catecholamines stimulates a suite of physiological adjustments (Perry *et al.*, 1989), including dilatation of the arterio-arterial pathway and constriction of the arterio-venous pathway through the gills, thereby causing an increase in perfusion of the gas-exchange surfaces and increased perfusion of the systemic circulation.

As the gills expose a large surface area to the environment, the branchial epithelium is a particularly sensitive site for the action of water-borne pollutants. Detergents are extremely common pollutants of the aquatic environment. Detergents are amphipathic molecules that decrease the surface tension of lipid structures. They are formed of a hydrophobic portion consisting of a long hydrocarbon chain (C_{12}-C_{18}) and of a hydrophilic portion which can be a positively charged (cationic detergent), negatively charged (anionic detergent) or polar group (non-ionic detergent). Detergents exert a toxic effect on fish, primarily through effects on the branchial epithelium (Abel and Skidmore, 1975; Abel, 1976). The phospholipids of the plasma membrane are amphipathic molecules resembling detergents. Thus, detergents can dissolve in the membrane, leading to increases in membrane fluidity and at high concentrations this results in the formation of mixed micelles and solubilization of the membrane (Helenius and Simon, 1975). ß-adrenergic receptors are located within the cell membrane, together with the adenylate cyclase sub-units with which they must interact to exert their cellular effects (Ori and Schramm, 1976; Limbird and Lefkowitz, 1977). This indicates that

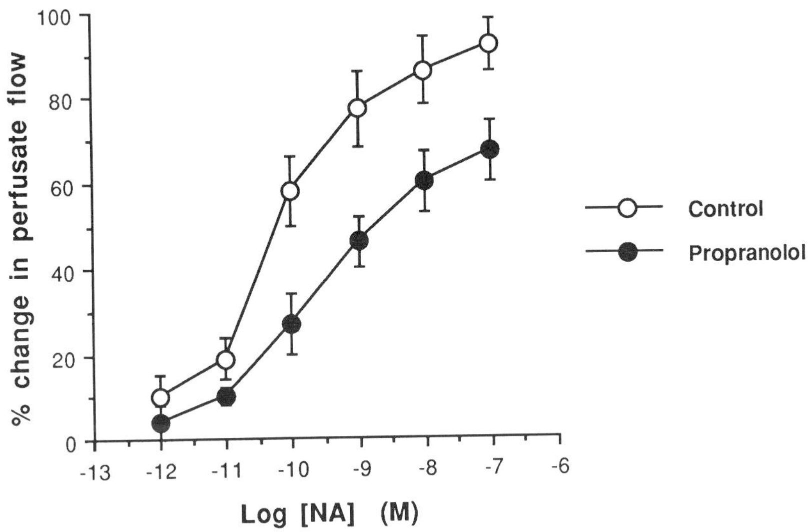

Figure 1: Concentration response curve for the increase in perfusate flow (expressed as % control flow) through eel (*Anguilla anguilla*) gills produced by noradrenaline in the presence and absence of 10^{-6}M propranolol. Values = mean ± S.E., n = 15.

detergents in the aquatic environmemt might interfere with adrenergic responses by the gill vasculature, by dissolving in the gill vascular epithelia and interfering with receptor:adenylate cyclase binding.

The isolated gill preparation developed by Rankin and Maetz (1971) is the major means by which the effects of detergents on adrenergic control of blood flow have been investigated. Briefly, this involves the placement of cannulae in the afferent and efferent branchial arteries of a gill excised from a freshly sacrificed animal.

Once in place, the cannulae are tightly ligatured to avoid fluid loss from accessory vessels and to occlude the arterio-venous pathway. The gill is then suspended in a water bath gassed with 95% O_2, 5% CO_2. The inflow cannula is attached to a reservoir containing a fish Ringer solution, placed at a height of about 40-50 cm above the gill

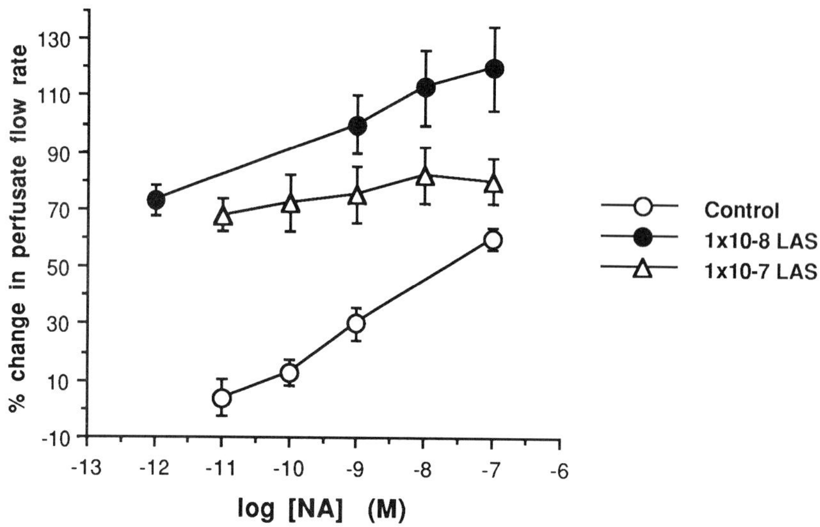

Figure 2: Concentration response curve for the increase in perfusate flow (expressed as % control flow) through isolated brown trout (*Salmo trutta*) gills produced by noradrenaline in the presence of two different concentrations of LAS in the perfusate. Values = mean ± S.E., n = 15.

preparation. The outflow cannula terminates in a drop counter attached to a potentiometric recorder which is placed at 20-30 cm above the gill preparation. Thus, the inflow and outflow pressures of the gill approximate those *in vivo*. The drop counter monitors flow of perfusate through the isolated gill preparation. The fish Ringer solution should be low in calcium (to prevent spontaneous contractions of the gill vasculature) and contain polyvinylpyrrolidone as an osmotic filler.

This preparation allows assessment of the effects of catecholamines on resistance to perfusate flow through the gill, and the effects of detergent on these adrenergic responses. Figure 1 shows a cumulative dose-response curve of the effects of NA on perfusate flow through the isolated gill of an eel, *Anguilla anguilla*, and the effects of propranolol on the vascular responses. Noradrenaline decreases vascular resistance (increases perfusate flow) in a dose-dependent

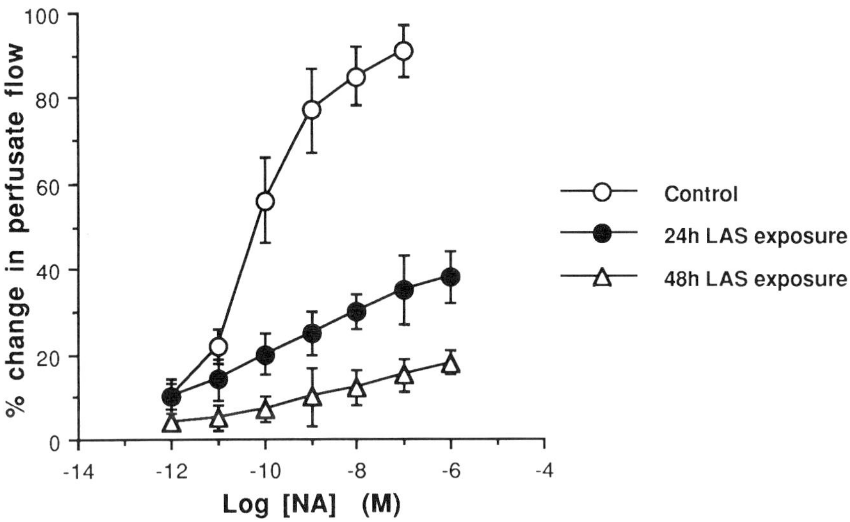

Figure 3A: Concentration response curves for the increase in perfusate flow (expressed as % control flow) produced by noradrenaline in gills from eels exposed for 24h and 48h to 1 mg/l LAS. Values = mean ± S.E., n = 15.

manner, and this effect is partially blocked by propranolol, indicating that it is effected via ß-adrenergic receptors. This is also true of gills isolated from brown trout, *Salmo trutta* (Bolis and Rankin, 1980). Figure 2 shows the effects of two concentrations of Linear Alkyl Sulphonate (LAS) in the perfusate on the response of an isolated brown trout gill to noradrenaline (Bolis and Rankin, 1980). LAS inhibited the effects of NA on perfusate flow. The effect of LAS on the vasoactive response of the gill to NA cannot be attributed to the vasodilatory effect of LAS itself, because addition of a different detergent, Sodium

Lauryl Sulphate (SLS), to the perfusate flowing through isolated eel gills caused vasoconstriction but still inhibited the vasomotor response to NA (Bolis and Rankin, 1975; Stagg *et al.*, 1981), indicating that the inhibition of NA vasoreactivity in isolated gills was not a function of vasoactive effects of the detergent itself. Addition of detergent to the

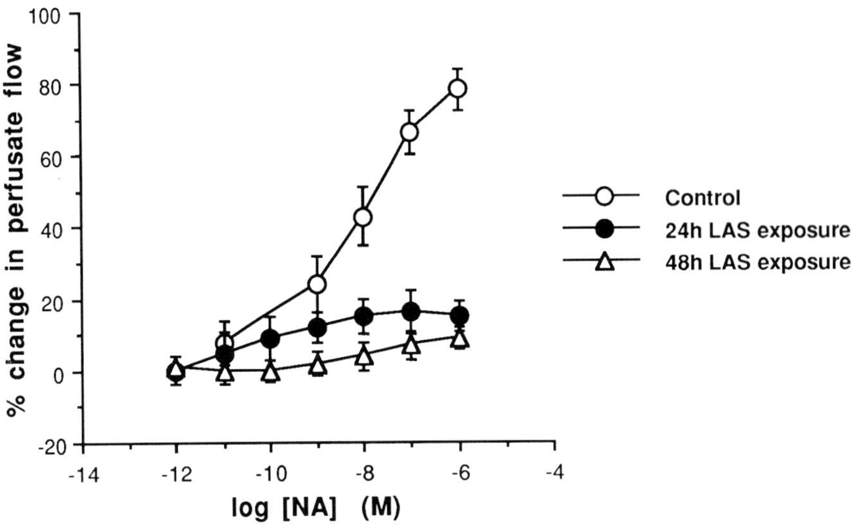

Figure 3B: Concentration response curves for the increase in perfusate flow (expressed as % control flow) produced by noradrenaline in gills from brown trout exposed for 24h or 48h to 1 mg/l LAS. Values = mean ± S.E., n = 15.

perfusate is an acceptable simulation of conditions *in vivo*, as detergent is known to cross the gill epithelium easily and accumulate in fish tissue (Tovell *et al.*, 1974; 1975). Uptake of detergent, however, is dependent on the external divalent ion concentration and on gill mucus accumulation (Tovell *et al.*, 1974), so addition of detergent to the gill bathing solution would lead to variable uptake rates and variability in results.

In order to assess the potential effects of detergents *in vivo*, eels and trout were exposed to 1 mg/L LAS in their holding tanks. Gills subsequently isolated from fish exposed to detergent in this way also showed a reduced responsiveness to NA. Figures 3A and 3B show the effects of NA on perfusate flow through gills isolated from eels (fig 3A) and brown trout (fig 3B) exposed for 24h or 48h to 1mg/L LAS.

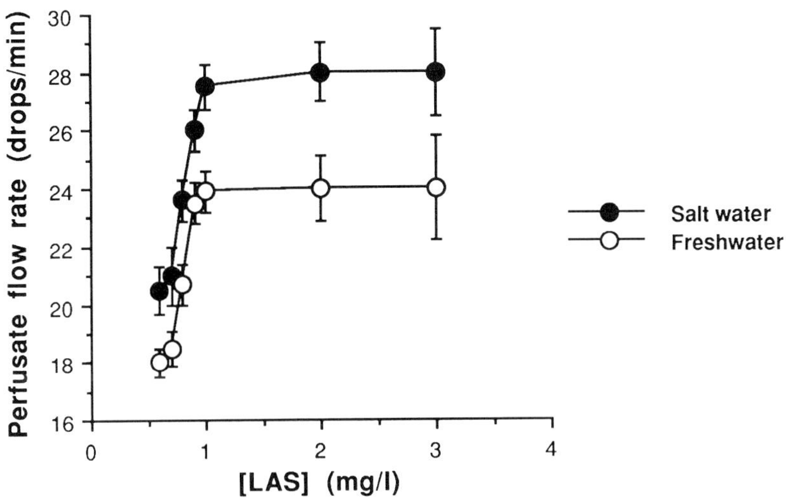

Figure 4: Concentration response curves for the increase in perfusate flow produced by LAS in gills of chum salmon (*Oncorhynchus keta*) from freshwater and seawater. Values = mean ± S.E., n = 8.

The effects of LAS on noradrenergic responses was related to time of exposure; fish at 48h showed a greater loss of sensitivity to NA than those exposed for 24h (Bolis and Rankin, 1980). Thus, under natural conditions, detergent pollution would compromise adrenergic control of gill blood flow. The extent to which this would occur is also dependent on environmental conditions as, e.g. the effects of LAS on chum salmon (*Oncorhynchus keta*) gills (fig 4) were potentiated in seawater as compared with freshwater (Bolis and Rankin, 1978).

Detergents do not inhibit responses by isolated gills to all vasoactive agents. Vasoactive intestinal peptide (VIP) has a vasodilatory effect in isolated *S. trutta* gills (Bolis *et al.*, 1984a). Exposure of trout to 1 mg/L LAS for 24 or 48h has little effect on vasoactive responses to VIP, although gills isolated from the same trout show a strong inhibition of NA sensitivity (Bolis *et al.*, 1984b). This

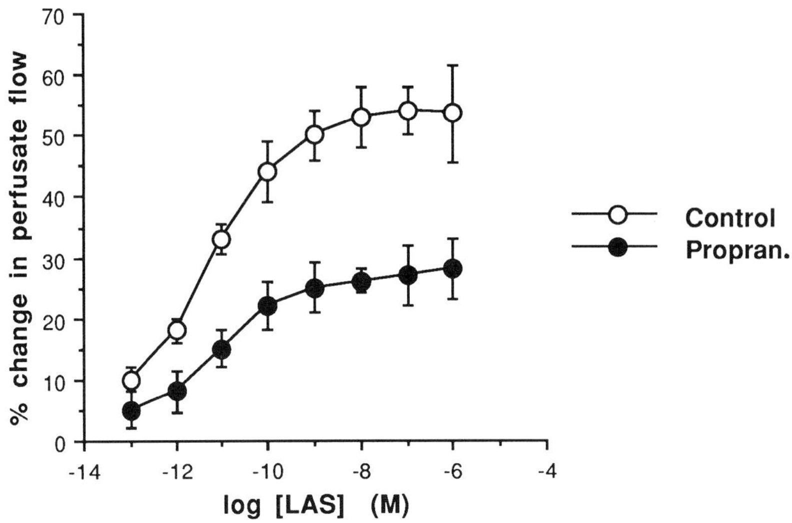

Figure 5: Concentration response curves for the increase in perfusate flow (expressed as % control flow) through eel gills produced by 1mg/l LAS in the perfusate in the presence or absence of propranolol. Values = mean ± S.E., n = 15.

indicates that the effects of detergents on responses to NA are mediated through effects on ß-adrenergic receptors.

It is possible that dissolution of LAS or SLS in the vascular epithelium inhibits ß-receptor-adenylate cyclase interactions by increasing membrane fluidity. Detergents also are reported to show high affinity binding to protein molecules, including adenylate cyclase (Neer, 1978). The vasodilatory action of LAS in the perfusate of salmon gills can be partially inhibited by ß-propranolol (fig 5), indicating that LAS may also have a direct effect on the ß-receptor (Bolis and Rankin, 1978). LAS is also known to cause a progressive loss of protein moieties from the gill epithelium (Roy, 1990). Thus, the effects of detergents on sensitivity of gills to NA is probably exerted through inhibition of receptor:adenylate cyclase interactions, direct effects of detergent on ß-receptors and also through gradual loss of membrane proteins in exposed fish.

The effects of detergents on NA sensitivity are reversible at low concentrations (Stagg *et al.*, 1981) indicating that fish can recover from mild detergent exposure, and effects appear to be inhibited by ß-endorphin (Bolis *et al.*, 1984b), suggesting that ß-endorphin release would ameliorate the effects of detergents on gill blood flow *in vivo*. The effects of detergent exposure on NA-mediated control of gill blood flow would be most serious under conditions of hypoxia or exercise, when fish would require increased perfusion of the gills in order to meet their requirements for O_2 uptake. Under these circumstances, the increase in blood flow through the gills elicited by release of circulating NA would be compromised in fish living in detergent-polluted water.

The effects of detergent on control of gill blood flow are visible at very low detergent concentrations (Stagg *et al.*, 1981) and the isolated gill preparation is inexpensive and relatively easy to prepare. Thus asessment of response to NA might be used as a sub-lethal indicator of detergent pollution, by e.g. comparing responses to NA in gills isolated from fish living above and below a sewage outfall.

References

Abel PD (1976). Toxic action of several lethal concentrations of an anionic detergent on the gills of brown trout (*Salmo trutta* L.). J. Fish Biol. 9:441-446

Abel PD and Skidmore JF (1975). Toxic effects of an anionic detergent on the gills of rainbow trout. Wat. Res. 9:759-765

Bolis L and Rankin JC (1975). Adrenergic control of blood flow through fish gills: Environmental implications. In: Comparative Physiology - Functional aspects of structural materials (Bolis L, Maddrell SHP and Schmidt-Neilsen K, eds). pp. 223-233. North-Holland, Amsterdam.

Bolis L and Rankin JC (1978). Vascular effects of acetylcholine, catecholamines and detergents on isolated perfused gills of pink salmon, *Oncorhynchus gorbuschka*, coho salmon, *O. kisutch*, and chum salmon, *O. keta*. J. Fish Biol. 13:543-547.

Bolis L and Rankin JC (1980). Interactions between vascular actions of detergent and catecholamines in perfused gills of the european eel *Anguilla anguilla* L. and the brown trout *Salmo trutta* L. J. Fish Biol. 16:61-74

Bolis L, Mandolfino M, Marino D and Rankin JC (1984). Vascular actions of vasoactive intestinal peptide and ß-endorphin in isolated perfused gills of the brown trout, *Salmo trutta* L. Mol. Physiol. 5:221-226.

Bolis L, Mandolfino M, Rankin JC and Trischitta R (1984). Effects of exposure to detergent on the vascular actions of vasoactive intestinal peptide, ß-endorphin and noradrenaline in isolated perfused gills of the brown trout *Salmo trutta* L. Mol. Physiol. 5:227-232.

Helenius A and Simons K (1975). Solubilisation of membranes by detergents. Biochim. Biophys. Acta 415:29-79.

Laurent P (1984). Gill internal morphology. In: Fish Physiology Vol. XA (WS Hoar and DJ Randall, eds.) pp 73-183. Academic Press, New York.

Laurent P and Dunel S (1976). Functional organisation of the teleost gill. I. Blood pathways. Acta Zool. 57:189-209.

Limbird LE and Lefkowitz RJ (1977). Resolution of ß-adrenergic receptor binding and adenylate cyclase activity by gel-exclusion chromatography. J. Biol. Chem. 252:799-802.

Neer EJ (1978). Size and binding of adenylate cyclase from bovine cerebral cortex. J. Biol. Chem. 253:1498-1502.

Nilsson S and Peterson K (1981). Sympathetic nervous control of blood flow in the gills of the Atlantic cod, *Gadus morhua*. J. Comp. Physiol. 144:157-163.

Ori YJ and Schramm M (1976). Coupling of catecholamine receptor from one cell with adenylate cyclase from another cell by cell fusion. Proc. Natl. Acad. Sci. 73:4410-4414.

Payan P and Girard JP (1977). Adrenergic receptors regulating patterns of blood flow through the gills of trout. Am. J. Physiol. 232:H18-H23.

Perry SF, Kinkead R, Gallaugher P and Randall DJ (1989). Evidence that hypoxemia promotes catecholamine release during hypercapnic acidosis in rainbow trout (*Salmo gairdneri*). Respir. Physiol. 77:351-364.

Rankin JC and Maetz J (1971). A perfused teleostean gill preparation: vascular actions of neurohypophyseal hormones and catecholamines. J. Endocrinol. 51:621-635.

Rankin JC and Bolis L (1984). Hormonal control of water movement across the gills. In: Fish Physiology Vol. XA (WS Hoar and DJ Randall eds.) pp 177-201. Academic Press, New York.

Roy D (1990). Detergent-induced changes in the protein constituents of various cell types of opercular epidermis of *Rita rita*. Biomed. Environ. Sci. 3:290-298.

Stagg RM, Rankin JC and Bolis L (1981). Effects of detergent on vascular responses to noradrenaline in isolated perfused gills of the eel *Anguilla anguilla* L. Env. Pollution 24:31-37.

Tovell PWA, Newsome C and Howes D (1974). Effect of water hardness on the toxicity of an anionic detergent to fish. Wat. Res. 8:291-296.

Tovell PWA, Howes D and Newsome CS (1975). Absorption, metabolism and excretion by goldfish of the anionic detergent sodium lauryl sulphate. Toxicology 4:17-29.

Wahlquist I (1981). Branchial vascular effects of catecholamines released from the head kidney of the Atlantic cod, *Gadus morhua*. Mol. Physiol. 1:235-241.

Wood CM (1974). A critical examination of the physical and adrenergic factors affecting blood flow through the gills of rainbow trout. J. Exp. Biol. 60:241-265.

List of Contributors

A. Belayew
Laboratoire de Biologie Moléculaire et de Génie Génétique
Université de Liège
Allée du 6 Août, B6
B-4000 Sart Tilman
BELGIUM

J.G. Bell
NERC Unit of Aquatic Biochemistry
School of Natural Sciences
University of Stirling
Stirling FK9 4LA
SCOTLAND

M.V. Bell
NERC Unit of Aquatic Biochemistry
School of Natural Sciences
University of Stirling
Stirling FK9 4LA
SCOTLAND

Gilles Boeuf
Station Ressources Vivantes
IFREMER
Centre de Brest
BP 70
29280 Plouzané
FRANCE

L. Bolis
Istituto di Scienze Farmacologiche
Università di Milano
via Balzaretti, 9
Milano 20133
ITALY

B. Breton
Laboratoire de Physiologie des Poissons
INRA
Campus de Beaulieu
35042 Rennes
FRANCE

N. Bromage
Institute of Aquaculture
University of Stirling
Stirling FK9 4LA
SCOTLAND

L. Byrnes
Department of Microbiology
National Diagnostic Center
University College Galway
Galway
IRELAND

P. Canciglia
Istituto di Fisiologia Generale
Università di Messina
via Tommaso Canizzaro, 278
Messina
ITALY

J.R. Cardwell
Department of Zoology
University of Alberta
Edmonton
Alberta
CANADA T6G 2E9

M. Carillo
Instituto de Acuicultura
Torre de la Sal
Ribera de Cabanes
12595 Castellon
SPAIN

D. Chourrout
Laboratory of Fish Genetics
INRA
78350 Jouy en Josas
FRANCE

C.B. Cowey
Department of Nutritional Sciences
University of Guelph
Guelph
Ontario
CANADA N1G 2W1

B. Davies
Institute of Aquaculture
University of Stirling
Stirling FK9 4LA
SCOTLAND

F. Deryckere
Department of Microbiology
National Diagnostic Center
University College Galway
Galway
IRELAND

J. Duston
Institute of Aquaculture
University of Stirling
Stirling FK9 4LA
SCOTLAND

G. Flik
Department of Animal Physiology
Faculty of Science
University of Nijmegen
6525 ED Nijmegen
THE NETHERLANDS

G. Flouriot
Laboratoire de Biologie Moléculaire
URA CNRS 256
Campus de Beaulieu
35042 Rennes Cedex
FRANCE

F. Gannon
Department of Microbiology
National Diagnostic Center
University College Galway
Galway
IRELAND

F.J. Gatesoupe
INRA/IFREMER
Centre de Brest
BP 70
29280 Plouzané
FRANCE

G. Hardiman
Department of Microbiology
National Diagnostic Center
University College Galway
Galway
IRELAND

N. Heisler
Department of Physiology
Max-Planck-Institut für experimentelle Medizin
D-3400 Göttingen
F.R. GERMANY

R.J. Henderson
NERC Unit of Aquatic Biochemistry
School of Natural Sciences
University of Stirling
Stirling FK9 4LA
SCOTLAND

R.K.H. Kinne
Max-Planck-Institut für Systemphysiologie
Rheinlanddamm 201
4600 Dortmund
F.R. GERMANY

E. Kinne-Saffren
Max-Planck-Institut für Systemphysiologie
Rheinlanddamm 201
4600 Dortmund
F.R. GERMANY

P.-Y. Le Bail
Laboratoire de Physiologie des Poissons
INRA
Campus de Beaulieu
35042 Rennes
FRANCE

F. Le Menn
Laboratoire de Biologie Marine
Institut de Biologie Animale et des Facultés
33405 Talence Cedex
FRANCE

Hong Lin
Department of Zoology
University of British Columbia
Vancouver
British Columbia
CANADA

G. Maisse
Laboratoire de Physiologie des Poissons
INRA
Campus de Beaulieu
35042 Rennes
FRANCE

J.A. Martial
Laboratoire de Biologie Moléculaire et de
Génie Génétique
Université de Liège
Allée du 6 Août, B6
B-4000 Sart Tilman
BELGIUM

D. McKenzie
Istituto di Scienze Farmacologiche
Università di Milano
via Balzaretti, 9
Milano 20133
ITALY

T. Mikolajczyk
Academy of Agriculture
Department of Ichthyobiology
30149 Krakow-Midliniki
POLAND

C. Pelissero
Laboratoire de Biologie Marine
Institut de Biologie Animale et des Facultés
33405 TALENCE Cedex
FRANCE

J. Pérez-Sanchez
Instituto de Acuicultura
Torre de la Sal (CSIC)
Ribera de Cabanes
12595 Torre de la Sal
Castellon
SPAIN

A.-C. Poncelet
Laboratoire de Biologie Moléculaire et de
Génie Génétique
Université de Liège
Allée du 6 Août, B6
B-4000 Sart Tilman
BELGIUM

W. Popek
Academy of Agriculture
Department of Ichthyobiology
30149 Krakow-Midliniki
POLAND

C. Randall
Institute of Aquaculture
University of Stirling
Stirling FK9 4LA
SCOTLAND

David J. Randall
Department of Zoology
University of British Columbia
Vancouver
British Columbia
CANADA

F. Rentier-Delrue
Laboratoire de Biologie Moléculaire et de
Génie Génétique
Université de Liège
Allée du 6 Août, B6
B-4000 Sart Tilman
BELGIUM

G. Salbert
Laboratoire de Physiologie des Régulations
URA CNRS 256
Campus de Beaulieu
35042 RENNES Cedex
FRANCE

J.R. Sargent
NERC Unit of Aquatic Biochemistry
School of Natural Sciences
University of Stirling
Stirling FK9 4LA
SCOTLAND

B. Sekkali
Laboratoire de Biologie Moléculaire et de
Génie Génétique
Université de Liège
Allée du 6 Août, B6
B-4000 Sart Tilman
BELGIUM

P.W. Sorensen
Department of Fisheries and Wildlife
University of Minnesota
St. Paul
Minnesota 55108
U.S.A.

N.E. Stacey
Department of Zoology
University of Alberta
Edmonton
Alberta
CANADA T6G 2E9

C. Storelli
Dipartimento di Biologia
Università di Lecce
Strada Prov.le Lecce-Monteroni
73100 Lecce
ITALY

D. Swennen
Laboratoire de Biologie Moléculaire et de
Génie Génétique
Université de Liège
Allée du 6 Août, B6
B-4000 Sart Tilman
BELGIUM

M. Thrush
Institute of Aquaculture
University of Stirling
Stirling FK9 4LA
SCOTLAND

D.R. Tocher
NERC Unit of Aquatic Biochemistry
School of Natural Sciences
University of Stirling
Stirling FK9 4LA
SCOTLAND

J.P. Truchot
Laboratoire de Neurobiologie et Physiologie
comparées
URA CNRS 1126
Université de Bordeaux I
place du Dr Peynau
33120 Arcachon
FRANCE

Y. Valotaire
Laboratoire de Biologie Moléculaire
URA CNRS 256
Campus de Beaulieu
35042 RENNES Cedex
FRANCE

T. Verri
Dipartimento di Biologia
Università di Lecce
Strada Prov.le Lecce-Monteroni
73100 Lecce
ITALY

A. Wagner
Department of Microbiology
National Diagnostic Center
University College Galway
Galway
IRELAND

S.E. Wendelaar Bonga
Department of Animal Physiology
Faculty of Science
University of Nijmegen
6525 ED Nijmegen
THE NETHERLANDS

K. Yao
Laboratoire de Physiologie des Poissons
INRA
Campus de Beaulieu
35042 Rennes
FRANCE

S. Zanuy
Instituto de Acuicultura
Torre de la Sal
Ribera de Cabanes
12595 Castellon
SPAIN